Cellular Actuators

AF760574

Cellular Actuators

Modularity and Variability in Muscle-inspired Actuation

Jun Ueda

Georgia Institute of Technology, 771 Ferst Drive, Atlanta, GA 30332

Joshua A. Schultz

The University of Tulsa, 800 South Tucker Drive, Tulsa, OK 74104

H. Harry Asada

Massachusetts Institute of Technology, 77 Massachusetts Avenue, MA 02139

ELSEVIER

Butterworth-Heinemann
An imprint of Elsevier
elsevier.com

Butterworth-Heinemann is an imprint of Elsevier
The Boulevard, Langford Lane, Kidlington, Oxford OX5 1GB, United Kingdom
50 Hampshire Street, 5th Floor, Cambridge, MA 02139, United States

Copyright © 2017 Elsevier Inc. All rights reserved

No part of this publication may be reproduced or transmitted in any form or by any means, electronic or mechanical, including photocopying, recording, or any information storage and retrieval system, without permission in writing from the publisher. Details on how to seek permission, further information about the Publisher's permissions policies and our arrangements with organizations such as the Copyright Clearance Center and the Copyright Licensing Agency, can be found at our website: www.elsevier.com/permissions.

This book and the individual contributions contained in it are protected under copyright by the Publisher (other than as may be noted herein).

Notices

Knowledge and best practice in this field are constantly changing. As new research and experience broaden our understanding, changes in research methods, professional practices, or medical treatment may become necessary.

Practitioners and researchers must always rely on their own experience and knowledge in evaluating and using any information, methods, compounds, or experiments described herein. In using such information or methods they should be mindful of their own safety and the safety of others, including parties for whom they have a professional responsibility.

To the fullest extent of the law, neither the Publisher nor the authors, contributors, or editors, assume any liability for any injury and/or damage to persons or property as a matter of products liability, negligence or otherwise, or from any use or operation of any methods, products, instructions, or ideas contained in the material herein.

Library of Congress Cataloging-in-Publication Data
A catalog record for this book is available from the Library of Congress

British Library Cataloguing-in-Publication Data
A catalogue record for this book is available from the British Library

ISBN: 978-0-12-803687-7

For information on all Butterworth-Heinemann publications
visit our website at https://www.elsevier.com

ELSEVIER Book Aid International
Working together
to grow libraries in
developing countries

www.elsevier.com • www.bookaid.org

Publisher: Joe Hayton
Acquisition Editor: Sonnini Yura
Editorial Project Manager: Mariana Kuhl
Production Project Manager: Mohana Natarajan
Designer: Greg Harris

Typeset by VTeX

Contents

List of figures

List of tables

Introduction

About this book

Actuators are one of the key components in robotic and mechatronic systems. Motions generated by actuators enable computer-controlled systems to interact with the physical world. This book introduces the authors' research efforts on cellular robotic actuators performed at the Massachusetts Institute of Technology and Georgia Institute of Technology. Inspired by unique characteristics of biological muscles, a system-level approach was taken to design, fabricate, control, and analyze new robotic actuators, resulting in a unique and powerful methodology for biologically inspired actuation. The book also introduces the concept of cellular piezoelectric actuation, where mechanical structures with an array of single amplified piezoelectric actuators operate in parallel or series according to the needs of the application. Such actuators are fast and provide smooth contraction, which is interesting for human-like movements in arms and eyes.

The authors have been performing research on robust robot control and human function modeling for more than 10 years. Studying both artificial systems and biological systems allowed us to realize that there is still a gap between the two areas, mainly due to a substantial difference in architecture. To address the challenges within these areas, the authors had extensive discussions with colleagues in the community at workshops and conferences. Conclusions drawn from those discussions include: (i) new actuation technologies have a huge potential to impact robotics research, (ii) every actuator material has its advantages and disadvantages; seeking a single perfect material applicable to any possible applications is not the right approach, and (iii) high performance can be achieved by studying the dynamics of the plant (i.e., component-level research), discovering how to push the plant safely toward its performance envelope, and designing control architecture accordingly (i.e., system-level research). This third point is one of the unique aspects of actuator research that motivated the authors to write this book.

The intended audience is graduate-level students as well as professional engineers who are new to this field of research. Regarding prerequisites to the content covered in this book, the authors have assumed that the readers have knowledge in mechanical engineering or related discipline at the level of a BS degree, preferably in System Dynamics and Control. Reference materials will be provided for readers who are new to this area as well as for advanced readers for further study.

Motivation for biologically inspired actuation

The argument that compliant, linear, fast-contracting, artificial muscle actuators (fundamentally different from rotary industrial electromagnetic servo motors) must be used to reproduce biological motion still needs justification. It is true at this point that electromagnetic servo motors are still the first choice to design a highly mobile humanoid. Dynamic limb movements can be generated by servo motors placed at joints, allowing different actuator configurations than those of humans. Considering the well-known airplane–bird analogy (or car–cheetah), an optimal engineering solution is not always mimicking biological mechanisms completely.

We should keep in mind that the choice of industrial servo systems for a new robotic system is supported by the fact that not only the actuators themselves, but also drivers and sensors are highly developed. More pointedly, industrial actuators are mature *as a system*. Model selection and installation of industrial servos are well-documented. Extensive effort at the system-level has been made to optimize the performance. In contrast, although some new actuator materials are promising and potentially superior to electromagnetic actuators, they are not necessarily mature as a system. This is one of the major research challenges that the community must address.

On the other hand, the potential of biologically inspired robotics and mechatronics should not be limited to currently available actuator choices. Current servo drives are well optimized for factory automation (FA). We can still "borrow" industrial servo drive technology for non-FA purposes; however the outcomes may be limited. One example is the noticeable difference between human motion and robot motion. It is known that time-optimal point-to-point control of an inertial system is bang-bang. However, human reaching motion does not follow the bang-bang control concept, but rather exhibits a smoother, i.e., bell-shaped profile, possibly by following the minimum jerk or minimum torque change hypothesis. A musculoskeletal system roughly consists of muscles and link mechanisms where we find a larger difference between muscles and servo motors than between the human skeleton and robot link mechanism. One cannot say definitively, but it is still reasonable to suspect that the difference in actuation may be the source of the disparity in motion.

Biological muscles and artificial muscle-type actuators

For the reasons mentioned above, our goal is to develop a new muscle-inspired actuator system that has common features observed in biological muscles. This book deals with miniature actuator systems that utilize piezoelectric actuator materials, in particular lead zirconate titanate (PZT). PZT-based actuators are widely used in industry for high-precision and high-speed positioning. This book has strategically chosen PZT over other actuator materials, such as electroactive polymers and shape memory alloys, to study the system dynamics in twitch-type fast contraction. Due to their established market, not only are their material properties well studied, but also reliable driver circuits are readily available; such maturity at the single actuator level is preferable when studying the system engineering aspect of biologically inspired actuation. The book will discuss the unique properties of PZT further in subsequent chapters.

Cellular architecture

Biological muscles are very different from industrial actuators. By itself, a single biological muscle fiber is a rather unimpressive actuator; it merely produces discrete translational contractions. In contrast, a single industrial actuator must be precise and reliable because of specific market needs to displace a linear or rotary joint in a mechanism. Inspired by anatomy and neuromuscular physiology, the concept of "cellular robot actuators" has been proposed. The main inspiration proceeds from the observation that a single biological muscle consists of a number of nearly uniform motor units as subunits. Applying this concept to robotic actuation, a new robot actuator would consist of a number of small actuator units, or an array of cellular actuators. It should be noted that such cellular architecture would not have been encouraged from the viewpoint of traditional mechanism design because of its complexity in fabrication and maintenance. However, recent advances in precision machining and additive manufacturing technology have enabled fabrication of complex mechanisms that were not possible a decade ago. Furthermore, biologically inspired robotics has drawn great attention from the research community and general public. These developments have motivated the authors to pursue a rigorous study of the cellular actuation concept and its defining features.

Precisely speaking, the cellular actuator concept that will be presented in this book is to study the following properties observed in biological muscles:

- Coordination of a number of actuator units
- Control of discrete and fast dynamic contraction
- Compensation of variability and hysteresis in actuation

The reason this book focuses primarily on PZT actuators in applications is because of their history of reliability in stand-alone actuator applications. This book is not intended to be a comprehensive reference on active materials used in actuation, but rather an introduction to the concept of cellular actuation. For this reason, *some* active material is needed to demonstrate the concepts, and PZT fits the bill nicely. For a comparative survey of active materials, the reader is directed to [161,198].

Although PZT is by no means a biological material, it mimics biological muscles in that it exhibits large amounts of hysteresis in actuation that must be compensated for. Note that this book does not aim to address other challenges associated with the cellular biology or electrochemistry in the muscle structure. For the readers' convenience, the book provides a basic introduction to piezoelectric effects. Readers interested in the materials-level physics should refer to books such as [279,262]. Another topic that this book has chosen to investigate is effective compensation of hysteresis in an alternative (but slower) material, shape-memory alloy (SMA). Cellular actuators made from SMA wire actuators have been developed, and some examples will be treated in this book as well.

An additional emerging non-traditional artificial muscle actuator material (though not treated in this book) is electroactive polymer (EAP). Readers interested in this actuator material can find well-written books such as [18].

Outline of this book

To produce a sufficient stroke from the PZT materials comparable to biological muscles in terms of strain, unique strain amplification mechanisms have been developed. Chapter 1 introduces design and fabrication of such PZT cellular actuators. The nested rhombus exponential strain amplifying mechanism hierarchically nests mechanical structures known as moonies to produce a large effective strain. The design is intended to generate a two-order-of-magnitude greater displacement than original small displacements of PZT actuators.

From the actuator design point of view, kinematics and dynamics of a small actuator unit should be modeled. When an array of small cellular actuators act as a single actuator, the properties of this array should be characterized. Chapter 2 presents modeling methods of a single cell, single actuator array, and antagonistic actuator arrays. The mechanical strain amplifier is modeled by a two-port network model based on elastic beam theory. Nesting of multiple amplifiers for exponential strain amplification is mathematically represented and its theoretical limit is shown. When many cellular actuator units form an actuator array, the number of ways in which the cellular units could be connected is vast. Graph theory is applied to represent the myriad of actuator topologies for the analysis. The graph-theoretic approach designed for the cellular actuator architecture is called the "finger-print method."

Chapter 3 describes how to efficiently control an actuator array. Methods should be developed to compensate for mechanical drawbacks such as hysteresis and structural flexibility. Segmented-binary control is a simple and effective method at the single actuator level. Discrete-switching actuation provides effective open-loop vibration suppression of an actuator array with flexibility. When the number of actuator units in an array is so large that sending actuation commands to individual units is no longer a reasonable option, broadcast control based on stochastic control theory is an alternative. Individual actuator units are controlled as a stochastic multi-agent system. Stability and robustness of broadcast control will be discussed.

Chapter 4 introduces the design and operation of several interesting robotic devices enabled by the piezoelectric cellular actuators. One of these is a cellular actuator array with variable stiffness tunable resonance frequencies. Contact modes between internal mechanical members resulted in discrete operation of PZT actuators exhibiting a variety of different spring–mass behaviors. The second mechanism produces discrete displacements from structural buckling induced by two PZT actuators facing each other. The buckling actuators are an example of the cellular actuator that is intrinsically discrete. The proposed buckling mechanical structure in which the piezoelectric stacks are placed regulates the buckling direction and produces extended bipolar motion. A micro robotic gripper utilizes both the direct and converse piezoelectric effects, enabling simultaneous actuation and sensing of robotic gripping. Sensing of both displacement and force at the end-point of the gripper is possible by calibrating the electromechanical properties of the gripper including its strain amplifiers. A camera positioning system operated by cellular actuators is capable of scanning the environment as fast as human saccades. The system demonstrates the potential of fast contracting artificial muscle actuators in studying the coordination of motion control and image processing in the human ocular system.

Acknowledgment

This material is based upon work supported by the National Science Foundation under Grants ECCS-0932208 (Ueda), CMMI-300019 (Ueda), CMMI-1000727 (Asada), and OISE-1427787 (Schultz). Any opinions, findings, and conclusions or recommendations expressed in this material are those of the authors and do not necessarily reflect the views of the National Science Foundation.

The authors would like to thank Dr. David MacNair (Section 2.4), Dr. Levi Wood (Section 3.4), Dr. Thomas Secord (Section 4.1), Dr. Devin Neal (Section 4.2), Mr. Timothy McPherson (Section 4.3), and Dr. Michael Kim (Section 4.4) for their contribution to the studies presented in this book.

The first author would like to thank the members of the Bio-inspired Robotics and Human-Modeling Laboratory at the Georgia Institute of Technology, in particular, Dr. Melih Turkseven, Rohan Katoch, Antonio Moualeu, Ellenor Brown, and Euisun Kim, for their support for preparing materials, editing chapters, and other miscellaneous assistance. The manuscript was finalized during the first author's short stay at the Ben-Gurion University of the Negev, Be'er Sheva, Israel in June 2016. The first author would like to express the gratitude to Dr. Yael Edan, Dr. Ilana Niskly, and other faculty and staff members of the ABC Robotics Program at the Ben-Gurion University of the Negev. Last but not least, the authors would like to express their thanks to Ms. Mariana Kuhl Leme, Ms. Sonnini Yura, and Mr. Andre Gerhard Wolff of Elsevier for their support, patience, and encouragement.

Historical overview

Soft robots for unstructured environments

A frontier in robotics and autonomous machines is that of operating in unstructured environments, where the robot does not have *a priori* information about the environment. This could refer to the size and shape of an object to be grasped, terrain over which the robot must maneuver, or obstacles with which the robot may come in contact, to name a few. Often the environment may not be very pliable, meaning the robot cannot reach a desired position in space without penetrating a solid object. Hogan [103] states a general principle for robotic manipulators interacting with the environment as follows:

> *An important consequence of dynamic interaction between two physical systems such as a manipulator and its environment is that one must physically complement the other: along any degree of freedom, if one is an impedance, the other must be an admittance and vice versa.*

He goes on to say that most often the environment will inevitably be an admittance (being rigid, it cannot accept displacement inputs) and therefore the robot must behave as the impedance (it accepts a displacement input and generates a force). He goes on to present the now widely-used impedance control formulation that allows the user to control the robot with the interaction in mind. The controller, in essence, makes the manipulator behave as if it is held against the environment by a spring.

Inspired by the soft features of motion generation in human and other organisms, the field of Soft Robotics has emerged. Although compliant behaviors can be implemented by controllers, *physical* compliance is seen to be beneficial for robot operation in uncertain environments [112,22]. Physical softness is not subjected to failures in controller implementation, and it is still present even when the robot is off. This has led to several interesting applications, including a walking robot based on differential expansion of multimaterial chambers [257], an autonomous rolling belt with embedded computation [47], and an octopus-like robot [35].

A robot does not need to be completely devoid of rigid parts, however, to capture the key interaction properties of soft robotics. If each actuator in the robot is compliant, the inherent compliance in each degree of freedom will provide for compliant interaction found in soft robotics and vertebrate organisms. The seminal work of Pratt and Williamson [200] showed that incorporating an elastic element in series with a traditional servo motor produced desirable properties, including more stable force control and lower reflected inertia. Since that time, numerous developments have been made. One notable development has been the ability to vary the stiffness of the joint using variable impedance actuators. Examples include the Mechanically Adjustable Compliance and Controllable Equilibrium Position Actuator (MACCEPA) [278], an actuator that couples the series elastic element with a movable pivot [115], and antagonistically-based actuators with nonlinear coupling [213,108].

Soft actuators *can contribute to* safety in the event of unexpected collisions, but the presence of soft material in the robot by itself is an insufficient criterion to presume a robot is safe. Large deformations of elastic material can still cause injury, meaning that bounds on the collision energy should be calculated before opening the workspace to humans. One formal method for assessing the potential for injury is presented by Bicchi and Tonietti [27] based on the head injury criterion, and compliance is shown to reduce the collision energy. Actuator compliance can also be exploited to produce motions that are not possible with rigid actuators, such as throwing a ball a great distance. This happens by storing energy in the actuator through one part of the trajectory and rapidly releasing it at another [7,29].

A question emerges with regard to soft robots or soft actuators: what is soft? One possible definition is the hardness of the material, but this does not lend insight with regard to the functionality of the device. One possible definition is this: an actuator can be considered soft if, when the input to the actuator is held constant, deformation in a degree of freedom is an order of magnitude greater than the deformation of rigid links in the constrained direction.

Robot actuators

The earliest robots were envisioned to be capable of positioning (and orienting) objects at arbitrary points in space. The pick and place tasks conducted were in a very structured environment with fences and fixtures, and little uncertainty with regard to the location and orientation of the part. For the robot to be capable of more subtle tasks humans perform every day, researchers realized something else was needed. Unstructured environments pose a challenge to roboticists to this day. Humans are adept at imposing forces in the appropriate directions to perform tasks. Robot force control is possible with robot hardware, of course, but to apply a force, one needs to make sure that the robot is in contact with the object first. Contact is a funny thing; it is unilateral in nature and discontinuous. The end effector is moving in free space, then instantaneously changes to being in contact with no warning.

Impedance control [103] sets actuator efforts at each instant in time to effect a virtual spring between the end effector and the goal position. That way if the robot comes in contact with a wall and the goal is beyond the wall, the servos will just output a position dependent force that keeps the goal in contact with the wall. This "software spring" works well in many applications; in fact, it will feel like a spring to human users who perturb the end effector. The drawback is that impedance control, like any software method in robotics, is only as good as the hardware implementing it. If the servo motors, drive electronics and control are low bandwidth, or if the motors have substantial cogging effects, these will affect the performance.

A physical spring will behave as spring-like without bandwidth limitations; this will lead to the development of physical, hardware series elastic elements being combined with motors [200,300]. Alongside this development came the introduction of more novel forms of actuation including shape memory alloy (SMA) [242,79], piezoelectric [262,45,183,36,60,61,92,116,182,186], and many others [210,34,97,196,226]. As these are active materials, the elasticity of the material itself can produce the required spring behavior. Many of these newer systems, however, still suffer from concerns such as small force or displacement capability and thus have limited applicability on their own [160].

Fig. I shows a qualitative comparison of various actuator materials such as dielectric elastomers, SMA, PZT, conducting polymer, electrostrictive elastomer, and natural muscle. Extensive quantitative data may be found in references such as [161] and [198], but such data shows large variance and a continually shifting mean as materials improve. The figure shows that each smart material performs well in certain categories but not in others. Thin film PZT has a high power density, high bandwidth, and high efficiency. PZT outperforms other actuator materials with respect to bandwidth. Its maximum stress is as large as that of SMA, and the efficiency is comparable to electrostrictive elastomers. Furthermore, PZT is the most reliable material that is usable in harsh environments. Polypyrrole conducting polymers, for example, degrade quickly, despite attractive expected features. On the other hand, the strain of PZT is much smaller than that of other materials.

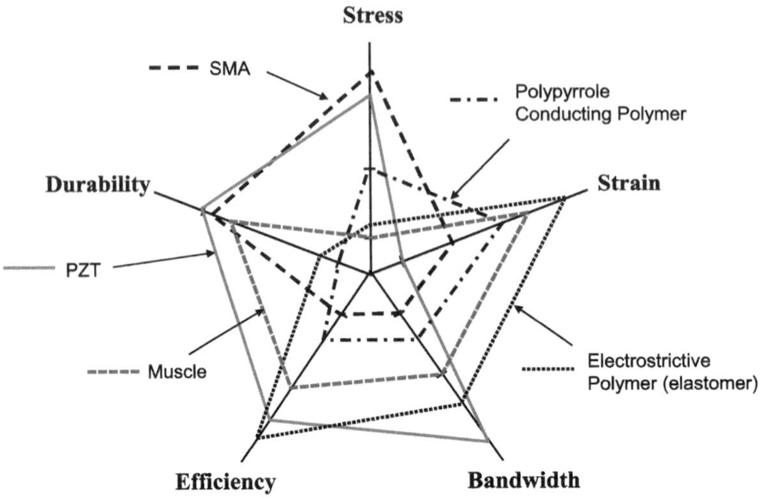

Figure I Qualitative comparison of actuator materials and natural muscle.

Redundant actuators

While the softness of muscles is a major part of what allows humans to function successfully in unstructured environments, less attention has been paid to the role that the discretization of muscle into cells plays in this endeavor. Few works exist that apply actuation with a "cellular structure" in the engineering context. The primary advantage of a cellular structure is that it is extremely unlikely that more than a few "cells" will fail at any given time. This means that rather than provoking a failure of a system to operate, cell failure produces a reduction in performance. This reduction in performance may be acceptable, such as walking with a slower gait or grasping less tightly [259]. This changes the perspective of putting redundancy into engineering systems, as failure can be mitigated by simply including extra "cells" rather than complete redundant systems [58]. The work of Du and collaborators, motivated by resiliency, looked at the consequences of failure systematically in an actuator consisting of a bank of servomotors [65,67,66]. Although not cellular in nature, Fillipini, Sen, and Bicchi discuss the consequences of failure in antagonist pairs with multiple belts, and characterize the residual functionality of the system in response to different kinds of failures [75].

The earliest known formal study of "actuators composed of sub-actuators" is the work of Huston et al. [110,109]. They introduce a few notational devices, and conduct some preliminary experiments using shape memory alloy, but they do not address properties of large systems or system dynamical properties. Secord and Asada [222] analyze a chain of actuation units where each behaves as a rigid body if inactive, and as fixed stiffness if active. They establish rigorously several intuitive results about how the resonant frequency behavior changes with the activation pattern and establish upper and lower bounds on the resonant frequency. MacNair and Ueda [158] provide an automated method to calculate the dynamic equations of motion for an interconnected set of units that follow the Hill model (with damping), joined by masses. They employ a non-boolean incidence matrix-like technique called the "fingerprint" method [159].

Discretization of muscle-like actuators has begun to catch on in recent years, and several platforms have been developed. Bryant, Meller, and Garcia developed a platform based on bundles of McKibben-style pneumatic actuators [32]. Davis and Calwell developed a muscle-like pneumatic actuator with multiple bladders and demonstrated basic healing capability by injecting latex foam in response to a puncture [51]. Mathijssen, Schultz, Vanderborght, and Bicchi developed a compact solenoid-based "actuation unit" that could be combined with other "actuation units" to form a muscle of the stroke and force desired [167,214]. This compact, modular platform incorporated the concept of strong and weak motor units in human muscle for greater control signal richness. Kianzad et al. used a thermally activated conductive polymer fiber approach, and incorporated the pennate nature of muscle into their prototype [127]. Kani et al. [125] created a platform whereby individual servomotors are coupled to a common drive shaft through elastic elements and developed a control architecture with variable impedance. This burst of recent works is exciting for the field, and many interesting properties of this architecture have begun to emerge.

Generation of natural movements

Another rationale for cellular muscle-like actuation is that it may produce more natural motions, benefiting the field from the movement economy and social robotics perspectives. Generation of natural movements, or the movements created by biological systems including humans and animals, has been one of the biggest scientific questions discussed in physiology for decades. Human natural movements are stereotypical in nature [49,281,280,33,202,56,184]. Excluding small individual differences, motions with the same objective, such as arm-reaching or walking, tend to be very similar among different people. Stereotypical adaptation processes are also observed in neurological patients [50,138,206]. Together these observations imply the existence of general rules for coordination of multiple muscles to generate natural movements. Humans can easily distinguish natural motions from artificially generated ones, even from limited visual information such as the trajectories of light points attached to moving limbs [118,260]. A robotic limb which "moves like a robot" gives a feeling of trepidation to those working nearby. As robotic movements become more natural, people interacting with the robot are able to more easily predict the robot's movements, even subconsciously, and feel more comfortable close to the robot. Such perceived realism of movement has been an interest in Computer Graphics [84], Psychology [102], and Social Robotics [189,57].

The definition of what causes a natural motion is still an open subject of research. Many research efforts searching for these underlying rules of natural movement focus on the variability in spike generation which results in variability of muscle forces [98, 43,119,245,255,253,93,187]. Harris and Wolpert showed that the standard deviation (SD) of the commanded signal varies proportionally with the mean of the command signal [98]. They argued that this signal-dependent noise plays a central role in motor control and that movements are organized to minimize the variance at the endpoint. When effectuating a point-to-point motion, for example, Harris and Wolpert argue that the optimal strategy is to reach the desired endpoint with the minimal error due to the noise in the motor signals. The trajectory, velocity, and acceleration for the motion are determined by solving this optimization problem. They discovered that the calculated trajectories according to this criterion are similar to stereotypical human motions, or natural movements [98]. Todorov et al. [255] suggested that this optimization strategy is equivalent to the optimality principle in muscle force generation [49,280,33,202]; that is, a minimization of the sum of muscular stress raised to a power, which is subject to the force/torque constraints of a given task. Simmons et al. has applied this concept to optimal control of a standard 2-DOF manipulator with non-redundant planar rotary joints [234]; however, signal-dependent noise was computed and merely added to the control signal to mimic a natural system.

Since the signal-dependent noise was computationally generated, and does not naturally occur as an artifact of AC/DC rotary motors, this method seems somewhat contrived. As will be seen in this book, with the cellular actuator architecture, signal-dependent noise is a naturally occurring, innate artifact of the muscle-like architecture, and can be exploited in the control architecture. To truly study the effect of neurologically-inspired control architectures in robotics, an actuation platform that operates in the same fashion as muscles (the destination of neural impulses) is necessary. Cellular actuator arrays provide a suitable physical platform for further research to understand muscle optimality criteria and the control signals used to control human muscles. The signal distribution methods presented later in this book demonstrate the same characteristics as the signal-dependent noise observed in biological muscle, and developing optimal control methods for the cellular actuator arrays will likely lead to greater understanding of the brain's control over biological muscle. Furthermore, since optimality criteria like signal dependent noise are believed to lead to natural motion, actuator arrays which follow this optimality criteria should be capable of achieving natural motion.

Cellular actuator concept

Inspiration from biological muscles

Muscles are "actuator rich" systems, or dynamical systems with many degrees of internal freedom and relatively few inputs and outputs. A skeletal muscle consists of five layers of hierarchical structure, starting with the lowest functional units called sarcomeres which contract to provide varying levels of displacement and stiffness [166].

Fig. II shows the molecular representation and actual structure, as seen under a light microscope, of a sarcomere. The M-lines and Z-lines of sarcomeres in biological muscle and the interaction between the myosin and actin fibers generates displacement and subsequent contractile force. The actuation of the muscle is provided by myosin interacting with actin in the presence of ATP to create contractile displacement between the Z-line and M-line on either side of the sarcomere. Additionally, the

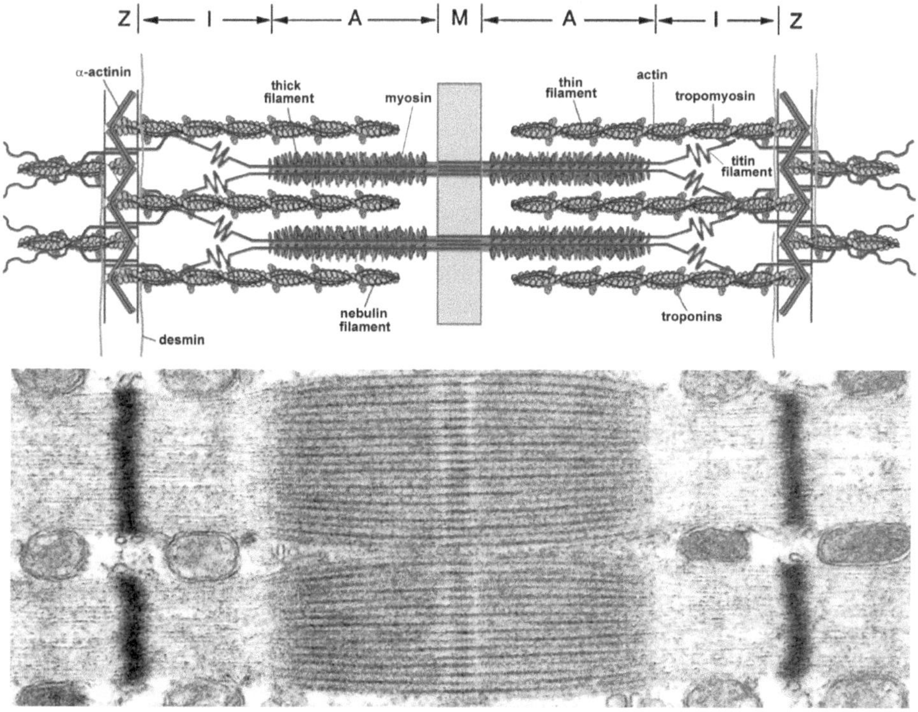

Figure II Molecular representation and structure of a sarcomere. Image taken from [188] and used with permission under the creative commons license.

strands of actin and the connecting titin fibers are flexible, allowing the contractile displacement of the sarcomere to be translated into contractile force when under the influence of an external load or blocking force. Individual sarcomeres have a consistent contraction length, and additional overall displacement, or similarly force, is built by recruiting or activating more sarcomeres in the muscle, not by further displacing a given sarcomere.

These sarcomeres are far more numerous than Golgi tendons and muscle spindles, the internal receptors used to measure force, velocity, and displacement in the muscle. Furthermore, the number of motor neurons entering a muscle is also much smaller than the number of sarcomeres. It is anticipated that the central nervous system is probably not aware of the full internal states of the muscles, nor can it specify the individual contractions of the sarcomeres. However, a smooth and accurate gradation of response can be obtained from a muscle. This implies that there is a certain mechanism coordinating a vast number of small functional units in such a way that a fraction of them are selectively activated to accommodate the aggregate output of the cellular units.

At the molecular level, studies reported that the activation of sarcomeres is not governed by a deterministic control, but a stochastic process due to the diffusion of calcium ions [247]. Other references argue that the actomyosin contraction process, the essential process of actuation, is a Brownian process [134]. At the macroscopic level, a skeletal muscle shows smooth motion although the muscle fibers are known to have either "ON" (producing tension) or "OFF" (relaxed) state [166], and they exhibit prominent hysteresis [136]. It is also notable that a muscle can function properly despite a significant fraction of the cellular units being fatigued or non-functional. The system is quite robust and stable.

Today's artificial muscle actuators, although similar in some aspects, are significantly different in structure and control from biological muscles. Simulating the anatomical structure and motor control architecture of a skeletal muscle, we could gain some insights as to how an artificial muscle can be built and controlled. This leads to an alternative to the design of today's artificial muscle actuators, named the "Cellular Actuator Concept," which is worth investigation for long-term research interests.

Binary cellular structure. Functional units lower in the muscle hierarchy take on a binary state, which can be modeled as ON–OFF finite state machines. Bistable ON–OFF control has salient features in coping with complex nonlinearities of actuator materials. Muscle fibers have prominent hysteresis as addressed by [136]. Most materials for artificial muscle actuators, too, have prominent hysteresis and state-dependent complex nonlinearities [241,162,97,226,44]. As shown in Fig. III, bistable ON–OFF control would not depend on these complex nonlinearities as long as the state of the material is pushed towards either ON or OFF state by using inputs u_{OFF} for OFF or u_{ON} for ON. Dynamic transition may be influenced by the varying nonlinearities. Nonetheless, the control problem becomes much simpler for ON–OFF control, as demonstrated by [225] for SMA and by [196] for dielectric elastomers.

The cellular architecture has another important feature with respect to speed of response if we take shape-memory alloy (SMA) as an example. As the size of cellular units decreases, the speed of response increases for those actuator materials that entail transport of matter. Activating sarcomeres entails diffusion of calcium ions, activating SMA needs heat transfer, and conducting polymers need ion migration. Common to all these actuator materials is the fact that the speed of response increases when the actuator materials are segmented into many small units or thin films, and the reservoir of ions or heat is closely located to the cellular units. For example, thin film SMA for a micro actuator [79] has a small amount of thermal capacitance compared to macroscopic SMA applications, thus the response time is substantially reduced.

Fig. IV shows an artificial muscle control system having a binary cellular structure. Instead of driving the whole actuator material in bulk, the actuator material is divided into many small segments, each controlled as a bistable ON–OFF finite state machine [225]. The displacement of the actuator is given by the aggregate sum of the binary outputs of all the cellular actuators. As the size of each cell decreases, the speed of response increases and the resolution improves. Further details of the binary cellular actuator control will be discussed in Section 3.3.

Broadcast control. Increasing the number of cellular units and reducing the size of each cell can bring about improved resolution and faster response. In such an "actuator-rich" system, as the number of cellular units increases, it is extremely challenging, or almost infeasible, to control all the cellular units directly by centralized

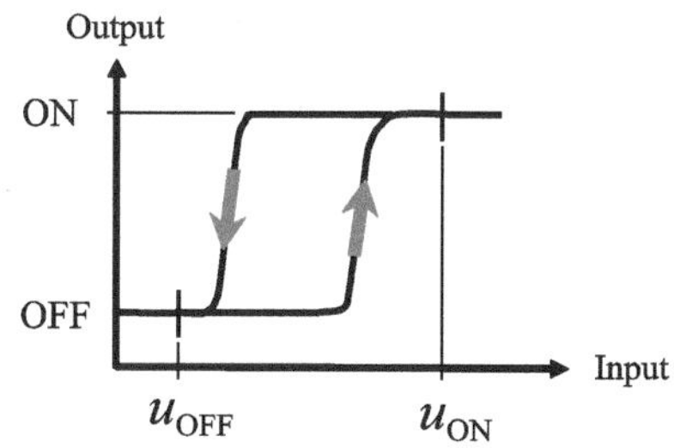

Figure III Bistable ON–OFF Control. © 2007 SAGE Publications

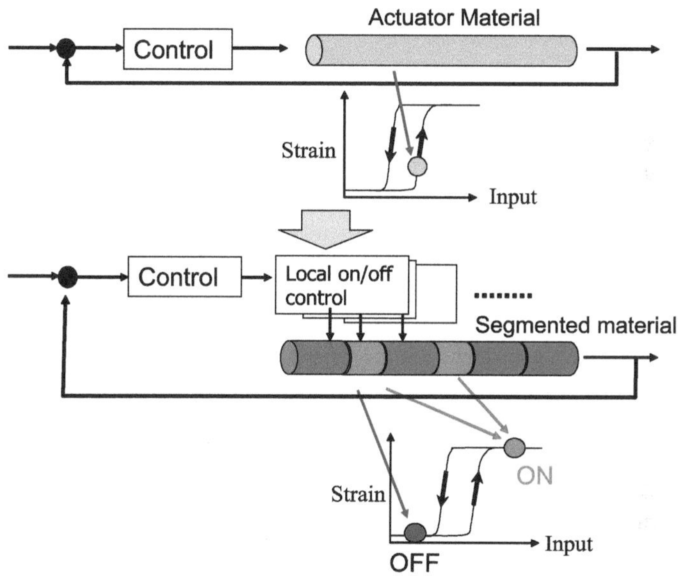

Figure IV Bistable ON–OFF control of actuator materials. © 2007 SAGE Publications, modified and reprinted with permission

control. Fig. VA illustrates a central controller directly controlling a number of individual cellular units through a communication line, e.g., a bus line. It is unlikely that biological muscle employs a traditional control architecture such as this. To each motor neuron, more than 1,500,000,000 sarcomeres (e.g., 150 fibers/neuron [243], 1000 myofibrils/fiber [261], and 10,000 sarcomeres/myofibril [166]; note that the numbers vary in the literature.) are connected, which are too many to communicate and control individually. Addressing all the cellular units for sending individual control commands entails long addressing bits, which eat up the channel capacity. Each motor neuron transmits a control signal from the central nervous system to a target muscle fiber. The control signal is then disseminated through a network of T tubules to a number of sarcoplasmic reticula, which activate a bundle of sarcomeres. From this observation, it can be hypothesized that the same information is "broadcasted" over a vast number of low-level units, at least to the level of sarcoplasmic reticula,

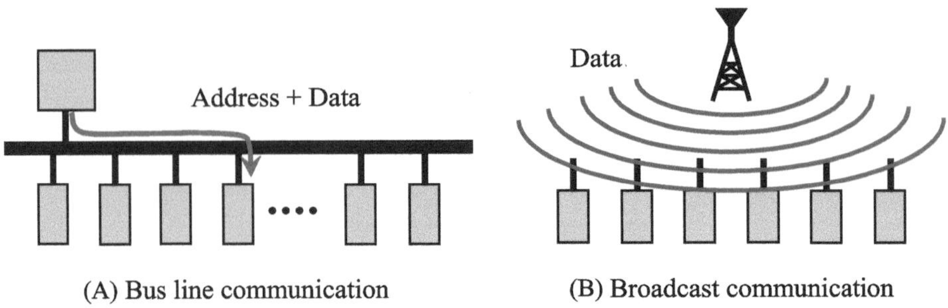

Figure V Communication between controller and cellular units.

rather than different information being delivered to individual contractile units. Thus, collections of muscle fibers can be grouped according to which motor neuron causes them to contract, rather than where they are geometrically located within a given fascicle. These are termed motor units [164]. Fig. VB illustrates the broadcast nature of communications between the motor neuron and the cellular units.

Decentralized stochastic recruitment. If the same information is broadcast to all the cellular units *and* each unit can take only the ON or OFF state, the consequence is that all the units turn ON or OFF at the same time. This would appear to contradict the fact in muscle physiology that an ensemble of sarcomeres can take multiple levels of excitation. This contradiction can be resolved if each cellular unit makes a stochastic decision in response to the broadcast information. Each sarcomere is activated with calcium ions through a diffusion process, which is a stochastic process. In other words, the sarcomere activation is stochastic, and the probability with which each sarcomere is activated depends on the ion density and diffusion characteristics. In the literature a number of groups have reported the stochastic nature of calcium release and recapture processes. Moreover, stochastic behavior can be observed at various motor control processes, ranging from motor unit firing [173] to actomyosin motors [134]. Molecular-level processes in particular, such as calcium release, breakdown of ATP, etc., are influenced by thermal noise resulting in stochastic behavior. Based on the above observations in the motor control, it may be hypothesized that the response of the cellular units are not uniform but have variations due to the stochasticity in the signal transduction processes. This indicates that even though the control command, or nerve impulse, is sent uniformly to all units, the response of all the units may not be the same. The model is depicted in Fig. VI. Stochastic decision-making at local units regulates the aggregate output of the ensemble units without deterministic coordination.

Binary control of an actuator array

The cellular PZT actuators introduced in Chapter 2 have an interesting artifact. By virtue of the layered amplification structure, each actuator will contain multiple PZT stacks. This is more than just a curiosity of manufacturing such an actuator. As will be seen, this architecture lends itself to a very convenient control method. Each piezo-

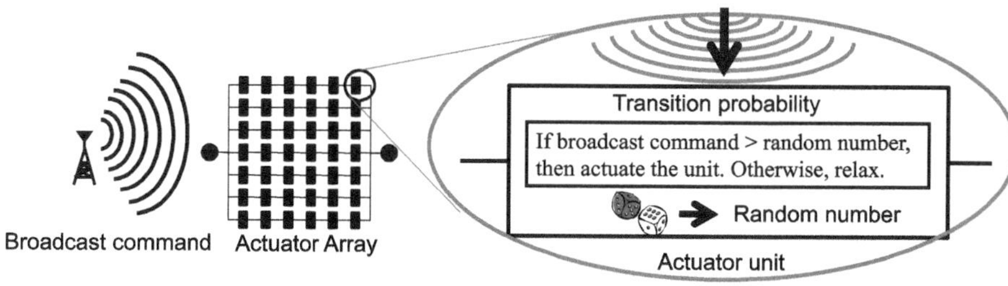

Figure VI Broadcast control and stochastic decision making.

electric stack can be activated completely independently of the others. This is a radical paradigm shift from most control systems; traditionally a control system receives a continuously variable analog signal as its command, corresponding to a valve spool displacement in a hydraulic system, or a motor current in a electrical servo system, for instance. In a cellular actuator architecture, the control input corresponds to how many PZT stacks should be in the "ON" state, which raises issues of quantization, and redundancy (commands with equivalent timings are non-unique).

Fig. VII shows the contrast in control paradigm. Each column shows progressively increasing levels of actuation effort and the size of the arrows shows the increase in force that results. Of course, a cellular actuator can be run just the same way as a traditional actuator is, simply by branching the control signal $u(t)$ to each cellular unit. This is represented by column A. A more biologically inspired paradigm is to take advantage of the cellular structure, and instead of using each and every unit partially, completely activate the number of units that produces the actuation effort desired. This is represented by column B. It does not take many units before the resolution is sufficient for most practical purposes [215]. This is how the human body produces muscle contractions; it activates *motor units* (collections of muscle fibers connected to the same neuron) completely to produce a contraction. In this oversimplified model, without rate coding or firing rate modulation, each motor unit can be "ON" or "OFF" – nothing in between. To increase the actuation effort, the nervous system must select another motor unit that is in the inactive state and bring it into the active state. This is known as *recruitment* [164,70].

Controlling the actuator through recruitment has several advantages, particularly with regard to piezoelectric cellular actuators. The actuators developed in Section 1.2 each contain several piezoelectric stacks, each of which can be activated independently, rendering it an architecture that works well within the recruitment paradigm. The first is with regard to what happens when the actuator sustains damage. The final row of Fig. VII shows a red × across one of the cells, indicating that it is damaged and is stuck in the inactive state. Using the traditional control paradigm, the control effort will always be less than commanded. Using recruitment, this is only a problem when that particular unit is the one recruited – the problem will be intermittent in nature. In the example, the damaged unit was not involved in the command (luckily) and the effect would not be noticed. There are other benefits to operating each motor unit (PZT stack) in ON–OFF mode only. It avoids hysteresis in the PZT material be-

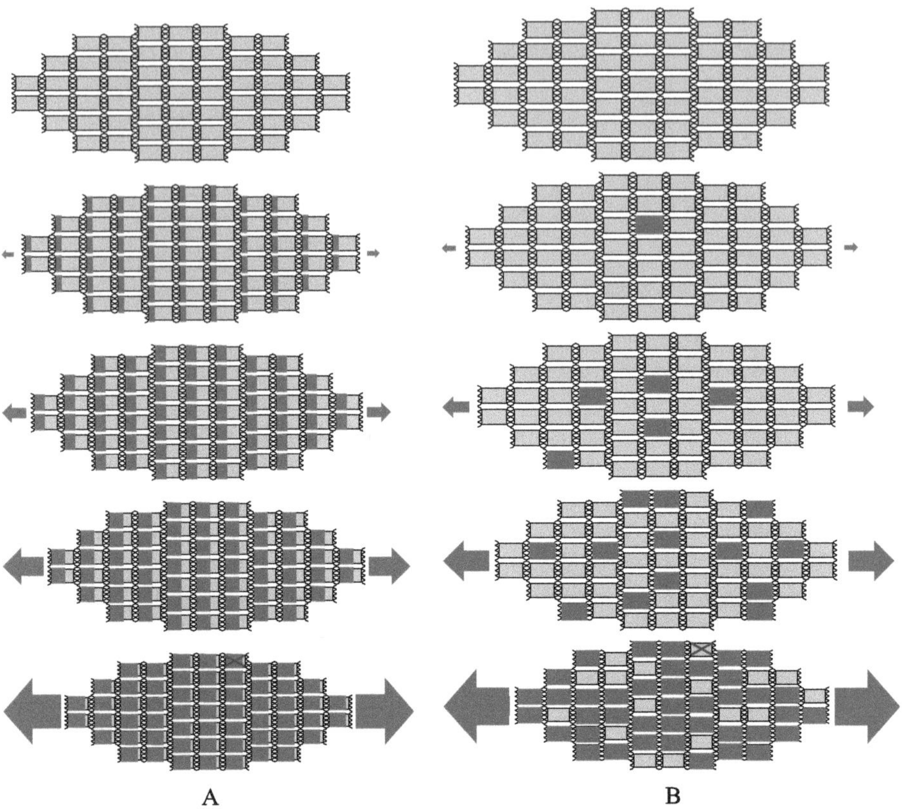

Figure VII Traditional control paradigm on a cellular actuator: (A) contrasted with a paradigm that operates by recruitment; (B) using recruitment takes advantage of the cellular structure. (For interpretation of the colors in this figure, the reader is referred to the web version of this Introduction.)

cause the only two points of the hysteresis loop used are the tips of the loop, which are unique as shown in Fig. III. In addition, ON–OFF operation simplifies the drive circuitry: each stack can be driven by a pair of switching transistors. There is no need for complicated drive electronics that operate in the linear region of the transistor. This may lead to energy savings in the drive electronics when compared to analog drivers.

Broadcast feedback with stochastic recruitment

Combining the above aspects inspired by skeletal muscles lead to the control architecture described here. Fig. VIII shows a novel concept of controlling a vast number of actuator units [263]. Instead of wiring many control lines to each individual cell, each cellular actuator has a stochastic local control unit that receives the broadcast signal from the central control unit and cycles its state in a simple ON–OFF manner. This stochastic coordination, called "stochastic recruitment," is inspired by the stochastic elements of muscle fiber (or motor unit) recruitment [134].

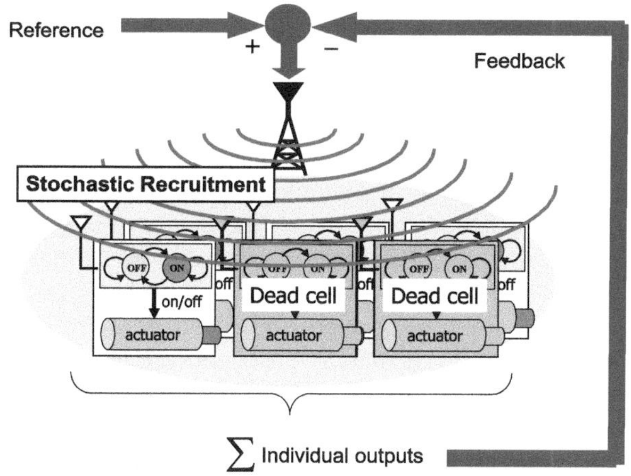

Figure VIII Broadcast feedback.

In the inspired actuator architecture, a control signal generated at a central controller is broadcasted from a broadcast station. Each cellular actuator has a decision-making unit that decides whether to accept or ignore the broadcast control command in a stochastic manner. The probability from one state to another can be determined by the external and internal factors, i.e., the amplitude of the broadcast command and/or current states. The aggregate outputs, i.e., muscle displacement, force, etc., are detected by sensors. Note that this system architecture is not intended to fully mimic true biological muscle control. Highly complex neurological and biochemical processes involved in the multi-layer muscle hierarchy are ignored; the whole system is reduced to just a two-layer distributed stochastic control system. Instead, this model is to manifest how the aggregate output of a number of cellular units can be controlled although the number of independent units is numerous. To verify that this control architecture functions properly and in a stable manner, a precise mathematical description and rigorous analysis are needed, which are the focus of Section 3.2. Further details of the binary cellular actuator control will be discussed in Sections 3.3 and 3.4.

Discussion

This chapter has presented new architecture for robot actuators inspired by muscle behavior, which in turn has the potential to be a novel approach to the control of a vast number of cellular units. A simple ON–OFF control is enough for individual cellular actuator units since the net effect upon the output displacement and force is merely the summation and average of the total PZT stack actuators. Therefore, the local control unit controls the cellular actuator unit in an ON–OFF manner, which overcomes the hysteresis of the material and simplifies the amplifier. This cellular architecture has the following advantages:

1. **Reconfigurability.** The number of stacks and bundles for the final layer can be configured relatively freely, thus enabling high reconfigurability. By applying broadcast control, complex rewiring is not necessary. This feature is discussed in Section 1.2.
2. **Adjustability.** As described in Section 1.3, the shape and size as well as force–displacement relationship are adjustable by tuning design parameters.
3. **Sustainability.** The cellular actuator is robust to the failure of cells as well as to non-uniformity of displacement and/or transition probabilities of individual cells. The robustness of the cellular control with stochastic control is presented in Section 3.2. Even in the presence of these problems, which are considered unavoidable in micro manufacturing, the aggregate output of the cellular units converge to a reference robustly by merely broadcasting the aggregate output error.
4. **Controllability.** Bistable ON–OFF control does not depend on hysteresis, as long as the binary control is applied. This will be discussed in Sections 3.3 and 3.4. In addition, as the size of cellular units reduces, the speed of response increases for those actuator materials that entail transport of matter. For example, thin film SMA for a micro actuator [79] has a small amount of thermal capacitance, thus the response time is substantially reduced.

Structure of cellular actuators

1

1.1 Strain amplified piezoelectric actuators

1.1.1 Piezoelectric materials

Piezoelectric materials have crystal structures such that an electric field is generated by mechanical deformation. Conversely, an applied electric field will cause a mechanical deformation. These effects are known as the direct and converse piezoelectric effects [126]. One of the most widely used models for piezoelectric actuators comes from IEEE Standard 176. The electromechanical coupling is described in tensor notation [126]. Electric displacement and strain are assumed to be linearly dependent on electric field strength and stress. However, the range in which the linear model is accurate is quite small, as piezoceramics are known to exhibit pronounced hysteresis. Subsequently, more complex models have been developed, such as the one proposed by Goldfarb and Celanovic that is now widely in use [87]. Since significant hysteresis has been observed in piezoelectrically driven actuators, this model is used instead of the simpler linear model.

A schematic representation of the model is shown in Fig. 1.1. The model describes the hysteric behavior of the actuator in addition to the electrical–mechanical interplay caused the direct and converse piezoelectric effects. The model is described by Eqs. (1.1) through (1.4):

$$q_{pzt} = \alpha x_{pzt} + C_{pzt} V_{pzt}, \tag{1.1}$$

$$V_H = H(q_{pzt}), \tag{1.2}$$

$$\frac{x_{pzt}}{F + \alpha V_{pzt}} = \frac{1}{k_{pzt}}, \tag{1.3}$$

$$V_{total} = V_{pzt} + V_H, \tag{1.4}$$

where q_{pzt} is electric charge, V_{total} is the input voltage supplied to the actuator, α is the transformer ratio of the actuator with units of N/V, C_{pzt} is the clamped capacitance of the actuator, F is the external force, and k_{pzt} is the short circuit stiffness of the actuator. $H(q_{pzt})$ is a hysteresis operator that can be calibrated and modeled by phenomenological models [140,139] introduced in Appendix A.1.

1.1.2 Strain amplification mechanisms

Actuators will require a basic active unit that contracts in response to an electrical stimulus like human muscle. Numerous active materials are available that contract under the application of an electric field. Lead zirconate titanate (PZT) piezoelectric actuators exhibit a two-way interaction with the load, like biological muscle. A piezoelectric ceramic generates a voltage in response to a mechanical stress (the direct piezoelectric effect) or a mechanical strain when an electric field is applied (the

Cellular Actuators, DOI: 10.1016/B978-0-12-803687-7.00001-6
Copyright © 2017 Elsevier Inc. All rights reserved.

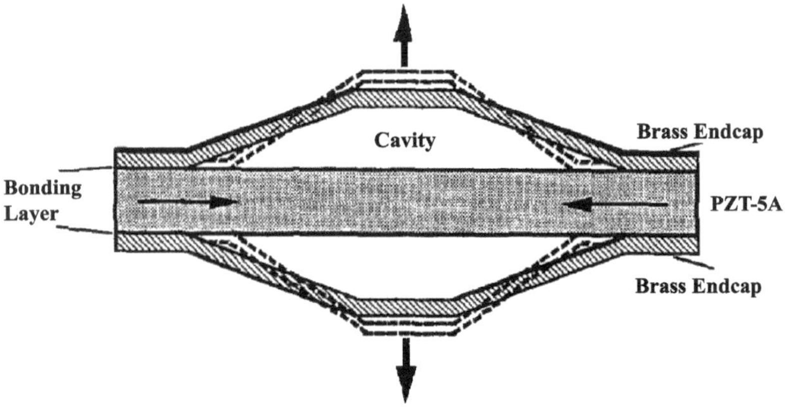

Figure 1.5 Cymbal actuator [60]. © 1997 IEEE, reprinted with permission

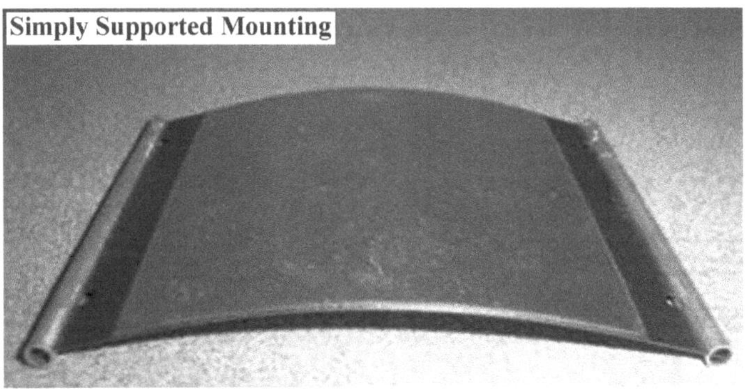

Figure 1.6 Thunder actuator [46]. Courtesy of Face International Corporation.

film PZT actuator is a better alternative since it has good compatibility with most MEMS processes along with their high energy density (1.2×10^7 J/m^3), compactness, fast dynamic response time, low input power, and low driving voltage. The small strain of PZT, however, has limited its wide use in many MEMS applications. A typical PZT actuator has a strain limit on the order of 0.1%. For a cellular actuator where a maximum strain of 10% might be desired, a PZT micro-actuator would need a means of amplification of its strain.

An extensile cellular PZT actuator has been developed by using a compact stroke amplifying mechanism [44] and shown in Fig. 1.7. The device was fabricated using an SU-8 structure as an amplifying mechanism for a thin-film PZT beam sandwiched between top and bottom electrodes. The amplifying structure is in the shape of a bow-tie so that the extension of the PZT member along the y-axis produces the extension of the amplification structure along the x-axis. Each actuator "cell" can be arrayed in series and/or in parallel, and the design can accommodate different force/displacement requirements. Fabricated devices consisting of three actuator cells in series showed a strain amplification ratio in excess of 10:1 per cell, which gave about 1% strain. The

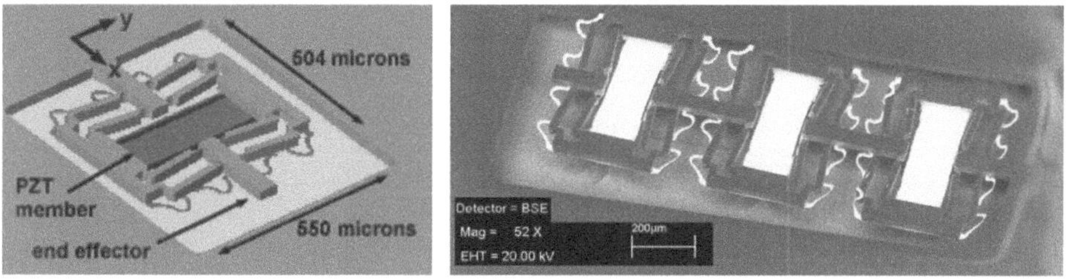

(A) (B)

Figure 1.7 Extensile PZT strain amplifier: (A) single-cell extending PZT actuator model, (B) fabricated three cell PZT actuator. © 2006 IEEE, reprinted with permission

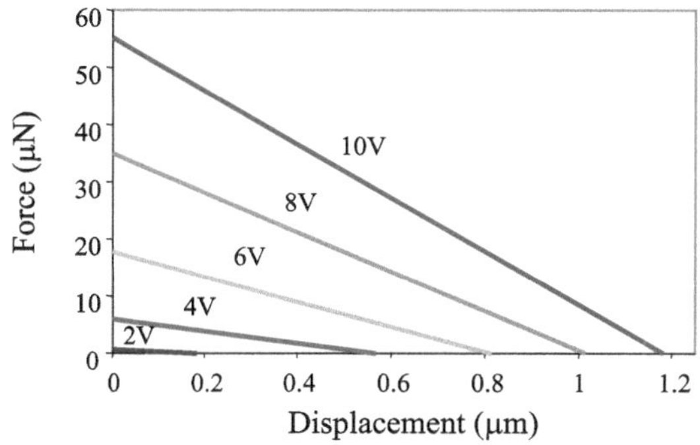

Figure 1.8 Force–displacement curve of the expanding PZT actuator. © 2006 IEEE, reprinted with permission

amplified strain ϵ is given as follows:

$$\epsilon = \frac{\Delta L}{l} = \frac{a\epsilon_{pzt}W}{l} \tag{1.5}$$

where ΔL is the displacement along the x-axis, ϵ_{pzt} is the strain achieved by PZT film, l (= 504 µm) is the length, and W (= 550 µm) is the width of the cell. a (= 10.0) is an amplification ratio by this amplifying mechanism, and W/l is the aspect ratio of the cell. Fig. 1.8 shows the measured force–displacement characterization, with a maximum blocked force of 55 µN and free displacement of 1.18 µm at 10 V. This actuator has a second-order resonance at 5.7 Hz.

An MEMS amplifier with a large stroke of strain of 10% or more for a thin-film PZT actuator can be designed (Fig. 1.9A) [271]. The required large strain is realized by modifying both the amplifying mechanism and the aspect ratio. The two side beams are bowed inward by 2 degrees, which will give displacements as the PZT beam activates at the center. A thin-film PZT beam (0.5 µm thick, 80 µm wide) contracts when an electric field up to 10 V is applied to top and bottom electrodes. The expected maximum strain of each cell is about 7% with the device. Fig. 1.9B shows a conceptual module which has serially connected and stacked PZT cells.

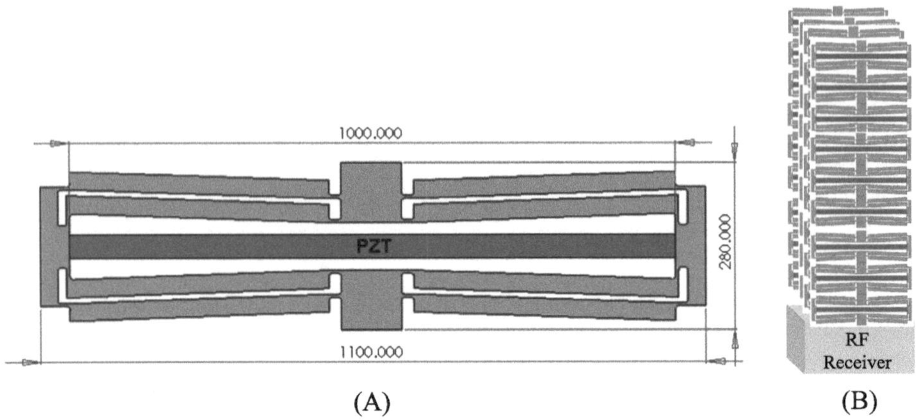

Figure 1.9 MEMS–PZT cellular actuator: (A) a large strain contracting PZT actuator cell design, (B) serially stacked and connected PZT actuator cells into a module. © 2006 IEEE

1.1.4 Discussion

As presented in this chapter, modular versions of amplified piezoelectric stacks using a rhomboidal architecture have been actively studied and some of them are commercially available. Given that natural muscle can produce more than 20% of strain, although there is extensive research on PZT material to amplify its strain, the presented designs fall short of the current macro-scale robotics applications; even with an amplification factor of 10–20 [186,45,60,116,182], the devices only displace a few microns. To break this barrier, the book introduces new designs of strain amplification mechanisms in Sections 1.2, 1.3, 4.1, and 4.2.

1.2 Nested rhombus exponential strain amplification

1.2.1 Large effective strain piezoelectric actuators

Despite their energy efficiency, piezoelectric stacks using multiple piezoelectric layers are of little use in robotics because they have a very small stroke length. A promising technique is the use of amplifying mechanisms to increase the displacement. In this approach, a piezoelectric stack imposes a force–displacement boundary condition on a strategically chosen point (the input) of a linearly elastic deformable body. The geometric properties of the body make the displacement at another point (the output) much larger than the displacement of the piezoelectric stack. Some of the electrical energy supplied to the piezoelectric stack is stored in the deformable body as strain energy. In some sense, this is viewed as a parasitic effect, but a properly-designed control system can utilize this stored energy to power subsequent motions.

The proposed strain amplification approach to amplify small displacements by placing several amplified stacks in series and amplifying the output of this series

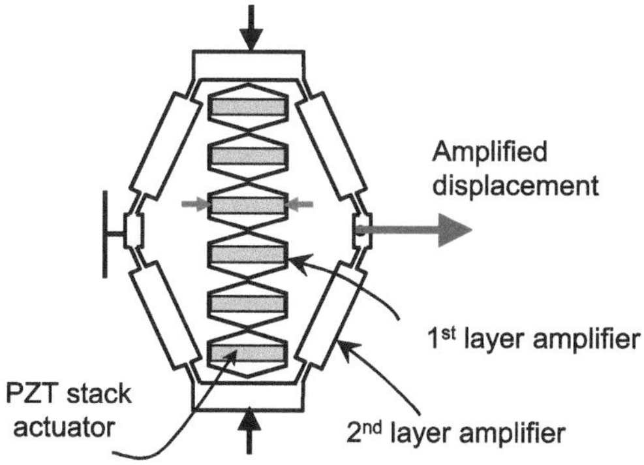

Figure 1.10 Nested structure for exponential strain amplification.

combination a second time [265,266,264,267]. This new structure, named a nested rhombus multi-layer mechanism [267], amplifies the displacement of PZT actuators to produce 20–30% effective strain, which is comparable to natural skeletal muscles. The key idea is to hierarchically enclose smaller flextensional actuators with larger amplifying structures, achieving exponential strain amplification. Note that this concept is fundamentally different from traditional layered structures, such as telescoping cylindrical units [183] or stacking multiple plates connected by actuator wires [3]. Unlike these traditional stacking mechanism, where the amplification gain is proportional to the dimension of the lever or number of stacks, the amplification gain of the new mechanism increases exponentially as the number of layers increases.

Fig. 1.10 shows such a device that shows a vertical column of PZT stacks with standard amplifiers in the "first layer." The "second layer" has an additional rhomboidal amplifying mechanism. This multi-stage amplification technique produces strains on the order of human muscle, approximately 22%. The proposed actuator is a modular device, and multiple units can be bundled, much like muscle cells. For this reason, actuators of this type are termed cellular actuators because many of them can be connected in series and parallel combinations to achieve desired actuation characteristics.

1.2.2 Rhombus strain amplification mechanisms

This section briefly describes the kinematics of traditional "Moonie" flextensional mechanisms [182] that will be used for the basis of our proposed "nested rhombus" multi-layer mechanism. As shown in Fig. 1.11A, the main part of the mechanism is a rhombus-like hexagon that contracts vertically as the internal unit shown in gray expands. The vertical displacement, that is, the output of the mechanism, is amplified if the angle of the oblique beams to the horizontal line is less than 45 degrees. Fig. 1.11B illustrates how the strain is amplified with this mechanism. Let h_1, w_1, and ϵ_0 respectively be the height, width, and strain of the internal unit. Also let d_1 be the initial gap between the surface of the internal unit and the apex of the rhombus mechanism. In this section it is assumed that all the joints are purely rotational and that all the beams are completely rigid. Also assume that the internal unit is extensible; it can be

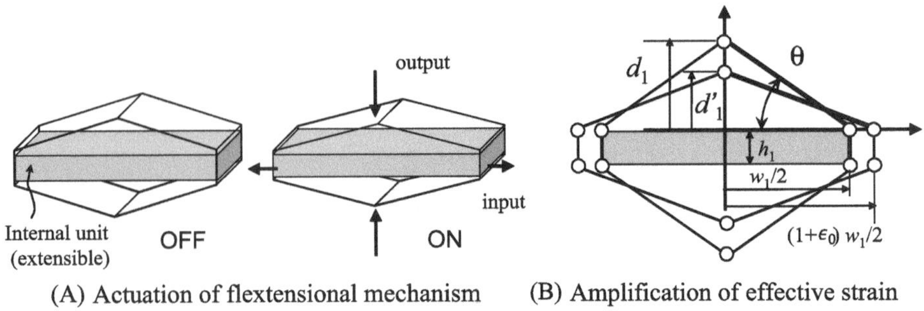

Figure 1.11 Amplification principle of flextensional mechanisms [182]. © 2010 IEEE, reprinted with permission

extended to a contractive case with minor changes to the following formulation. As the internal unit, e.g., a PZT stack actuator, expands, the gap d_1 contracts to d'_1:

$$d'_1 = \sqrt{d_1^2 - (\epsilon_0^2 + 2\epsilon_0)w_1^2/4}. \tag{1.6}$$

Then, the amplification gain a_1 of the displacement is given by

$$a_1 = \frac{2\Delta x_1}{\epsilon_0 w_1} \tag{1.7}$$

where $\Delta x_1 \triangleq d_1 - d'_1$. For small ϵ_0 this can be approximated to $a_1 = \frac{w_1}{2d_1} = \cot\theta$ where θ is the angle of the oblique beam to the horizontal line, as shown in the figure. Note that this instantaneous amplification gain does not apply to large strain because of the nonlinearity in (1.6). A smaller value for the angle of the oblique beams θ gives a larger amplification gain. However, the angle θ needs to be carefully determined to avoid buckling of the beams due to unexpected external forces. As will be explained below, buckling of the beams cannot be neglected in the mechanical design since the working direction alternates from layer to layer in the proposed nested structure. Typically this amplification gain alone can increase displacement to only $3 \sim 5$ times larger.

The initial length of the rhombus mechanism measures $2d_1 + h_1$ along the output axis. Since the displacement created in this output direction is $2\Delta x_1$, the "effective strain" along the output axis can be defined as

$$\epsilon_1 \triangleq \frac{2|\Delta x_1|}{2d_1 + h_1}. \tag{1.8}$$

Note that the lateral size, perpendicular to the output direction, is not included in the definition of the effective strain. Comparing this to the input strain ϵ_0 yields the strain amplification gain given by

$$\kappa_1 \triangleq \frac{\epsilon_1}{\epsilon_0} = a_1 \frac{w_1}{2d_1 + h_1} \tag{1.9}$$

where $w_1/(2d_1 + h_1)$ is the ratio of the width to the height of the rhombus, i.e., the aspect ratio of the mechanism. Note that both the displacement amplification and the aspect ratio of the mechanism contribute to the resultant strain amplification κ_1. Strictly speaking, the aspect ratio is not a strain amplifier. However, since (i) the effective strain amplification is defined to be the ratio of output displacement to the natural body length in the same direction as the output, and (ii) the direction of input strain and that of the output displacement are perpendicular to each other, the effective gain κ_1 is apparently amplified by the aspect ratio. Increasing the aspect ratio increases the strain amplification gain κ_1. However, space constraints must be considered since a larger aspect ratio increases the lateral size of the actuator that would affect on force capacity. Typically κ is from 3 to 20 for practical applications.

1.2.3 Nested rhombus structure

The above mechanism for amplifying small displacements of PZT actuators have already been developed both in macro [182] and micro scale [45] and have been applied to commercial products [2]. Our method is to extend this technique to: (i) gain an order-of-magnitude larger strain amplification, and (ii) build a modular structure that is flexible and extensible.

Fig. 1.12 illustrates a new mechanism, called a "nested rhombus" strain amplifier, which consists of a multitude of rhombus mechanisms arranged in a hierarchical structure. The inner-most unit, i.e., the building block of the hierarchical system, is the standard rhombus mechanism, or conventional flextensional mechanism, described above. These units are connected in series to increase the output displacement. Also these units can be arranged in parallel to increase the output force. The salient feature of this hierarchical mechanism is that these rhombus units are enclosed with a larger rhombus mechanism that amplifies the total displacement of the smaller rhombus units. These larger rhombus units are connected together and enclosed with an even larger rhombus structure to further amplify the total displacement. Note that the working direction alternates from layer to layer, i.e., the second layer rhombus extends when the inner-most first layer units contract as shown in Fig. 1.11B.

As this enclosure and amplification process is repeated, a multi-layer strain-amplification mechanism is constructed, and the resultant displacement increases exponentially. Let K be the number of amplification layers. Assuming that each layer amplifies the strain κ times, the resultant amplification gain is given by κ to the *power* of K:

$$\kappa_{total} = \kappa^K. \tag{1.10}$$

For $\kappa = 15$ the gain becomes $\kappa_{total} = 225$ by nesting two rhombus layers and $\kappa_{total} = 3375$ with three rhombus layers. The nested rhombus mechanism with this hierarchical structure is a powerful tool for gaining an order-of-magnitude larger amplification of strain. As described before, our immediate goal is to produce 20% strain. This goal can be accomplished with $\kappa = 15$ and $K = 2$ as $0.1\% \times 15 \times 15 = 22.5\%$.

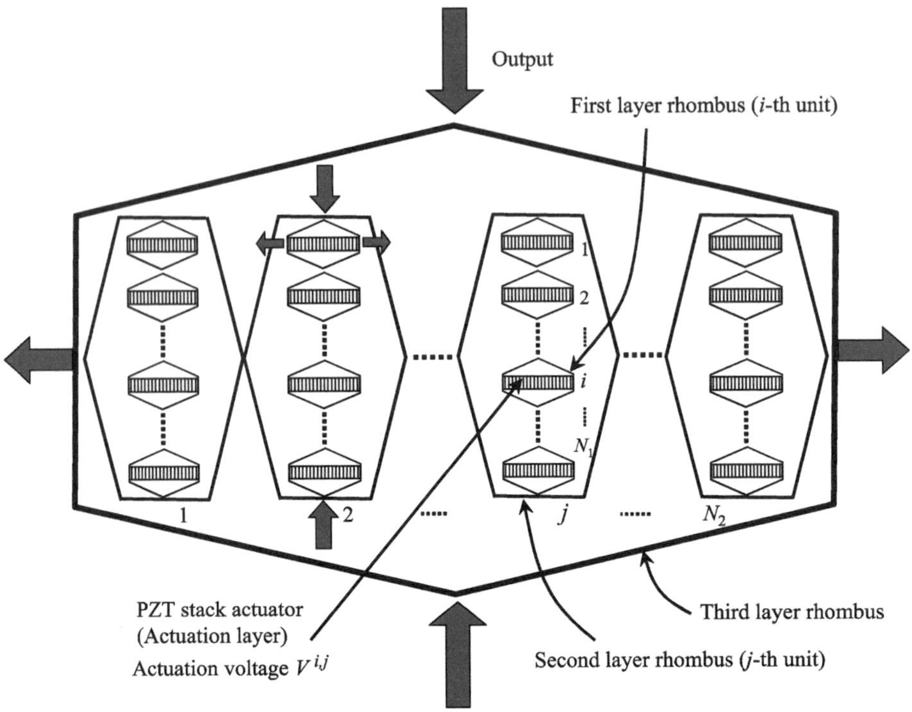

Figure 1.12 Generalized nesting for exponential strain amplification. The strain is amplified by three layers of rhombus strain amplification mechanisms, with the first layer, called an actuator layer, consisting of the smallest rhombi directly attached to the individual PZT stack actuators. © 2010 IEEE, reprinted with permission

This nested rhombus mechanism has a number of variations, depending on the numbers of serial and parallel units arranged in each layer and the effective gain in each layer. In general the resultant amplification gain is given by the multiplication of each layer gain, $\kappa_{total} = \prod_{k=1}^{K} \kappa_k$ where $\kappa_k \triangleq \epsilon_k/\epsilon_{k-1}$ is the kth layer's effective gain of strain amplification computed recursively with the following formula:

$$\epsilon_k = \frac{2d_k - \sqrt{4d_k^2 - (\epsilon_{k-1}^2 + 2\epsilon_{k-1})w_k^2}}{2d_k + h_k} \quad (k=1, \ldots, K). \tag{1.11}$$

Another important feature of the nested rhombus mechanism is that two planes of rhombi in different layers may be arranged perpendicular to each other. This allows us to construct three-dimensional structures with diverse configurations. For simplicity, the schematic diagram in Fig. 1.12 shows only a two-dimensional configuration, but the actual mechanism is three-dimensional, with output axes being perpendicular to the plane. Three-dimensional arrangement of nested rhombus mechanisms allows us to densely enclose many rhombus units in a limited space. For example, Fig. 1.13 illustrates a 3-dimensional structure. Note that the serially connected first-layer rhom-

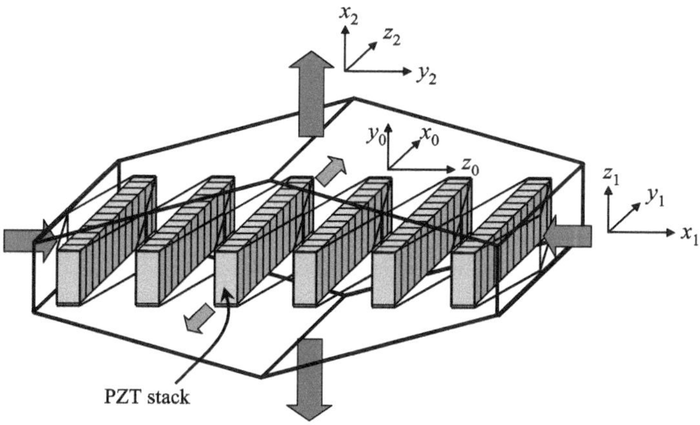

Figure 1.13 Three-dimensional nesting for 20% strain. © 2010 IEEE, reprinted with permission

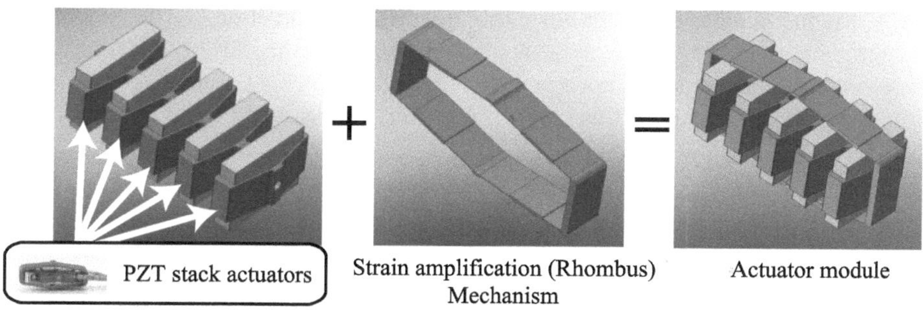

Figure 1.14 Schematic assembly of nested rhombus multi-layer mechanism. © 2013 Springer, reprinted with permission

bus units are rotated 90 degrees about their output axis x_1. This makes the rhombus mechanism at the second layer more compact; the length in the x_2 direction is reduced. Namely, the height h_1 in Fig. 1.11B, which is a non-functional dimension for strain amplification, can be reduced. These size reductions allow us not only to pack many PZT units densely but also increase the effective strain along the output axis, ϵ_1, since h_1 is involved in the denominator of (1.8). A schematic assembly process of the proposed structure is shown in Fig. 1.14. A series of piezoelectric actuators with the "Moonie"-type strain amplification mechanisms are connected (in Fig. 1.14, five actuator units are serially connected) and nested in a larger rhombus amplification mechanism.

Furthermore, it should be noted that diverse configurations can be made simply by changing the serial and parallel arrangements of the same building blocks. This modular design is a powerful method for building diverse actuators with different strokes and forces matched with the load impedance.

1.2.4 Properties of ideal nested rhombus PZT actuators

In the nested rhombus PZT actuator, displacements of the individual PZT actuators are aggregated and transmitted through the multiple layers of strain amplification mechanisms, resulting in an output displacement at the final layer. Similarly, the output force is the resultant force of many PZT actuators. In this section, these aggregate force and displacement are analyzed in relation to the individual PZT actuator outputs based on an ideal kinematic and static model of the nested rhombus mechanism.

Consider a PZT stack actuator shown in Fig. 1.15. Let l_{pzt}, w_{pzt}, and h_{pzt} be the length, width, and height of a PZT stack actuator, respectively. The x-axis is defined as the actuation direction. The choice of y- and z-axes is arbitrary. For descriptive purposes, the y-axis is chosen in the direction of w_{pzt} as shown in Fig. 1.15. The displacement of this PZT stack actuator when no load is connected to the actuator is given by

$$\Delta x_{pzt} = N_{film} \cdot d_{33} \cdot V, \tag{1.12}$$

where N_{film} is the number of PZT films along the actuation direction, d_{33} is piezoelectric coefficient, and $V(> 0)$ is voltage applied to each PZT film. Strictly speaking, the piezoelectric coefficient d_{33} is not a constant; according to [171], it may vary significantly as strain gets larger. In this chapter, however, it is assumed to be constant. The inherent stiffness of this actuator is given by $k_{pzt} = \frac{E_{pzt} h_{pzt} w_{pzt}}{l_{pzt}}$, where E_{pzt} is the elastic modulus of PZT material.

The no-load displacement given by (1.12) results from the balance between the net force f_{pzt} produced by the PZT and the restoring force due to the stiffness k_{pzt}, which is proportional to Δx_{pzt}. Unlike standard electromagnetic actuators, e.g., DC and AC motors, PZT and other actuator materials cannot produce force independent of their displacement. Due to the inherent structural stiffness, the net output force of these actuator materials gets substantially lower when producing a displacement at the same time.

Consider the following force–displacement relationship where the force generated by the PZT stack actuator while producing displacement Δx_{pzt} is given by

$$f_{pzt} = k_{pzt}(\chi V - \Delta x_{pzt}) \tag{1.13}$$

with $\chi = N_{film} d_{33}$. As this PZT stack is imbedded in a first rhombus mechanism, the force is reduced to $1/a_1$ and the displacement is amplified a_1 times, i.e., $f_1 = f_{pzt}/a_1$ and $\Delta x_1 = a_1 \Delta x_{pzt}$. Assuming that the rhombus mechanism is lossless and that the beams are completely rigid and are connected with free joints, the force–displacement relationship at the output axis of the first layer rhombus mechanism is given by

$$f_1 = \frac{f_{pzt}}{a_1} = \frac{k_{pzt}}{a_1} \chi V - \frac{k_{pzt}}{(a_1)^2} \Delta x_1. \tag{1.14}$$

Note that the equivalent stiffness of the PZT stack viewed from the output side of the rhombus mechanism is attenuated by a factor of $1/(a_1)^2$.

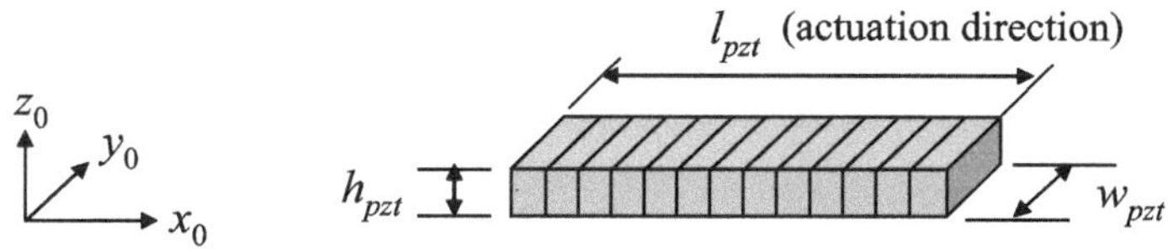

Figure 1.15 Actuator coordinate system of PZT stack actuator. © 2010 IEEE, reprinted with permission

Suppose that N_1 units of this first layer are connected in series and are enclosed with a second-layer mechanism. Each unit is numbered from 1 to N_1. Parallel connection in a layer is not considered since it forms a closed kinematic chain for ideal rhombus mechanisms and solving the kinematic chain problem is not essential. Let V^i, f_1^i, and Δx_1^i $(i = 1, \ldots, N_1)$ respectively be the voltage, force, and displacement of the ith unit in the serial connection of the first layers. The force is common to all the N_1 units:

$$f_1^1 = f_1^2 = \cdots = f_1^{N_1} \triangleq f_1^{com}. \tag{1.15}$$

From (1.14), we have

$$\frac{k_{pzt}}{a_1}\chi V^i - \frac{k_{pzt}}{(a_1)^2}\Delta x_1^i = f_1^{com} \quad (i = 1, \ldots, N_1). \tag{1.16}$$

Suppose that the second rhombus mechanism amplifies displacement and attenuates force a_2 times. The resulting displacement at this layer is given by

$$\Delta x_1^1 + \cdots + \Delta x_1^{N_1} = \frac{\Delta x_2}{a_2}. \tag{1.17}$$

From (1.16) and (1.17), the relationship between the output force and displacement for the second layer is given by

$$f_2 = \frac{f_1^{com}}{a_2} = \frac{\chi k_{pzt}}{N_1 a_1 a_2}\sum_{i=1}^{N_1} V^i - \frac{k_{pzt}}{N_1 (a_1)^2 (a_2)^2}\Delta x_2. \tag{1.18}$$

Repeating the same process yields a relation between the aggregate displacement and force along the Kth layer output axis given by

$$\begin{aligned} f_K &= \frac{\chi k_{pzt}}{\prod_{k=1}^{K} a_k} \cdot \frac{1}{\prod_{k=1}^{K-1} N_k} \sum_{1}^{N_{K-1}} \cdots \sum_{j=1}^{N_2} \sum_{i=1}^{N_1} V^{i,j,\cdots} \\ &\quad - \frac{k_{pzt}}{\prod_{k=1}^{K-1} N_k \prod_{k=1}^{K} (a_k)^2} \Delta x_K. \end{aligned} \tag{1.19}$$

In total, $N_{K-1} \cdot N_{K-2} \cdots N_1$ PZT units are involved in the system, and $V^{i,j,\cdots}$ in the above equation represents the voltage applied to each individual PZT actuator. See

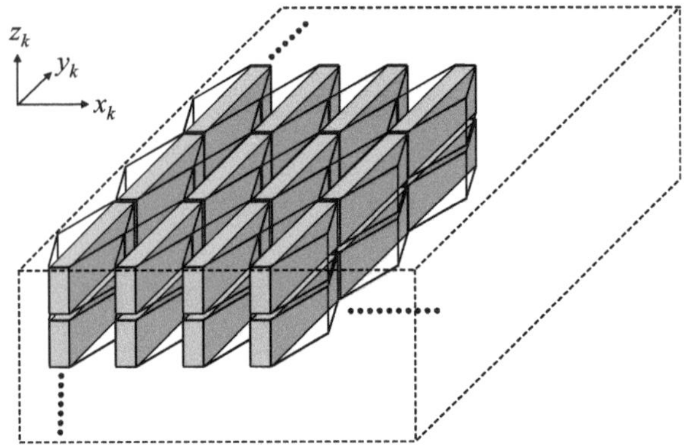

Figure 1.16 Three-dimensional stacking of actuator units. © 2007 IEEE, reprinted with permission

Fig. 1.12 for $K = 3$ where $V^{i,\,j}$ is applied to the ith PZT unit in the first layer involved in the jth unit ($j = 1, \ldots, N_2$) of the second layer.

As shown in Fig. 1.16, each layer can be configured by connecting the units in serial and parallel. The nesting level K has to be chosen in terms of desired direction of actuation, e.g., K needs to be odd for contractive actuation if the actuation layer is extensible.

From the above results we can note that:

1. Given applied voltages, the maximum of the aggregate displacement is obtained when no force is generated, i.e., under free load. This aggregate free-load displacement Δx_K^{free} is proportional to the total sum of the inputs $\sum_1^{N_{K-1}} \cdots \sum_1^{N_2} \sum_1^{N_1} V^{i,\,j,\cdots}$ amplified by a factor of $\prod_{k=1}^{K} a_k$.

2. The maximum of the aggregate force is obtained when the output displacement is totally blocked. This aggregate blocking force f_K^{block} is proportional to the average of the entire inputs, namely $\frac{1}{\prod_{k=1}^{K-1} N_k} \sum_1^{N_{K-1}} \cdots \sum_1^{N_2} \sum_1^{N_1} V^{i,\,j,\cdots}$.

If the total number of PZT actuators is very large, the individual PZT stack actuators can be driven with simple ON–OFF controls [270,263], since the net effect upon the output displacement and force is the sum and average of many PZT actuators. Expensive analogue drivers and controllers are unnecessary for the cellular actuators. As the number of PZT actuator units increases, discretization error becomes small and smooth output displacement and force can be expected.

1.2.5 Feasibility check for 20% effective strain

Fig. 1.13 also illustrates a design example aiming at 20% effective strain. In this design example, 6 first-layer rhombus units are connected in series. As described in the

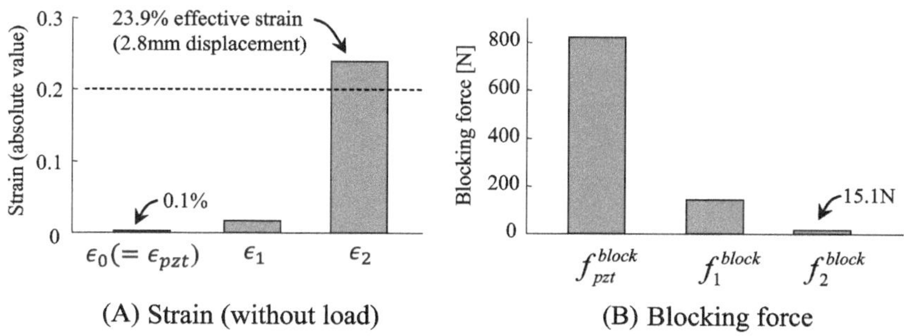

Figure 1.17 Idealized analysis. Compliance of the amplification mechanism is not considered. © 2010 IEEE, reprinted with permission

previous section, the 3-dimensional structure plays a key role for large strain. The serially connected units are rotated 90 degrees and inserted into the second layer rhombus. By this design, the second layer rhombus extends when the PZT stack actuators are turned ON since they are extensible and the number of amplifying layers is 2.

Let the size of the PZT stack actuator be 12.8 [mm] (l_{pzt}) $\times$ 6 [mm] (w_{pzt}) $\times$ 2.5 [mm] (h_{pzt}). The initial gap, d_1, between the surface of the PZT stack actuator and the apex of the first rhombus mechanism is 1.1 [mm]. We apply typical values of PZT-ceramics for Young's modulus and strain, i.e., $E_{pzt} = 55.0$ [GPa], and $\epsilon_{pzt}(= \epsilon_0) = 0.1\%$. These dimensional parameters have been determined according to a commercially available PZT actuator, Cedrat APA50XS [2], as the first layer unit. The size of the second layer is 12.0 [mm] (length, actuation direction) $\times$ 28.2 [mm] (width) $\times$ 12.8 [mm] (height). The thicknesses of the amplifying mechanisms and connection parts between the units have been neglected for simplicity.

From iterative calculations of (1.14) and (1.18), the amplified strain and reduced blocking force are obtained as shown in Fig. 1.17. The prospective displacement is 2.8 mm for the actuator length of 12 mm, which is equivalent to $\epsilon_2 = 23.9\%$. This result implies that over 20% stain is feasible by the proposed nested structure. Also, the resultant blocking force is predicted to be 15.1 N.

1.2.6 Discussion

This chapter has presented a nested rhombus structure for PZT actuators. The nesting notational convention and design parameters have been defined. The idealized analysis has been given for fundamental design of the nested structure. A design example has shown that over 20% of effective strain is feasible for PZT stack actuators.

The proposed strain amplification method is essentially a trade-off free displacement and blocked force. The compliance of the amplification mechanisms must be considered, otherwise a designed structure would not produce an indented strain. This issue is discussed in the following chapters.

The modular design of the PZT cellular actuator allows a wide variety of configurations using the designed actuator as a building-block. Also as shown in Fig. 1.18,

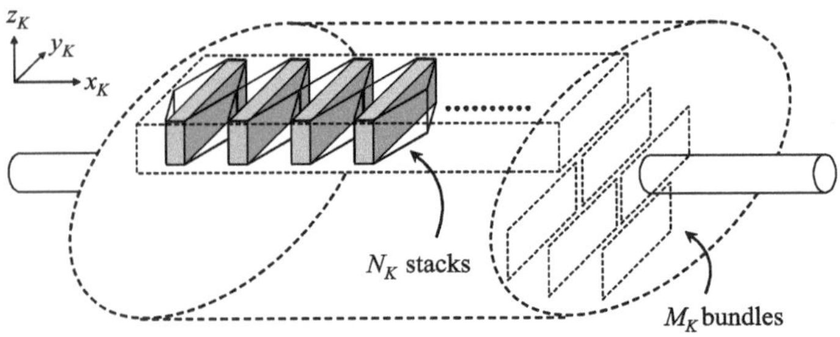

Figure 1.18 Final layer connection. © 2007 IEEE, reprinted with permission

(A) Actuator Array: 6 nested structures are connected in series

(B) 12 stacks and 4 bundles

(C) 6 stacks and 7 bundles

Figure 1.19 Reconfigurability of the cellular actuators. © 2007 IEEE, reprinted with permission

the final actuator may be configured by connecting N_K units of the final layers in serial and M_K units in parallel based on required specifications of performance as well as spacial constraints. Fig. 1.19A shows a prototype actuator array of 6 units, which simply increases the displacement. The number of stacks and bundles are determined according to a specific application. Fig. 1.19B shows an example configuration where

12 cells are connected in series and 4 arrays are connected in parallel. These arrays are easily reconfigurable by changing the connectors; Fig. 1.19C shows a configuration for larger force and shorter displacement.

1.3 Design of nested-rhombus cellular actuators

1.3.1 Nested rhombus mechanisms with structural flexibility

The initial design presented in Section 1.2 was developed based on the ideal kinematic model having rigid beams and free joints at the strain amplification mechanism. Actual mechanisms, however, inevitably have some compliance at the structure, which may degrade the aggregate force and displacement. Not only the compliance of piezoelectric stack actuators but also the compliance of the amplification structures need to be taken into account in designing the nested strain amplification mechanism to minimize its adverse effect.

This section deals with static analysis of the nested rhombus mechanisms in order to investigate fundamental characteristics between the strain and stiffness. The effect of mass distribution and damping, which plays essential role in dynamic characteristics, will be neglected. Dynamic analysis and synthesis are reported in [221]. The main goal of this section is to understand physical limitations and design trade-offs of the nested rhombus mechanisms. Meeting a static performance goal is the primary concern in designing a strain amplifier of this kind.

1.3.1.1 Effects of joint stiffness and beam compliance

Consider a spring load serially connected to a PZT stack actuator, as shown in Fig. 1.20. Let k_{load} be a spring constant of the load, and Δx_{pzt} be the displacement of the load. The following equations hold:

$$f_{pzt} = (k_{load} + k_{pzt})\Delta x_{pzt}, \tag{1.20}$$

$$f = k_{load}\Delta x_{pzt}, \tag{1.21}$$

where f is the actuator output force applied to the load. The free-load displacement Δx_{pzt}^{free} is calculated by letting $k_{load} = 0$.

Eliminating Δx_{pzt} yields

$$f = \frac{k_{load}}{k_{pzt} + k_{load}} f_{pzt}. \tag{1.22}$$

Note that the actuator output force f becomes significantly lower than the original PZT force f_{pzt} when k_{pzt} gets larger or k_{load} gets lower. Similarly, the output displacement Δx_{pzt} also gets attenuated:

$$\Delta x_{pzt} = \frac{k_{pzt}}{k_{pzt} + k_{load}} \Delta x_{pzt}^{free}. \tag{1.23}$$

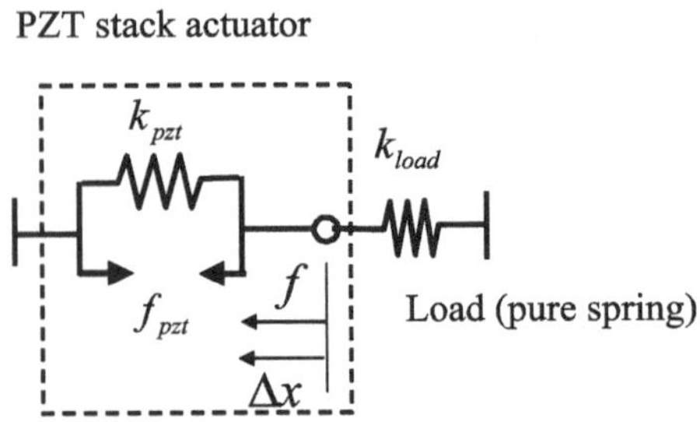

Figure 1.20 Model of PZT stack actuator connected to a spring load.

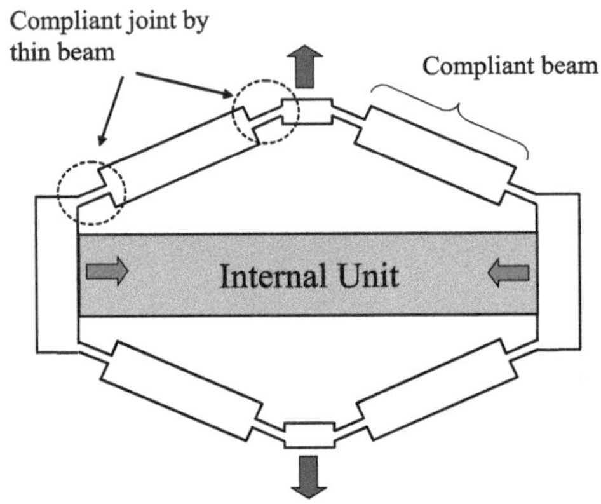

Figure 1.21 Embodiment of a rhombus mechanism.

The simple model described above shows that both the output force and displacement are attenuated due to the compliance of the connected load as well as the stiffness of the actuator itself. When this PZT stack actuator is connected to a rhombus strain amplification mechanism, an external load having properties similar to the above k_{load} and k_{pzt} will be imposed on the PZT actuator. As many layers of the amplification mechanism are attached to the PZT stack actuator, these structural effects will be even more prominent. In the ideal mechanism shown in Fig. 1.11B, it is assumed that the four beams of the rhombus are completely rigid and that all the joints are free to rotate and purely revolving. However, these assumptions do not hold in real structures.

Note that fabrication of free joints is difficult in small scale due to mechanical tolerance and play. For the first and second layers, in particular, where the displacement is extremely small, the displacement created by the PZT is likely to diminish due to play at the joints. Therefore, flexural pivots and flexible beams [45,2,116] have been used for amplifying PZT displacement. Fig. 1.21 shows an example embodiment of the rhombus mechanism. These flexural joints and beams inevitably bring undesirable properties to the system. There are three types of undesirable properties:

1. First, the joints are no longer free, but they impose a spring load that the PZT has to overcome. Fig. 1.22 depicts this parasitic effect of joint stiffness; some fraction of the PZT force is wasted for coping with the joint stiffness. This results in a reduction

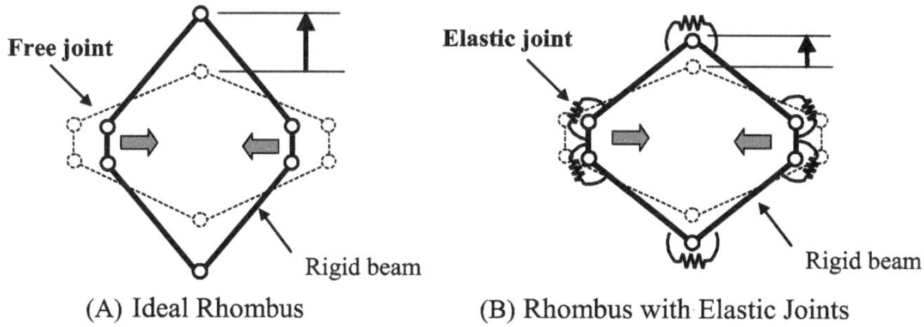

Figure 1.22 Effect of joint compliance on free-load displacement.

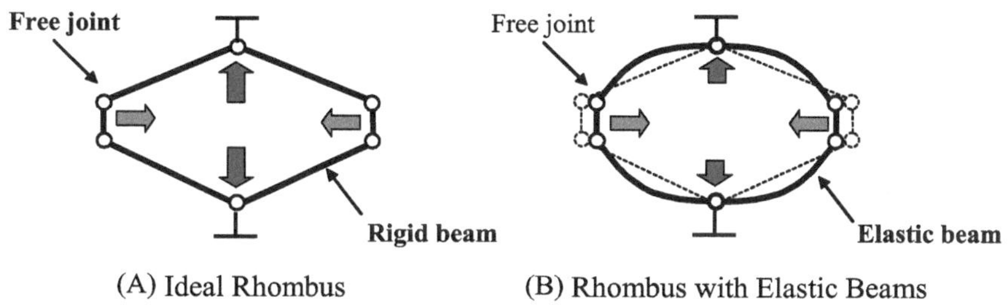

Figure 1.23 Effect of beam compliance on blocking force.

of free-load displacement. This implies that the joint stiffness has an equivalent effect to that of the PZT stiffness k_{pzt} in Fig. 1.20. The stiffness of the joints brings increased stiffness k_{pzt} for the PZT to overcome.

2. Second, flexibility at the beams may attenuate the displacement and force created by the PZT. Consider the case where the output displacement is blocked as shown in Fig. 1.23. As the PZT generates a displacement, the beams are deformed and thereby the transmitted force becomes lower; at least it does not reach the same level as that of the rigid beams. Similarly, if the output axis is coupled to another compliant load, the output force and displacement will be prorated between the load compliance and the beam compliance. As the beam stiffness becomes lower, the output force and displacement decrease.

3. Third, flexural joints not only create pure rotational displacements, but also often cause unwanted translational displacements. These elastic deformations at the joint along the direction of the beam incur the same problem as the beam compliance; the force and displacement created by the PZT tend to diminish at the joints.

It is important to distinguish two different types of compliance in the above cases: one is to take place in the kinematically admissible space of the ideal rhombus mechanism, and the other is in the orthogonal complement to the former, termed the constrained space [11]. The joint stiffness described in item 1 is in the admissible motion space, while items 2 and 3 are in the constrained space. Curved beams, such as seen in Moonies, contain compliance in both constrained and admissible spaces. The distributed compliance can be approximated into the two types of lumped compliant elements. To minimize the adverse effects of the nested rhombus mechanism, the stiffness in the admissible space must be minimized and the stiffness in the constrained space must be maximized. As multiple layers of strain amplification mechanisms are used, the compliances in the admissible and constrained spaces become more intricate. In the following sections the kinematic and static characteristics of multi-layer compliant rhombus mechanisms will be analyzed.

1.3.1.2 Modeling of single-layer flexible rhombus mechanisms

Consider the case shown in Fig. 1.24A where a rhombus mechanism, including Moonies, is connected to a spring load. k_{load} is an elastic modulus of the load, and k_{pzt} is an elastic modulus of the internal PZT stack actuator. Δx_{pzt} is the displacement of the internal unit, and f_{pzt} is the force applied to the amplification mechanism from the internal unit. f_1 is the force applied to the load from the actuator, and Δx_1 is the displacement of the load. In this figure, the internal unit is assumed to be contractive for later convenience.

The rhombus strain amplification mechanism is a two-port compliance element, whose constitutive law is defined by

$$\begin{bmatrix} f_I \\ f_O \end{bmatrix} = \mathbf{S} \begin{bmatrix} \Delta x_{pzt} \\ \Delta x_1 \end{bmatrix} \tag{1.24}$$

where $\mathbf{S}^{2\times 2} = \begin{bmatrix} s_1 & s_3 \\ s_3 & s_2 \end{bmatrix}$ is a stiffness matrix. f_I is the net force applied to the mechanism from the internal unit, and f_O is the reaction force from the external load. Note that the stiffness matrix $\mathbf{S}$ is non-singular, symmetric, and positive-definite; $s_1 > 0$, $s_2 > 0$, and $s_1 s_2 - s_3^2 > 0$. The symmetric nature of the stiffness matrix follows Castigliano's theorems. When the input port is connected to a PZT stack actuator producing force f_{pzt} with inherent stiffness k_{pzt} and the output port is connected to a load of stiffness k_{load}, we have

$$f_I = f_{pzt} - k_{pzt}\Delta x_{pzt} = s_1 \Delta x_{pzt} + s_3 \Delta x_1, \tag{1.25}$$

$$f_O = -f_1 = -k_{load}\Delta x_1 = s_3 \Delta x_{pzt} + s_2 \Delta x_1. \tag{1.26}$$

Eliminating Δx_{pzt} from the above equations yields

$$f_{pzt} = -\left(\frac{k_{pzt} + s_1}{s_3} k_{load} + \frac{s_2(k_{pzt} + s_1) - s_3^2}{s_3} \right) \Delta x_1. \tag{1.27}$$

Defining

$$\tilde{f} \triangleq \frac{-s_3}{k_{pzt} + s_1} f_{pzt}, \tag{1.28}$$

$$\tilde{k} \triangleq \frac{s_2(k_{pzt} + s_1) - s_3^2}{k_{pzt} + s_1} = \frac{s_2 k_{pzt} + \det\mathbf{S}}{k_{pzt} + s_1} > 0, \tag{1.29}$$

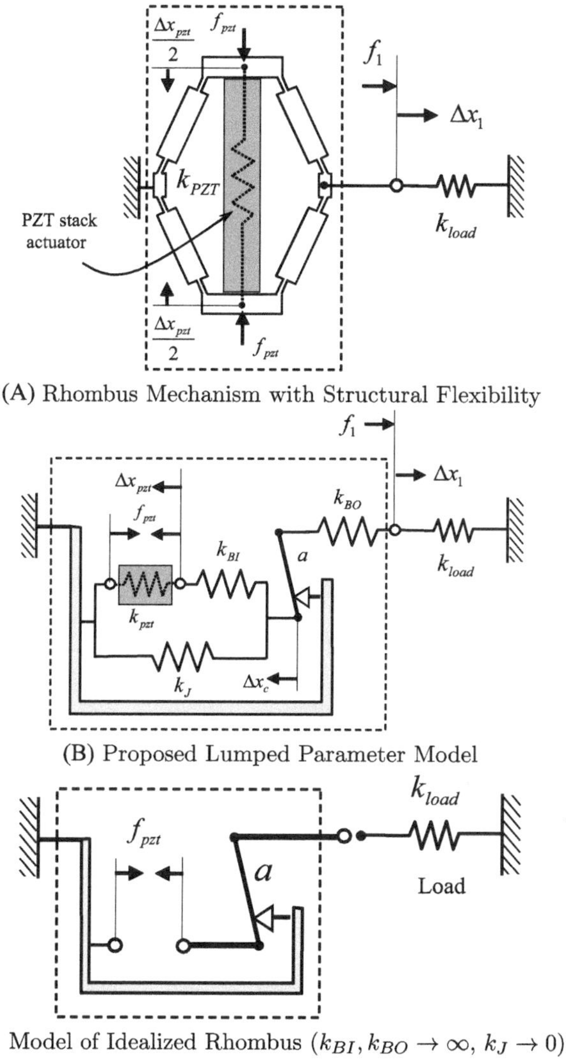

Figure 1.24 Model of Rhombus Mechanism with Flexibility.

the above equation (1.27) reduces to

$$\tilde{f} = (k_{load} + \tilde{k})\Delta x_1. \tag{1.30}$$

Force $\tilde{f}$ and stiffness $\tilde{k}$ represent the effective PZT force and the resultant stiffness of the PZT stack, all viewed from the output port of the amplification mechanism.

A drawback with the above two-port model representation is that it is hard to gain physical insights as to which elements degrade actuator performance and how to improve it through design. In the previous section, two distinct compliances were introduced, one in the admissible motion space and the other in the constrained space. To improve performance with respect to the output force and displacement, the stiffness in the admissible motion space must be minimized, while the one in the constrained space must be maximized. To manifest these structural compliances, we propose a lumped parameter model shown in Fig. 1.24B with 3 spring elements, k_J, k_{BI}, and k_{BO}, and one amplification leverage a. As the spring constants, k_{BI} and k_{BO}, tend to infinity, the system reduces to the one consisting of all rigid links, where the output Δx_1 is directly proportional to the input displacement Δx_{pzt}. Stiffness k_J impedes this rigid body motion, representing the stiffness in the admissible motion space. Elastic deformation at k_{BI} and k_{BO} represent deviation from the rigid body motion.

From Fig. 1.24B,

$$f_{pzt} + k_{BI}(\Delta x_c - \Delta x_{pzt}) - k_{pzt}\Delta x_{pzt} = 0, \tag{1.31}$$

$$a \cdot k_{BO}(a \cdot \Delta x_c - \Delta x_1) + k_J \Delta x_c + k_{BI}(\Delta x_c - \Delta x_{pzt}) = 0, \tag{1.32}$$

$$f_1 = k_{load}\Delta x_1 = k_{BO}(a \cdot \Delta x_c - \Delta x_1), \tag{1.33}$$

where Δx_c is the displacement at the connecting point between the leverage and springs; however, this point is virtual, and Δx_c does not correspond to a physical displacement. This model is applicable to a wide variety of "rhombus-type" amplification mechanisms, including Moonies.

Consider the blocking force when the PZT stack actuator generates its maximum force, $f_{pzt\ \max}$, given as follows:

$$f_1^{block} = \frac{a k_{BI} k_{BO}}{(a^2 k_{BI} k_{BO} + k_{BI} k_J) + k_{pzt}(a^2 k_{BO} + k_J + k_{BI})} f_{pzt\ \max}. \tag{1.34}$$

Similarly, the free-load displacement for this rhombus mechanism, where $k_{load} \to 0$, is given by

$$\Delta x_1^{free} = \frac{a k_{BI}}{k_{pzt}(k_{BI} + k_J) + k_J k_{BI}} f_{pzt\ \max}. \tag{1.35}$$

As addressed above, these equations imply that the blocking force will be maximized as $k_{BI}, k_{BO} \to \infty$. Similarly, $k_J \to 0$ maximizes Δx_1^{free}.

The other advantage is that the 3-spring model is able to represent the ideal rhombus shown in Fig. 1.23A as a special case as shown in Fig. 1.24C. See (1.47) and confirm that the stiffness matrix $\boldsymbol{S}$ cannot be defined for both $k_{BI}, k_{BO} \to \infty$ and $k_J \to 0$. As described in Section 1.3.2, the number of unknown parameters becomes 4, as the rigid amplification leverage is explicitly included, which makes the calibration problem ill-posed; however, this amplification leverage is necessary to include the ideal case. In addition, 3 lumped springs are considered minimum to satisfy the input–output bidirectionality, which is a basic requirement of Castigliano's theorems.

1.3.1.3 Model simplification

From Eqs. (1.31) to (1.33), the relationship between f_{pzt} and Δx_1 is given by

$$\begin{aligned}(a\ k_{BI}k_{BO})f_{pzt} = \\ \Big[k_{load}\{a^2k_{BI}k_{BO} + k_{BI}k_J + k_{pzt}(a^2k_{BO} + k_J + k_{BI})\} \\ + k_{BO}(k_{BI}k_J + k_{pzt}k_J + k_{BI}k_{pzt})\Big]\,\Delta x_1.\end{aligned} \tag{1.36}$$

The above equation can be written as

$$\tilde{f}_1 = (k_{load} + \tilde{k}_1)\Delta x_1 \tag{1.37}$$

where

$$\tilde{k}_1 \triangleq \frac{k_{BO}(k_{BI}k_J + k_{pzt}k_J + k_{BI}k_{pzt})}{(a^2k_{BI}k_{BO} + k_{BI}k_J) + k_{pzt}(a^2k_{BO} + k_J + k_{BI})}, \tag{1.38}$$

$$\tilde{f}_1 \triangleq \frac{ak_{BI}k_{BO}}{(a^2k_{BI}k_{BO} + k_{BI}k_J) + k_{pzt}(a^2k_{BO} + k_J + k_{BI})}f_{pzt}, \tag{1.39}$$

implying that the proposed lumped parameter model shown in Fig. 1.24B can be further simplified to Fig. 1.25. Note that the direction of $\tilde{f}_1$ is opposite to f_{pzt} due to the amplification leverage. This simplified model has a similar form as in (1.30). As will be described in the following section, this simplification enables performance evaluation of complex nested mechanisms simply by nesting a simplified model of lower layers into a higher lumped parameter model. As a result, the performance of the overall mechanism such as aggregate displacement and force can be predicted in a recursive manner without developing a full finite element method (FEM) model [233] or solving continuum elastic equations [232]; parameter calibration at the layer-level is sufficient.

1.3.1.4 Recursive formula of aggregate force and displacement of flexible nested mechanisms

The goal of this section is to describe a recursive formula to obtain an equivalent model for a general nested mechanism. Fig. 1.26 shows a nested rhombus structure. As addressed in the previous sections, each nested layer can be represented by its equivalent model, which enables us to describe the force–displacement property for the nested structure in an iterative manner. Let K be the number of nesting layers. Also, let k_{Jk}, k_{BIk}, k_{BOk}, and N_k be the joint compliance, beam compliances, and the number of serial connection for the kth ($k = 1, \dots K$) layer. This mechanism involves $N_{K-1} \cdot N_{K-2} \cdots N_1$ PZT stack actuators.

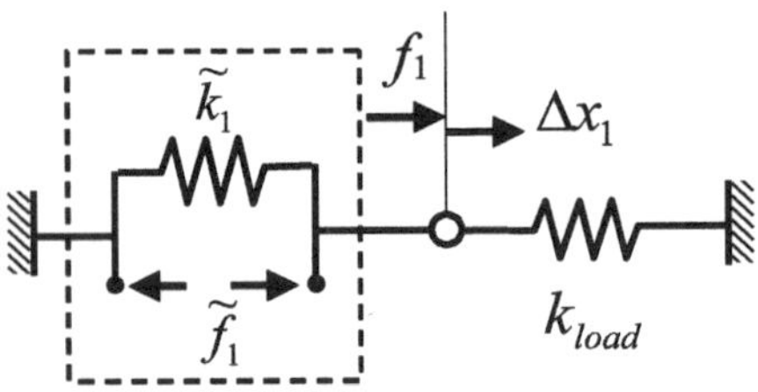

Figure 1.25 Simplified Representation of Lumped Parameter Model.

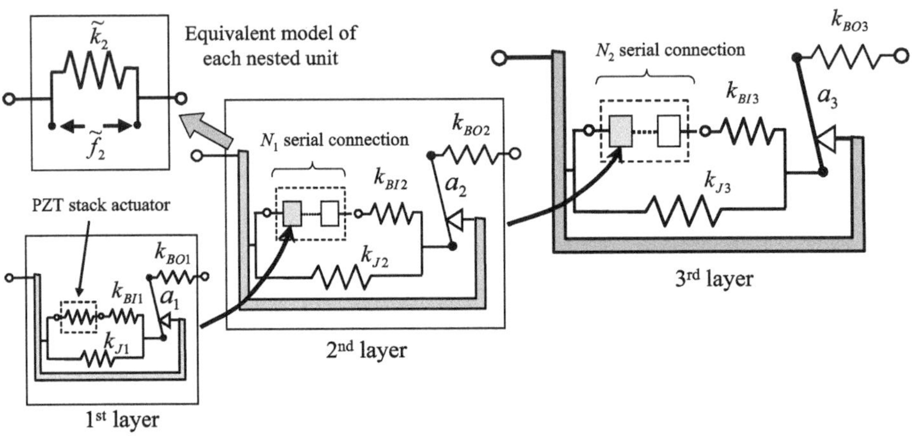

Figure 1.26 Recursive formula for a nested rhombus model.

By applying (1.38) and (1.39), the equivalent model for the kth layer can be represented by

$$\tilde{k}_k = \frac{k_{BOk}(k_{BIk}k_{Jk} + \frac{\tilde{k}_{k-1}}{N_{k-1}}k_{Jk} + k_{BIk}\frac{\tilde{k}_{k-1}}{N_{k-1}})}{(a_k^2 k_{BIk}k_{BOk} + k_{BIk}k_{Jk}) + \frac{\tilde{k}_{k-1}}{N_{k-1}}(a_k^2 k_{BOk} + k_{Jk} + k_{BIk})}, \tag{1.40}$$

$$\tilde{f}_k = \frac{a_k k_{BIk}k_{BOk}}{(a_k^2 k_{BIk}k_{BOk} + k_{BIk}k_{Jk}) + \frac{\tilde{k}_{k-1}}{N_{k-1}}(a_k^2 k_{BOk} + k_{Jk} + k_{BIk})} \times \frac{1}{N_{k-1}} \sum_{i=1}^{N_{k-1}} \tilde{f}_{k-1}^i, \tag{1.41}$$

where $\tilde{f}_{k-1}^i$ is the equivalent force of the ith unit in the $(k-1)$th layer. Recall that $\tilde{f}_k$ is proportional to the average of the entire forces at the $(k-1)$th layer as described in the previous section.

Assume that all actuators in the $(k-1)$th layer are controlled in a binary manner [263], i.e., $V_{k-1}^i = \{V_{\max}\ (\text{ON}),\ 0\ (\text{OFF})\}$. Also assume that all units are uniform and that each unit generates $\tilde{f}_{k-1}^{block}$ as its blocking force. Therefore, when N_{k-1}^{ON} units out of N_{k-1} actuators are ON, the last term of (1.41) can be replaced as

$$\frac{1}{N_{k-1}} \sum_{i=1}^{N_{k-1}} \tilde{f}^i_{k-1} \to \frac{N^{ON}_{k-1}}{N_{k-1}} \tilde{f}^{block}_{k-1}. \tag{1.42}$$

The free-load displacement changes accordingly. Both the aggregate free-load displacement and the blocking force are proportional to the number of ON units.

1.3.2 Verification and calibration of 3-spring lumped parameter model

The validity of the proposed lumped parameter model is confirmed by FEM. Consider the two amplification mechanisms shown in Fig. 1.27, for example. The size of the both mechanisms is: 40 [mm] (length, actuation direction) × 96 [mm] (width) × 5 [mm] (thickness). Brass (Young's modulus = 100.0 [GPa]) is used. The four structural lumped parameters, i.e., a, k_{BI}, k_{BO}, and k_J, are calibrated by the displacements and forces from two different conditions; one case is "blocked case" where the output displacement is totally constrained (or letting $k_{load} \to \infty$), and the other one is "free-load case" (or letting $k_{load} = 0$).

By applying an input force, f_{pzt}, Δx^{block}_{pzt}, and f^{block}_1 are measured for the blocked case, and Δx^{free}_{pzt} and Δx^{free}_1 are measured for the free-load case. From (1.31) to (1.33), we have

$$\frac{f^{block}_{pzt}}{\Delta x^{block}_{pzt}} = \frac{a^2 k_{BI} k_{BO} + k_J k_{BI}}{a^2 k_{BO} + k_J + k_{BI}} \triangleq X_1, \tag{1.43}$$

$$\frac{f^{block}_1}{\Delta x^{block}_{pzt}} = \frac{a k_{BI} k_{BO}}{a^2 k_{BO} + k_J + k_{BI}} \triangleq X_2, \tag{1.44}$$

$$\frac{\Delta x^{free}_1}{\Delta x^{free}_{pzt}} = \frac{a\, k_{BI}}{k_J + k_{BI}} \triangleq X_3, \tag{1.45}$$

$$\frac{f^{free}_{pzt}}{\Delta x^{free}_1} = \frac{k_J}{a} \triangleq X_4. \tag{1.46}$$

Note that the actual number of independent equations described above is 3, which can be confirmed by $X_1 = X_3(X_2 + X_4)$. This implies that the calibration of the four structural parameters, $[a, k_{BI}, k_{BO}, k_J]$, is an ill-posed problem. This can be confirmed by the two-port model representation in (1.24). The stiffness matrix $\boldsymbol{S}$ is generally given as

$$\boldsymbol{S} = \begin{bmatrix} X_1 & -X_2 \\ -X_2 & \frac{X_2}{X_3} \end{bmatrix}. \tag{1.47}$$

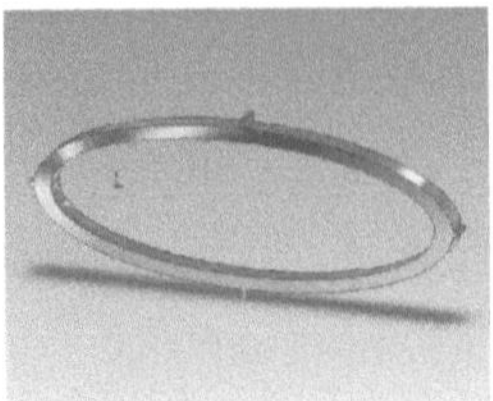

(A) Structure 1

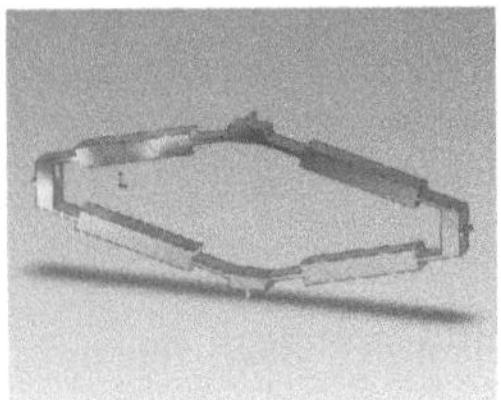

(B) Structure 2

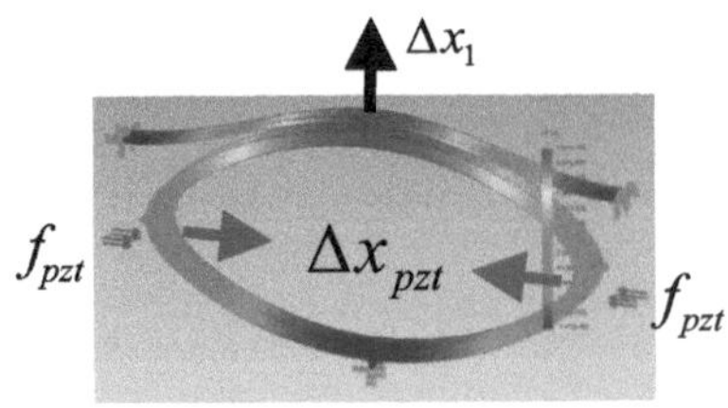

(C) Displacement when connected to a fixed beam

Figure 1.27 Example amplification mechanisms.

Recall that $\boldsymbol{S} = \boldsymbol{S}^T$ and that it fully represents the relation between the displacements and forces. Therefore, the number of independent elements is 3 by calibrating $\boldsymbol{S}$. Unlike the ideal rhombus mechanism consisting of all rigid links, the displacement amplification gain a cannot be defined uniquely as long as the stiffness in the constrained space is finite, i.e., $k_J > 0$. Note that the choice of a does not change $\boldsymbol{S}$ or the characteristics of the estimated model; however, a nominal gain $\hat{a}$ should be determined to have a physically feasible lumped parameter model, that is, $k_{BI}, k_{BO}, k_J > 0$. One way of determining $\hat{a}$ is based on free-displacement characteristics and kinematic characteristics of the structure such as the angle of the oblique beam θ, i.e., $X_3 < \hat{a} < \cot\theta$, to satisfy the requirement. X_3 can be assumed as a lower bound of $\hat{a}$ since X_3 is always lower than the actual a if k_J is positive. In addition, $\cot\theta$ can be assumed as an upper bound of $\hat{a}$ since this gain is realized only when $k_J = 0$. The following steps estimate the remaining parameters: $\hat{k}_J = \hat{a}\ X_4$, $\hat{k}_{BI} = \frac{X_3\hat{k}_J}{\hat{a}-X_3}$, and $\hat{k}_{BO} = \frac{(\hat{k}_{BI}+\hat{k}_J)X_2}{\hat{a}\hat{k}_{BI}-\hat{a}^2X_2}$.

Table 1.1 shows the observed values from FEM when applying f_{pzt}=10 [N]. The structural lumped parameters are calculated as shown in Table 1.2. The nominal amplification gains are determined accordingly based on the observed X_3's and kinematic characteristics to keep all spring constants positive. As shown in Table 1.1, structure 2 provides approximately 13 times larger free-load displacement than structure 1, while the blocking forces of the two structures are almost the same magnitude. This observation suggests that structure 2 has a more favorable structure than structure 1 as an amplification mechanism. This can be explained based on the estimated lumped parameters: The effective stiffness in the constrained space $\bar{k_B}$ viewed from the input port is calculated as

$$\bar{k_B} \triangleq \frac{1}{\frac{1}{k_{BI}} + \frac{1}{a^2 k_{BO}}}. \tag{1.48}$$

This expression gives $\bar{k_B} = 6.18\text{e+}06$ for structure 1 and $\bar{k_B} = 1.23\text{e+}07$ for structure 2. As a result, structure 2 has a smaller stiffness in the admissible space, k_J, and a larger stiffness in the constrained space, $\bar{k_B}$, compared with structure 1. Although structure 2 provides relatively good performance, it could also have problems in development due to its complex shape and in strength due to stress concentration at thin

Table 1.1 Observed values from FEM.

	Structure 1	Structure 2
Δx_1^{free} [m] (@ $f_{pzt} = 10$)	1.95e–05	2.61e–04
f_1^{block} [N] (@ $f_{pzt} = 10$)	2.73	2.51
X_1	6.64e+06	1.25e+07
X_2	1.81e+06	3.14e+06
X_3	2.88	3.89
X_4	5.13e+05	3.84e+04

Table 1.2 Estimated lumped parameters.

	Structure 1	Structure 2
$\hat{a}$	3.2	3.9
$\hat{k}_J$ [N/m]	1.64e+06	1.50e+05
$\hat{k}_{BI}$ [N/m]	1.46e+07	4.25e+07
$\hat{k}_{BO}$ [N/m]	1.05e+06	1.13e+06
Maximum stress [N/m^2]	5.35e+06	2.75e+07

Table 1.3 Estimated displacements.

		Structure 1			Structure 2		
Thickness [mm]	Disp. [µm]	FEM	Model	Error [%]	FEM	Model	Error [%]
1	Δx_{pzt}	6.48	6.49	0.035	36.76	37.20	1.184
	Δx_1	18.40	18.42	0.172	141.5	143.1	1.163
2	Δx_{pzt}	5.10	5.11	0.187	9.47	9.57	1.098
	Δx_1	13.28	13.34	0.467	34.09	34.46	1.098
3	Δx_{pzt}	3.56	3.56	0.051	3.65	3.67	0.492
	Δx_1	7.61	7.63	0.301	11.23	11.30	0.267
Ave. [%]				0.202			0.884

sections having large deformation. The maximum stress when producing the free-load displacement is shown in Table 1.2.

The validity of the calibrated models is confirmed by examining Δx_{pzt} and Δx_1 when connecting the mechanism to a spring load realized by a fixed beam shown in Fig. 1.27C. The length of the beam is L =100 [mm], and brass is used as material. There is a small protrusion on the top of the actuator so that the amplification mechanism and the fixed beam are assumed to be in point contact. Three different thicknesses, 1, 2 and 3 [mm], are used to vary the stiffness. Table 1.3 shows a comparison of the estimated displacements from the proposed lumped parameter model and the true values from FEM analysis. As can be observed in the table, the estimated values agree well with the true values, confirming the validity of the model.

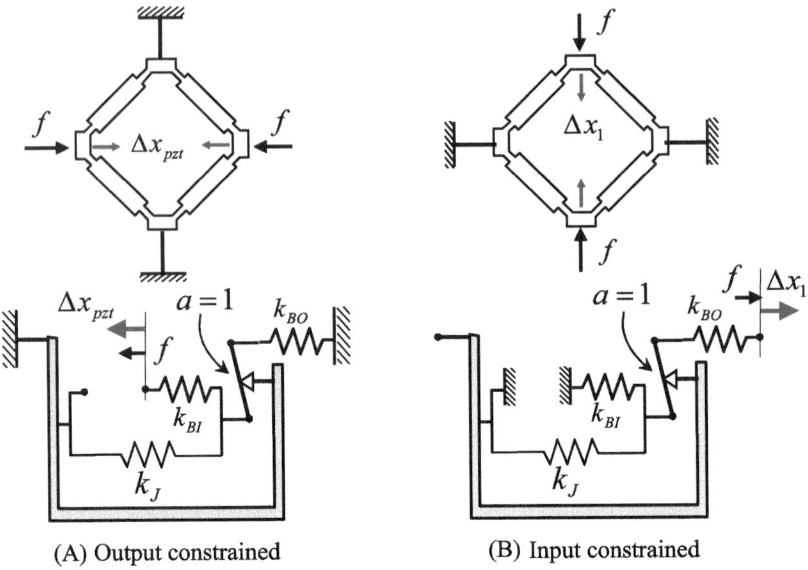

Figure 1.28 Requirement of input–output bidirectionality.

Consider two constrained cases shown in Fig. 1.28A–B to confirm the necessity of 3 lumped springs. For simplicity, the amplification mechanism is designed to have a square shape resulting in $a = 1$. $\Delta x_1 = \Delta x_{pzt}$ must hold for the displacements when applying the same magnitude of force f, which is a basic requirement of Castigliano's theorems. By using the proposed lumped parameter model, these constrained cases are represented by the figures shown below. The condition $\Delta x_1 = \Delta x_{pzt}$ is satisfied if $k_{BI} = k_{BO}$. This can also be confirmed in (1.47) where the off-diagonal elements are the same. This lumped parameter model has a redundancy in parameter calibration; however, 3 spring elements are minimally required to satisfy this condition.

1.3.3 Prototype two-layer actuator unit

1.3.3.1 Mechanical design

A prototype nested actuator with over 20% effective strain is designed based on the structural compliance analysis. Consider a nested structure with 2 amplification layers as shown in Fig. 1.13. The APA50XS "Moonie" piezoelectric actuators developed by CEDRAT, Inc. [2] are adopted for the first layer. According to the preliminary design in Section 1.2.2, over 20% of effective strain can be obtained by a two-layer mechanism; $K = 2$ and $\kappa = 15$. By stacking 6 APA50XS actuators for the first layer, i.e., $N_1 = 6$, this large strain may be achieved with a proper design of the second layer. From Table 1.4, we have $\tilde{k}_1 = 0.225 \times 10^6$ [N/m], and $\tilde{f}_1^{block} = 18.0$ [N] for the first layer units. The remainder of this section focuses on the design of the second-layer rhombus mechanism. From (1.40) and (1.41), an equivalent model is obtained for the

Table 1.4 Characteristics of the APA50XS actuator [2] for the first layer (definition of the dimensions has been modified).

Displacement	80 [µm]
Blocking force $\tilde{f}_1^{block}$	18.0 [N]
Stiffness $\tilde{k}_1$	0.225 [N/µm]
Voltage range	−20–150 [V]
Length (output actuation direction)	4.7 [mm]
Width (PZT stack actuation direction)	12.8 [mm]
Height	9.0 [mm]
Mass	2.0 [g]

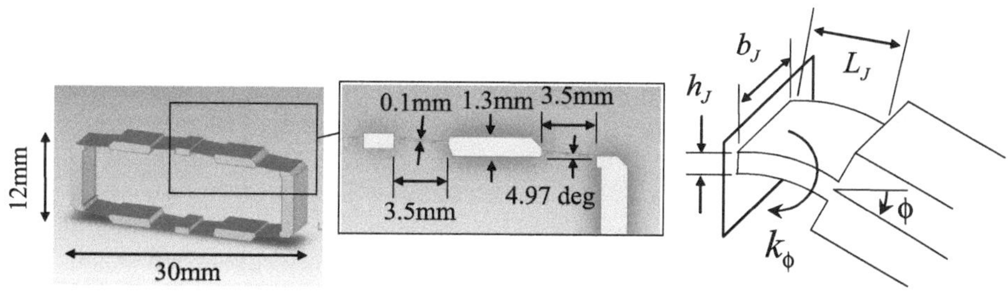

Figure 1.29 Design of a rhombus mechanism for the 2nd layer.

second layer by substituting (1.38) and (1.39), which provides a design guideline in terms of k_{BI2}, k_{BO2}, and k_{J2} for the target effective strain, i.e., 20%.

As described in the previous section, the stiffness in the admissible space, i.e., k_{J2}, must be minimized. The compliant joint may be represented as shown in Fig. 1.29. The rotational stiffness of this structure is given by $k_\phi = \frac{E b_J h_J^3}{12 L_J}$ where E is Young's modulus of the material. In order to reduce this stiffness, either the width b_J or thickness h_J must be reduced, or length of the gap L_J must be increased. Note that the reduction of h_J is the most effective for reducing k_{J2} since it is proportional to h_J^3; however, the thickness must be carefully determined considering manufacturing process. The maximum stress must be lower than the yield stress of material. In addition, in order to increase the stiffness in the constrained space, i.e., k_{BI2} and k_{BO2}, the oblique beam need to have a sufficient thickness except the thin part for the compliant joint.

Fig. 1.29 shows the designed rhombus mechanism for the second layer. Phosphor bronze (C54400, H08) is used for the material. The length of the mechanism in the actuation direction is 12 mm, and the width is 30 mm. The thickness of 0.1 mm is given to h_J for electrical discharge machining. The thickness of 1.3 mm is given to the oblique beams for sufficient stiffness. The oblique angle of the beams is 4.97 degrees that gives the displacement amplification ratio of 11.5, assuming the mechanism is ideal.

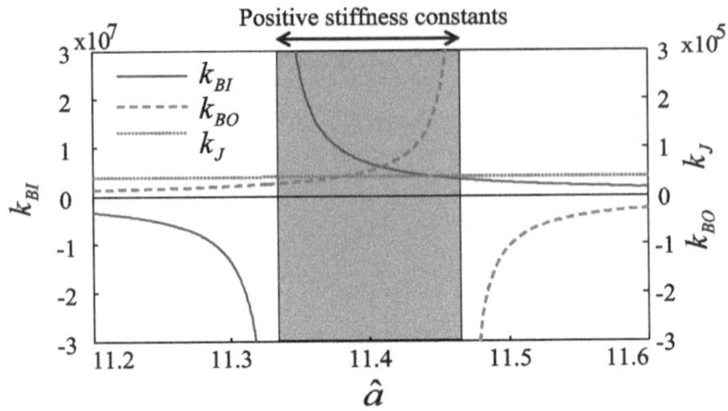

Figure 1.30 Choice of $\hat{a}$ for positive spring constants.

For the second layer mechanism shown in Fig. 1.29, $X_1 = 3.39 \times 10^6$, $X_2 = 2.95 \times 10^5$, $X_3 = 11.33$, and $X_4 = 3.50 \times 10^3$ are obtained from FEM analysis. The range of a_2 that makes all spring constants positive is shown in Fig. 1.30. Finally, $\hat{a}_2 = 11.4$ is chosen, which is between X_3 and $\cot\theta$ ($= 11.5$). Finally, the parameters for the lumped parameter model are calculated as $\hat{k}_{BI2} = 6.72 \times 10^6$ [N/m], $\hat{k}_{BO2} = 5.21 \times 10^4$ [N/m], $\hat{k}_{J2} = 3.98 \times 10^4$ [N/m] by determining the amplification gain as $\hat{a}_2 = 11.4$. The analysis using the lumped parameter model predicts that the maximum free-load displacement is 2.64 [mm], which is equivalent to 22% effective strain. Also, the maximum blocking force is 1.56 [N].

1.3.3.2 Performance evaluation

Fig. 1.31B shows the developed rhombus mechanism for the second, and Fig. 1.31C shows the assembled actuator with 6 first-layer units. The second layer mechanism weighs approximately 3 grams, resulting in the total weight of 15 grams. The connected first-layer units shown in Fig. 1.31A were inserted in the second layer mechanism and the both ends were manually bonded to the inside walls. Note that we noticed that the tolerances on the interfaces were important for satisfactory actuator performances. Fig. 1.32 shows snapshots of free-load displacement where 2 rhombus mechanisms are connected in series. This actuator extends since the first layer units are contractive.

The performance of this prototype is evaluated by measuring free-load displacement and blocking force. Fig. 1.33A shows the maximum free-load displacement measured using a laser displacement sensor (Micro-Epsilon optoNCDT 1401) when all 6 units are ON by applying 150 [V] actuation voltage. The measured displacement is 2.53 [mm], which is equivalent to 21.1% effective strain. Fig. 1.33B shows the blocking force where a sinusoidal wave input ranging from 0 to 150 [V] is applied. The maximum blocking force measured using a compact load cell (Transducer Techniques MLP) is 1.69 [N]. As shown in Table 1.5, the experimental values by using the lumped parameter model agree well with the estimated values, which confirm the validity of the approach.

Fig. 1.34 shows the aggregate displacements when ON–OFF control is applied to the internal 6 units with a constant actuation voltage when ON. For convenience, the

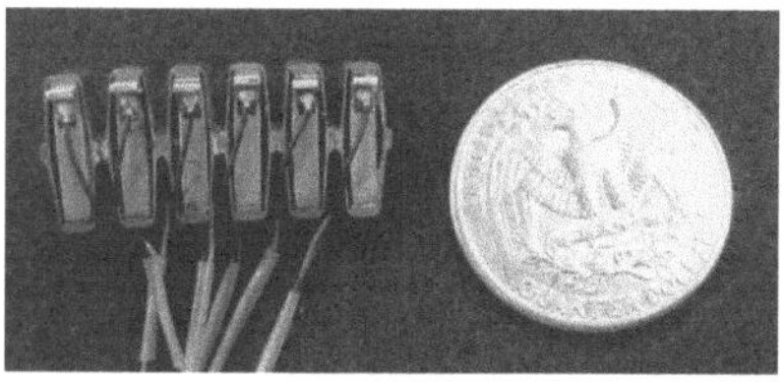

(A) Connected first layer units

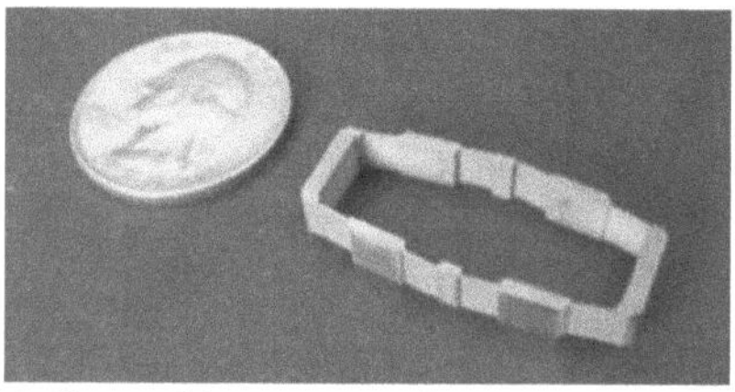

(B) Second layer amplification mechanism

(C) Assembled actuator

Figure 1.31 Prototype actuator: 6 CEDRAT actuators are used for the first layer.

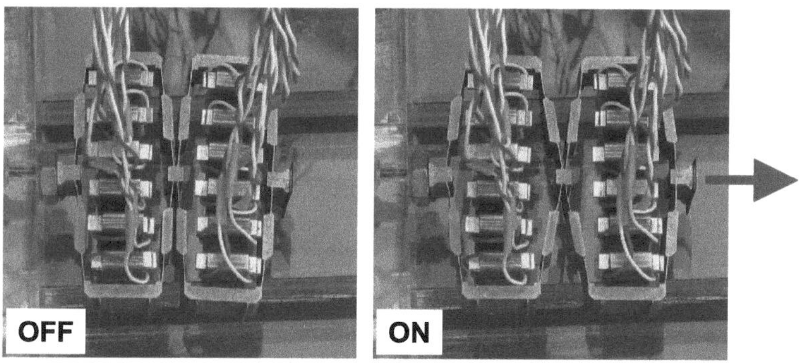

Figure 1.32 Snapshots of free-load displacement: Two nested rhombus mechanisms are connected in series. Each unit generates approximately 21% effective strain compared with its original length. © 2008 ASME, reprinted with permission

measured displacements are normalized by the maximum displacement when all 6 units are turned on. As described in Section 1.3.1.4, the distribution of the ON units in a layer does not theoretically affect on the aggregate displacement if an amplification mechanism encloses serially-connected internal units. As can be observed in Fig. 1.34, the measured displacement is not largely affected by the combination of ON units. For example, there are 20 ($= {}_6C_3$) combinations when the number of ON units is 3; however, the standard variation is at most 0.007, showing a sufficient repeatability. Another observation is that the increments of the displacement are slightly non-uniform, which is considered due to the nonlinearity of (1.6). In Section 1.2.2, this issue was neglected for simplicity; however, this characteristic should be further reflected to the design and control if rhombus mechanisms are used for creating a larger strain.

Table 1.5 Comparison between idealized model, proposed lumped parameter model, and experimental measurement.

	Analysis		Experiment
	Idealized model	**Lumped model**	
Effective strain	23.9 [%]	22.0 [%]	21 [%]
Free displacement	2.8 [mm]	2.64 [mm]	2.53 [mm] (std 5.9×10^{-3})
Blocking force	15.1 [N]	1.56 [N]	1.69 [N] (std 3.25×10^{-3})

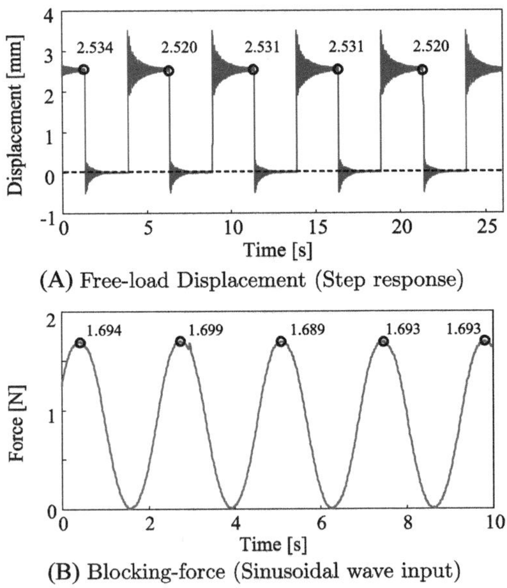

Figure 1.33 Experimental result.

A comparison between the idealized model and experimental measurement shown in Table 1.5 suggests that the aggregated force has been considerably attenuated, while the aggregated displacement or strain is as large as predicted by the idealized analysis. Note again that the result in Fig. 1.17 is never achieved, unless completely rigid beams and purely rotational joints without play are utilized. One of the difficulties in mechanical design is that physical structural parameters are intricately related to lumped parameters. For example, the increase of the gap L_J in Fig. 1.29 contributes to reducing the joint stiffness, but it also reduces the beam stiffness by having a long thin gap in the longitudinal direction. This gap may be reduced if the design focus is more on producing a larger blocking force.

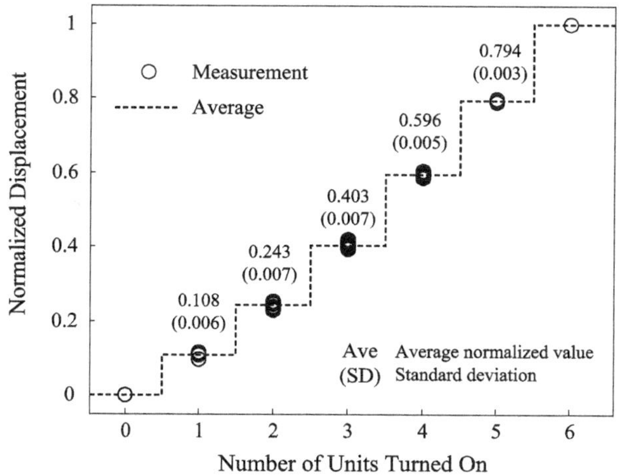

Figure 1.34 Binary ON–OFF Control Experiment.

(A) (B)

Figure 1.35 Micro gripper. © 2012 Springer, reprinted with permission

1.3.3.3 Micro manipulator design

Two types of miniature manipulators have been developed. Fig. 1.35 shows a design of a small gripper. A piece of metal that acts as a robotic "finger" was attached to each end of the outer rhombus mechanism as shown in Fig. 1.35A. The displacement of the tip is approximately 2.5 [mm] that is sufficient to pinch a small-outline integrated circuit as shown in Fig. 1.35B. Fig. 1.36 shows another design of a manipulator. A relatively long rod was attached to one end of the actuator unit that performs a pushing operation of small objects.

1.3.4 Contractile two-layer mechanism design

Fig. 1.37 shows a contractile two-layer nested rhombus PZT cellular actuator. Compared with the design shown in Figs. 1.31 and 1.32, this actuator with a "bow tie"

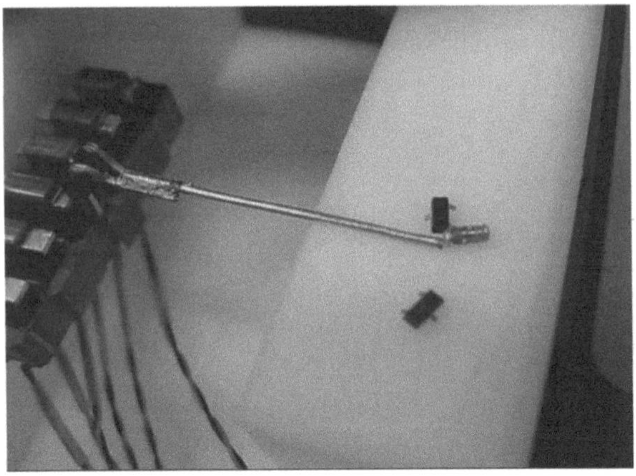

Figure 1.36 Micro manipulator. © 2012 Springer, reprinted with permission

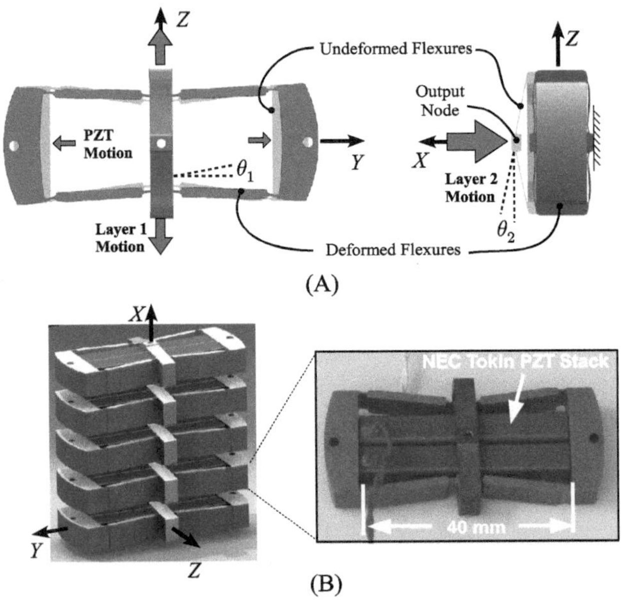

Figure 1.37 (A) Working principle showing deformed and undeformed flexures in two planes. (B) A 5 cell artificial muscle actuator based on PZT-driven flexures. The physical prototype uses two NEC Tokin PZT Stacks. © 2010 IEEE, reprinted with permission

shaped structure is contractile when the nested PZT actuators are ON [221]. Each cellular unit consists of a PZT stack and a special strain amplification flexure called a double-layer nested flexure.

Fig. 1.37A shows planar views of the cell design. From Fig. 1.37A, observe that as a voltage is applied to the PZT stacks, the first layer flexure is pushed outward along the Y-direction, which results in an outward amplified displacement in the Z-direction

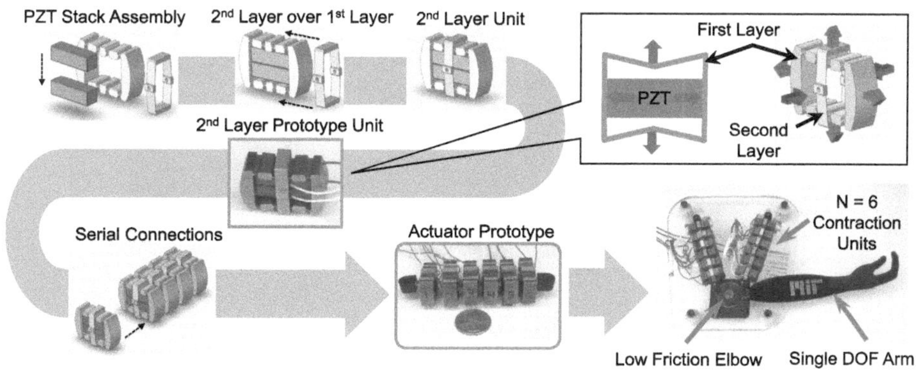

Figure 1.38 Fabrication of the cellular actuator arrays and 1-DOF robotic arm.

provided that the underformed beam angle θ_1 is small. The second layer flexure is then pushed outward in the Z-direction by the first layer, which results in a further amplified displacement in the inward X-direction when $\theta_2 \ll 1$. Therefore, a contraction force is generated along the X-axis output as the PZT stacks are activated. This contractile double-layer flexure design provides motion along the X-axis and allows the connection of multiple units in series without buckling of the flexures. The system exhibits small friction, natural axial compliance, and backdrivability.

A serial connection of 5 cells is shown in Fig. 1.37B where each prototype cell shown to the right has a 20 N blocking force and 1.5 mm free displacement (which is approximately 15% of its body length along the X-axis output). The 5 cell serial connection then produces 7.5 mm free displacement. In Section 4.1, the application of this design to actuation with tunable stiffness is discussed. Fig. 1.38 illustrates the fabrication process. The bow tie shaped structure for the first layer generates extending actuation. The rhombus shaped second-layer structure transforms it to contractive. Two 6-unit actuator arrays are developed and implemented into a one-degree-of-freedom robotic arm in an antagonistic arrangement for the bidirectional motion of 50 degrees at 150 V. Fig. 1.39A shows that the resonance frequency at 10 Hz is excited by the PZT actuation, and Fig. 1.39B shows the end-point compliance introduced by the structural flexibility.

1.3.5 Tweezer-style piezoelectric end-effector

1.3.5.1 Piezoelectric end-effector for robotic surgery and intervention

A prototype of a tweezer-style piezoelectric end-effector with nested-rhombus multilayer strain amplification mechanisms has been designed and developed [142,170, 94]. Fig. 1.40 shows an overview of the developed device. This robotic end-effector is a proof-of-concept prototype for a telerobotic system for surgery and intervention guided by magnetic resonance imaging (MRI). PZT piezoelectric actuators (APA35XS, CEDRAT, Inc.) were adopted. Five of the PZT actuators were connected in series and nested in a larger amplification mechanism made by phosphor bronze.

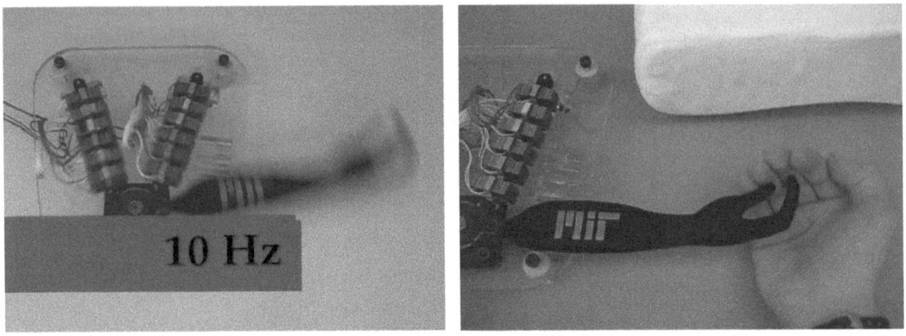

(A) Antagonistic arrangement of cellular actuators exciting the resonance frequency at 10Hz

(B) End-point compliance

Figure 1.39 Motion of the PZT-driven 1-DOF robotic arm.

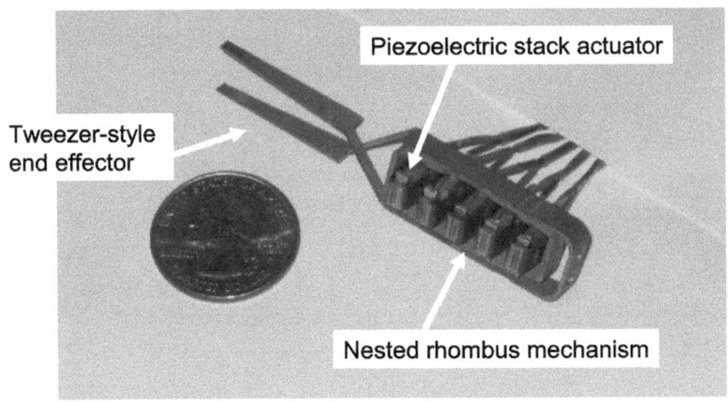

Figure 1.40 Assembled end-effector nesting actuator module.

A schematic assembly process is shown in Fig. 1.41. The first layer is made up of five commercially available Cedrat APA35XS "Moonie"-type piezoelectric actuators. A fixture shown in the figure may be used for precise assembly. A series of the five PZT actuators will be nested in a larger amplification mechanism made by phosphor bronze. This mechanism will be further nested in a tweezer-shape structure in which the lever action of the tweezer arms themselves provide a third layer of amplification. The length of the final assembly will be 70 mm in the longitudinal direction and the width will be 14 mm.

Note that this particular prototype contains small pieces of ferromagnetic metal in the first layer mechanisms and was not completely MRI compatible. This issue can easily be resolved by using a non-magnetic model of APA35XS that is also available from CEDRAT, Inc. The development of a refined device with MRI compatibility is in progress.

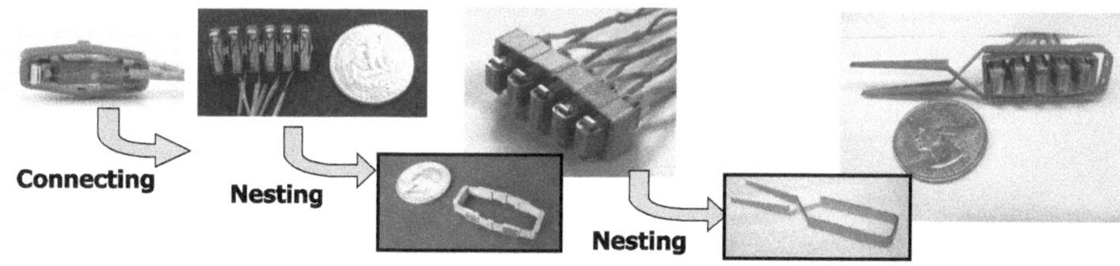

Figure 1.41 Assembly of a tweezer-style piezoelectric end-effector.

1.3.5.2 Modeling and design

The desired specifications of the robotic end-effector were determined based on a standard surgical clip. The goal is to achieve 1.0 [N] of force and 10 [mm] of displacement at the tip. The rhombus mechanism with PZT actuators is nested into a tweezer-shape structure with a reverse action mechanism. In this section, a Bernoulli–Euler beam model of the tweezer-shape structure is considered. A model of the structure is shown in Fig. 1.42. The displacement at point A, δ_A, where the force from the actuator module is applied, and the displacement at the tip δ_B can be written as

$$\delta_A = \left(\frac{C_{A2}}{2EI_1} + \frac{C_{A4}}{2EI_3}\right) \cdot W_A + \left(\frac{C_{A1}}{2EI_1} + \frac{C_{A3}}{2EI_3}\right) \cdot W_B, \tag{1.49}$$

$$\delta_B = \left(\frac{C_{B2}}{2EI_1} + \frac{C_{B5}}{2EI_3}\right) \cdot W_A + \left(\frac{C_{B1}}{2EI_1} + \frac{C_{B3}}{2EI_2} + \frac{C_{B4}}{2EI_3}\right) \cdot W_B, \tag{1.50}$$

where W_A, W_B are the forces at points A and B. W_B is a reaction force from an object. E is Young's modulus of the phosphorus bronze, and I_1, I_2, I_3 are the second moments of area of the sections shown in Fig. 1.42. $C_{A1} \sim C_{A4}$ and $C_{B1} \sim C_{B5}$ are structural coefficients obtained from Castigliano's theorem by assuming a Bernoulli–Euler beam model of the tweezer-shape structure (see Appendix A.2). Eqs. (1.49) and (1.50) are used to determine the dimensions of the end-effector to achieve desired performances. Let $\hat{W}_A$ be the force at the point A that needs to be applied to achieve the desired tip displacement when the tip of the end-effector is free (i.e., $W_B = 0$). From (1.49) we have

$$\hat{W}_A = \left(\frac{C_{B2}}{2EI_1} + \frac{C_{B2}}{2EI_3}\right)^{-1} \cdot \hat{\delta_A}. \tag{1.51}$$

The dimensions of the end-effector need be determined such that $\hat{W}_A$ and $\hat{\delta_A}$ do not exceed the maximum force and displacement of the actuator unit nested in the tweezer-shape structure.

Fig. 1.43 shows the final dimensions of the structure determined through trial and error by using finite element software. The length of the final assembly is 70 mm in the longitudinal direction and the width is 14 mm. The simulation result achieved 1.1 [N] of force and 9.0 [mm] of displacement at the tip, which are sufficiently close to the desired values. The stress analysis of the designed end-effector was conducted. The maximum von Mises stress when the actuator module exerts the maximum force was

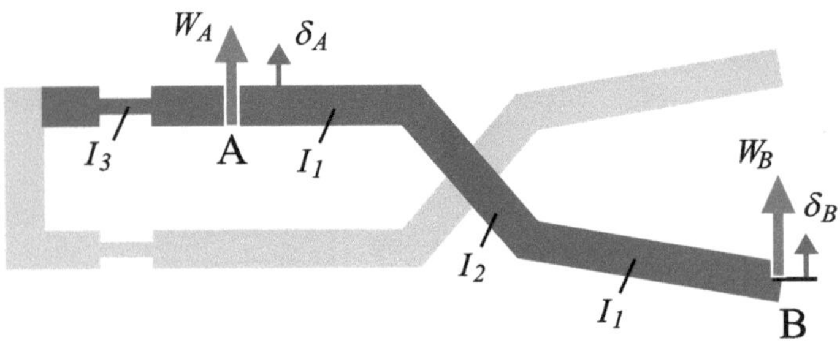

Figure 1.42 Schematic model of the end-effector structure. © 2010 JSME, reprinted with permission

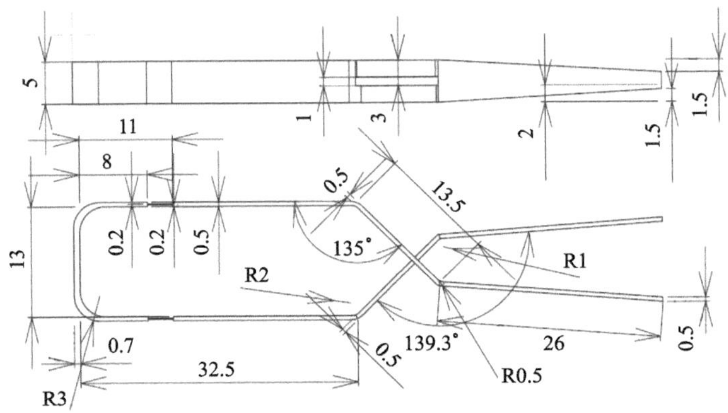

Figure 1.43 Drawings of the tweezer-style end-effector. © 2010 JSME, reprinted with permission

274 [MPa]. The designed end-effector has a sufficient strength since the yield stress of phosphorus bronze is 528 [MPa].

1.3.5.3 Fabrication and performance test

The fabricated tweezer-shape structure made by phosphorus bronze is shown in Fig. 1.44. The length of the structure is 68.6 [mm], the height is 5.0 [mm], and the width is 13.25 [mm], respectively. Fig. 1.45 shows the free displacement of the assembled end-effector. The prototype end-effector produced 8.8 [mm] of displacement and 1.0 [N] of static pinching force by applying 150 [V] input voltage to each of the five PZT actuators. The performance of the assembled end-effector was evaluated by using a force sensor and a laser displacement sensor. Fig. 1.46 shows the displacement and force profile when the input voltage changes from 0 to 150 [V] and from 150 to 0 [V]. Table 1.6 summarizes the results of the simulation and experiment.

In Fig. 1.47A, the developed end-effector was attached on a rotary motor and manipulated a small object. Fig. 1.47B shows a pinching of a rubber tube. The natural frequency of the fabricated end-effector was 36 [Hz]; uncompensated bandwidth of actuation up to 35 [Hz] was confirmed as shown in Fig. 1.47C. This actuation bandwidth is greatly wider than existing robotic micro grippers driven by tendons or

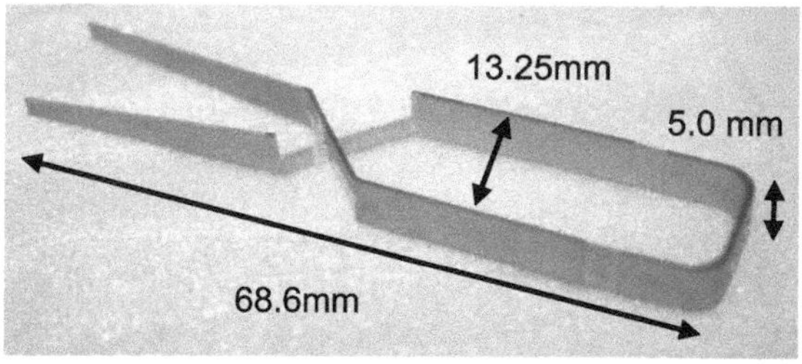

Figure 1.44 Fabricated end-effector.

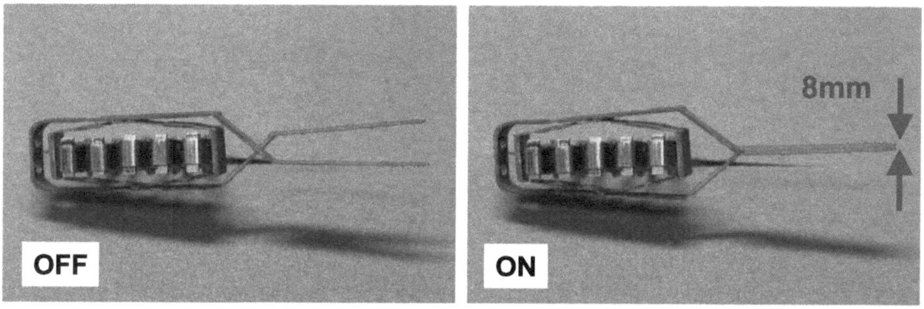

Figure 1.45 Motion of the end-effector. The developed end-effector has a reverse action mechanism; the tips close when the actuators are energized.

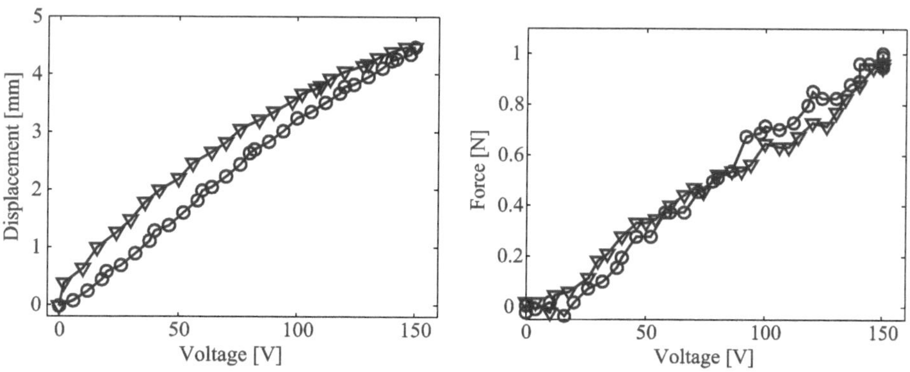

(A) Displacement profile (B) Force profile

Figure 1.46 Displacement and force performance. Circles are forward (from 0 to 150 [V]) and inverse-triangles are backward (from 150 to 0 [V]) directions; note that (A) shows the absolute displacement of one of the end-points. The total displacement is twice of this measurement. © 2012 IEEE, reprinted with permission

Table 1.6 Performance of the assembled end-effector.

	Displacement [mm]	Force [N]
Simulation	9.0	1.1
Experiment	8.8	1.0

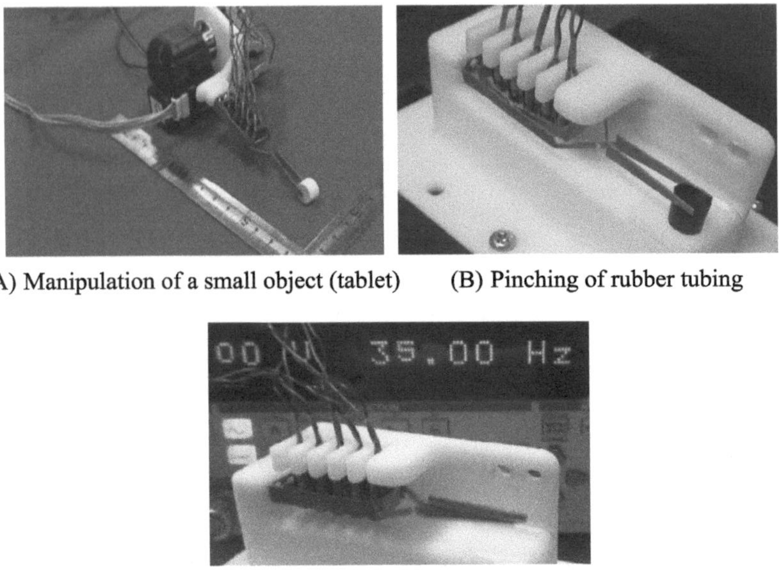

(A) Manipulation of a small object (tablet) (B) Pinching of rubber tubing

(C) High-speed operation (up to 35Hz)

Figure 1.47 Manipulation using the tweezer-style robotic end-effector.

MRI-compatible manipulators driven by fluid actuators. The bandwidth could be further improved if a vibration compensation controller is implemented.

In Section 4.3, a calibration procedure to model the hysteresis of the piezoelectric materials, as well as structural parameters, is discussed for sensor-less actuation of this mechanism.

1.3.6 Three-layer rhomboidal mechanism design and its application to a camera positioning mechanism

Another approach to develop a contractile cellular actuator is to design a three-layer structure with rhombus shaped amplification mechanisms. The majority of commercially available PZT stack actuators are extending actuators when ON. A single layer with a rhomboidal amplification mechanism switches the direction from extensile to contractile, or the reverse. Therefore, an odd number of amplification layers with rhomboidal mechanisms can configure an extending cellular actuator. Fig. 1.48A shows the schematic of the three-layer mechanism, and Fig. 1.48B shows a prototype. The three-layer actuator arrays are applied to a 1-degree-of-freedom camera positioning mechanism [217,219,132] as shown in Fig. 1.49.

A miniature image sensor (from Logitech C270 HD webcam) is used for image acquisition. The moving platen is connected to two rods that transmit the force from antagonistic pairs of cellular actuators located on both sides. The axis of the rod is positioned perpendicular, but not orthogonal to the pivot axis. When the rod is pushed

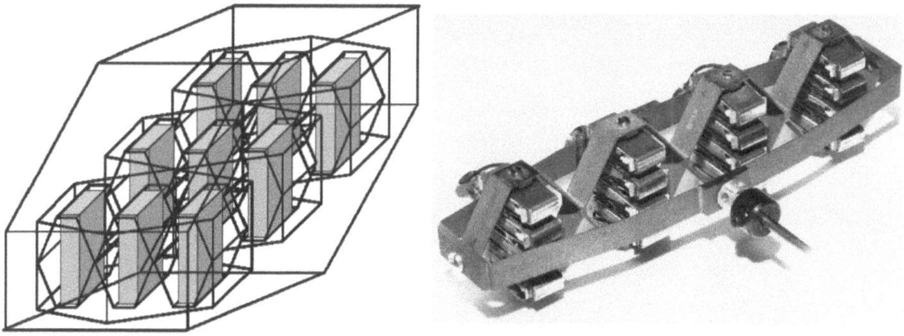

(A) Three-layer design for contractive actuation

(B) Prototype three-layer actuator

Figure 1.48 Three-layer rhomboidal amplification mechanism.

or pulled by the PZT actuators, a moment is applied to the moving platen, resulting in a panning (purely rotational) motion of the camera as shown in Fig. 1.50.

A total of 32 APA50XS (CEDRAT, Inc.) PZT actuators are used to position the camera. The cellular actuator can be controlled continuously by adjusting the input voltage from 0 to 150 V. However, binary control for each PZT unit, either 0 or 150 V, is applied. This is similar to physiological bang-bang type of inputs given in human saccades [69]. The PZT driver circuits operate discrete switching in the actuators. See Appendix A.3 for more details on the developed PZT driver circuits and real-time control system.

Fig. 1.51 shows the operation range of the camera orientation system. Since no gears or sliding mechanisms are used in the structures, this mechanism exhibits zero backlash and noiseless operation while extremely fast movement is achieved. The nested compliant mechanism has been reported to resemble characteristics of biological muscles [268]. This approach is similar to physiological bang-bang type of inputs given in human saccadic eye movement [69]. Since each cellular actuator has a total of 16 PZT actuators, the camera angle range is ± 16 PZT inputs or a tilting angle of approximately ± 13 degrees. The resolution of the discrete actuation of the camera orientation system was confirmed by using the same procedure presented in [219]. The camera positioner created 16 discrete angles that were measured by an encoder. The resolution of 0.8089 degrees is obtained by averaging all incremental movements where a standard deviation of 0.056 degrees is obtained. No significant nonlinearity of the angular displacement was observed in the overall activation since the standard deviation of the resolution was low. In addition, a high repeatability of the cellular actuator was confirmed in [219]. This camera positioning system is used for reproducing representative saccadic eye movements, which are discussed in Section 4.4.

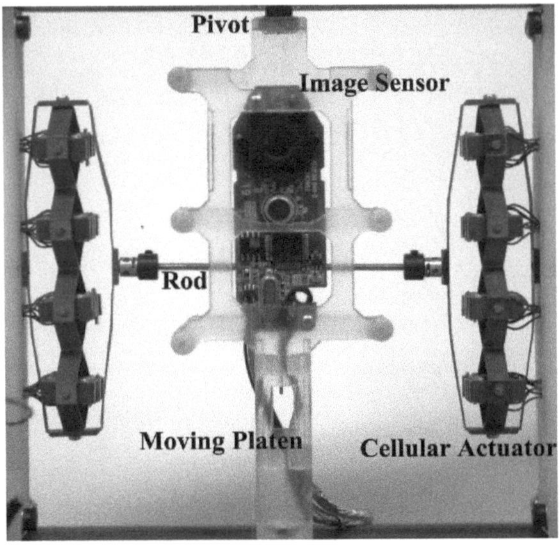

(A) Camera orientation system [219]

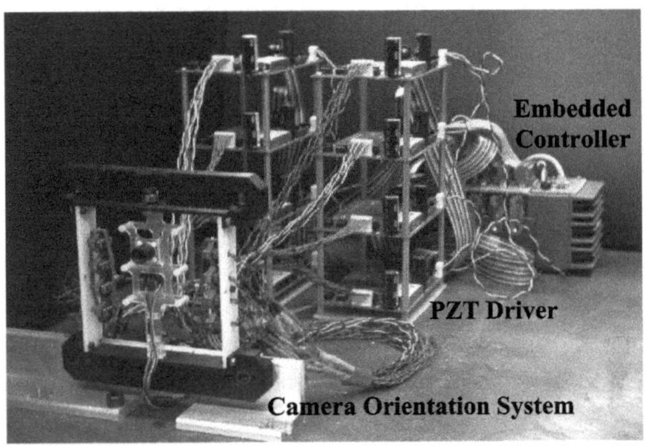

(B) System overview [132]

Figure 1.49 Robotic vision system. © 2015 SAGE, reprinted with permission

1.3.7 Discussion

A static lumped parameter model to represent and investigate fundamental characteristics of the strain and stiffness of nested rhombus mechanisms has been introduced. A recursive formula allows for modeling of an actuator unit with multiple amplification layers. Two-port network representation of the static lumped parameter model enables the calibration of structural parameters. Several designs of two- and three-layer amplification mechanisms and their application to a micro gripper and camera positioning mechanism have been presented. Mechanical designers can choose the shape of the amplification mechanism, either rhombus or bow tie, and the number of

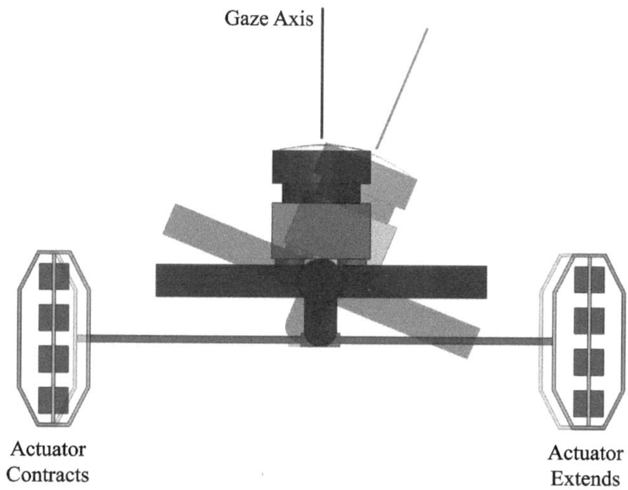

Figure 1.50 Pictorial representation of the camera orientation system [219]. The camera positioner driven by an antagonistic pair of cellular actuators.

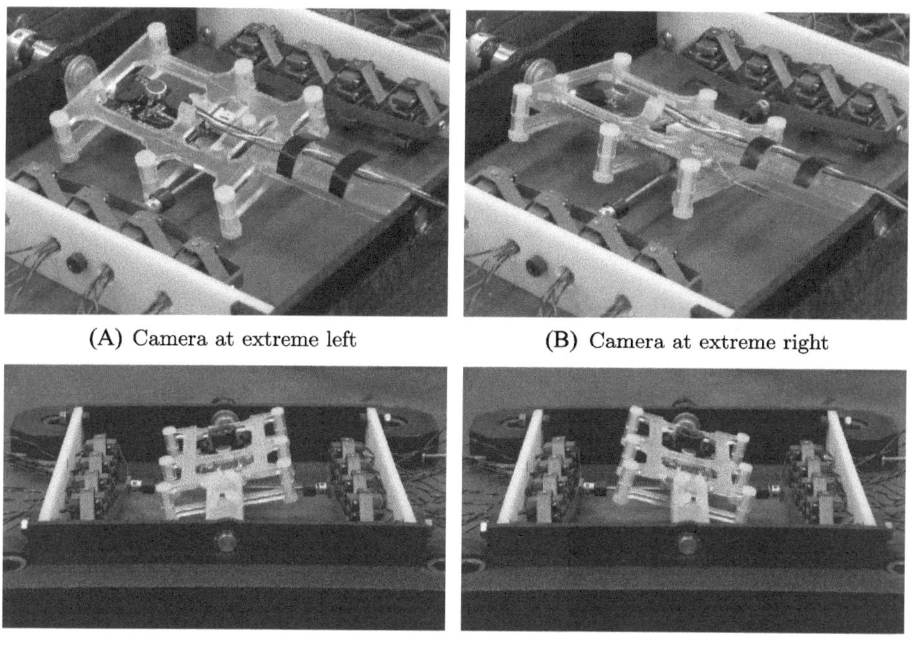

(A) Camera at extreme left

(B) Camera at extreme right

(C) Camera at extreme left

(D) Camera at extreme right

Figure 1.51 Single degree-of-freedom motion of the camera orientation system. © 2013 IEEE, reprinted with permission

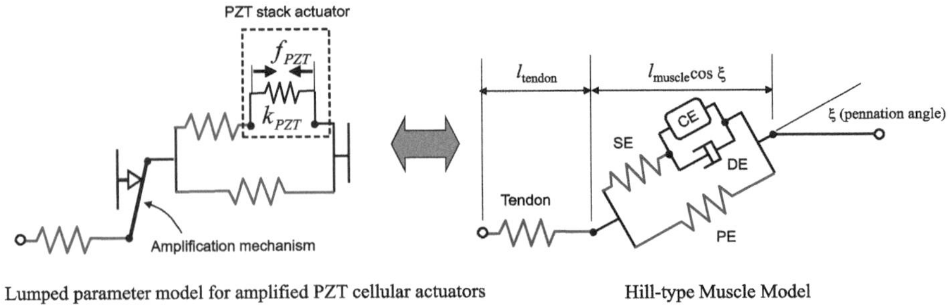

Figure 1.52 Muscle-like compliance of the amplification mechanism: (Left) lumped parameter model of the actuator; (Right) Hill's muscle model [294] with contractile element (CE) that produces force, series element (SE), parallel element (PE), and damping element (DE). l_{tendon} is the length of the tendon, l_{muscle} is the length of the muscle, and ξ is the pennation angle of the muscle.

amplification layers, considered up to 3 in this book, to configure extensible or contractile cellular actuators. The modeling presented in this chapter assumed that the linear characteristic in (1.13) holds for simplicity.

Although PZT ceramic itself is not compliant, the amplification mechanism introduces compliance as shown in the lumped parameter model in Fig. 1.24 as well as Fig. 1.39B. It should be noted that the lumped parameter model is similar to a well-known muscle model called Hill-type model as shown in Fig. 1.52 with minor differences. Recall that the 3-spring lumped parameter model was obtained to represent the characteristics of PZT actuators enclosed by compliant structures. Hill's model is also a lumped parameter model of skeletal muscles where no discrete mechanical elements are identified in the real muscle. In other words, essential force-producing elements are distributed and surrounded by compliant muscle tissues. While it is debatable and inconclusive, Hill's muscle model may be a general representation of compliant actuators with force generators and flexible structures that are characterized by means of two-port model such as (1.47).

Introducing compliance similar to biological muscles is crucial from the engineering point of view, where a robotic device directly interacts with humans, and the material of the actuator itself is not important; the use of biological tissues is not stipulated to obtain "muscle-like" compliance; PZT stack actuators perform equally well, in fact, surpassing biomaterials in terms of the reliability and the speed of response, provided the amplification structures are designed in an appropriate manner.

Dynamic analysis has been reported in [221]. How to control lightly damped actuator mechanisms by using open-loop switching commands is also discussed in Section 3.1. While the hysteresis compensation of piezoelectric materials is not the main aim of this chapter, Appendix A.1 describes empirical modeling of hysteresis that is applicable to self-sensing technique introduced in Section 4.3.

Modeling of cellular actuators

2

2.1 Two-port networks for single cell modeling

2.1.1 *Why a more involved model is necessary*

Prior sections introduced the amplification property of a rhomboidal mechanism. The longer the major (input) diagonal of the rhomboid compared to the minor (output) diagonal, the more "amplification" experienced, and this was encapsulated by the amplification ratio a. By arranging these rhomboidal mechanisms such that the output of each is the input to the subsequent one, (according to this view) one could expect to produce extremely large displacements from a single PZT stack just by including enough layers! Most engineers would naturally be suspicious, and suspect that somewhere there is a point of diminishing returns – that something has been omitted by this view.

Consider the rigid rhomboid with ideal frictionless hinges a second time (shown in Fig. 2.1). Let us say that this is the blocked case, and the output does not move. The model of the rhomboid from Section 1.12 would say that the force provided to the load is the blocked force of the PZT stack F_{block}, scaled down by the amplification factor a, or

$$F_{out} = \frac{1}{a} F_{block}. \tag{2.1}$$

This model, which assumes that the blocked force of the PZT is scaled down and transmitted to the load, is extremely optimistic, as we will see.

The rigid model of the rhomboid treats the angled beams between the hinges as levers; in the blocked case force is transmitted from input to output without any input deformation at all. What is really happening physically is that the PZT deforms the rhomboid at the input, creating a strain which in turn generates a stress. This stress propagates throughout the rhomboid, causing a stress and strain at the output that respects the boundary condition there. In reality it is impossible to transmit force to the output, even in the blocked state, without some deformation at the input to generate the stress in the material. This means that the PZT stack undergoes this same deformation at the input and is no longer in the blocked state, and the force applied at the input drops. Because the input force drops, the output force also drops below that predicted by (2.1) and this "amplification model" drastically overpredicts the force. In fact, the "beam and hinge" model does extremely poorly unless the impedance of the rhomboid is insignificant relative to the PZT in the free case, and the impedance of the PZT is insignificant relative to the rhomboid in the blocked case. These cannot both be true for the same rhomboid. In the case of a nested amplification mechanism, the impedances of each layer will generally be significant relative to one another and a more complete model is needed that is accurate across all ranges of operation. With only a modest increase in complexity, the two-port network model from circuit theory performs much more adequately in this regard. A force–current and displacement–voltage analysis

Cellular Actuators, DOI: 10.1016/B978-0-12-803687-7.00002-8
Copyright © 2017 Elsevier Inc. All rights reserved.

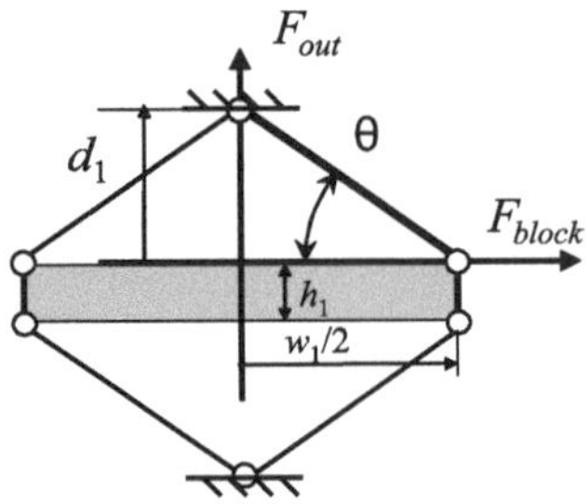

Figure 2.1 An overly simplistic understanding of the amplification principle of a rhomboidal strain amplifying mechanism that does not account for deformation in the piezoelectric material and the rhomboid itself.

allows the material deformation properties of the rhomboid to be included, resulting in a much more accurate displacement and force prediction.

2.1.2 Two-port models of strain amplifying compliant mechanisms

To produce a more accurate model of the behavior of a rhomboidal strain amplification mechanism, we borrow the notion of two-port models from network theory. Analogies are made between voltage and current and the key mechanical quantities of interest: force and displacement of the input to the amplification mechanism and its output flange that drives the load. The two-port model is not drastically more complex than the idealized amplification model, so why does it do so much better? It captures two phenomena not captured by the idealized model: first, it acknowledges that some input deformation is necessary to produce a force at the output and second, it encapsulates the bilateral nature of the elastic material – loads at the output will drastically affect how the input will behave.

2.1.2.1 Background to two-port models

In his book, Choma [42] devotes a chapter to two-port network theory, a simple way to model an electrical network when the mathematical models of the underlying components are either unknown, or are inordinately cumbersome. The two-port concept allows the designer to analyze the network as a "black box" and focus on its input–output behavior, while ignoring the constitutive relationships interior to the network. The idea is that the "black box" contains voltage-controlled current sources or current-controlled voltage sources and passive impedances. Two-port networks are represented by a square with two pairs of terminals emerging from it, as shown in Fig. 2.2. Each port has a voltage across the terminals and a current, which by convention is directed into the two-port network.

Whether a voltage-controlled current source or a current-controlled voltage source is appropriate depends on what quantities are chosen to be inputs and which are chosen to be outputs. The voltages and currents at each port make for a total of four quantities;

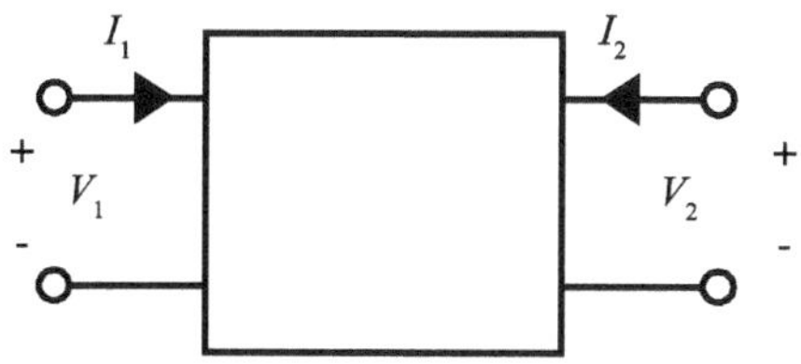

Figure 2.2 A representation of a two-port network model. © 2013 IEEE, reprinted with permission

the two-port formalism stipulates that two will be inputs, and two will be outputs. Any 2-input 2-output combination is allowable – the various input–output choices lead to a family of 2 × 2 *immittance* matrices. Some examples are:

- The currents at both ports are chosen as the inputs, and voltages are chosen as the outputs. The immittance matrix will have units of impedance.
- The voltages at both ports are chosen as the inputs, and the currents as the outputs. The immittance matrix will have units of admittance.
- The voltage and current at the left-hand port are chosen as the inputs, and the voltage and current at the right-hand port are chosen as the outputs. The immittance matrix is a matrix with mixed units known as the *forward transmission* matrix.

It is important to note that the physical system under consideration is the same regardless of the choice of inputs and outputs, so as one might expect, the various immittance matrices for a particular circuit or mechanical system are related to one another and can be derived in terms of each others' elements. Therefore, identification of one immittance matrix by analysis or experiment is sufficient to describe all the other immittance matrices. This will be analyzed in detail in Section 2.1.5.

The idea of using a two-port network to represent rhomboidal strain amplifiers was introduced in Section 1.3, which speaks of minimizing the stiffness in the admissible space, and maximizing it in the constrained space. Admittedly, this does not give much guidance on what rhomboids will perform well.

With the appropriate mechanical–electrical analogies, two-port networks can also be used to model rigid body or flextensional mechanical systems. This makes two-port networks a simple but powerful analysis tool, one that is particularly useful for describing the two-way interactions characteristic of biologically inspired actuation. Abdalla et al. [4] model not only the compliant mechanism as a two port network, but also the piezoelectric stack actuator itself. Because the electromechanical transduction of the PZT ceramic involves four quantities: voltage, current, displacement, and force, it also lends itself to a two-port network description.

The approach of this section is inspired in many ways by that of Abdalla et al. [4], but it has a different perspective and a different use case. There are two main differences: the first is that Abdalla et al. assume that the designer begins the design with knowledge of the load impedance, which is seldom the case in robotic systems in an unstructured environment. The piezoelectric cellular actuator must work well across a wide variety of output loads and operating conditions. Abdalla et al. are concerned with the tradeoff between geometric advantage (or displacement amplification factor)

$$a = \frac{\delta_{out}}{\delta_{in}}, \tag{2.2}$$

and efficiency (i.e., the amount of energy transferred to the load vs. how much is stored in internal deformation) of the mechanism. They show that these two quantities vary inversely. The second difference is in their use of the topology optimization approach, which can be said to represent the "other school of thought" in the design of compliant mechanisms from that taken in this book, as will be seen in the description to follow.

2.1.2.2 *Topology optimization vs. synthesis from a library*

The literature in the smart structures community on compliant mechanisms is vast. As one might imagine, there are numerous applications where movement of one type needs to be converted into movement of another type, and the mechanism doing so must be manufactured monolithically. Needless to say, some process is necessary to take a block of material and arrive at some material removal solution that has thin cross sections in the appropriate places so that movement of the actuator, most often thermal or piezoelectric pinches, pushes, presses, or lifts at some other parts.

There are two general categories of ways to do this. The first is topology optimization. Frecker provides a good overview of the techniques within this category [78]. In general, the design of a compliant mechanism has two distinct steps, a topology synthesis, followed by a dimensional synthesis [137]. The process begins with a "ground structure" [4,106], most often a uniform monolithic truss with uniform members. Subsequent steps are computational. The exact method may vary, but some algorithm begins to thicken or thin certain segments of the ground structure in response to a search direction expected to minimize an optimization criterion. As one might imagine, certain members become very thin, and when the thickness falls below some minimum criterion, they are removed (and never reinserted, regardless of the search direction). Removal of members eventually results in an optimal topology. The dimensional synthesis phase refines the thickness of individual members, but the connectivity of remaining material between the input and the load remains unchanged. The other procedure is the synthesis of a compliant mechanism from a set of candidate libraries with settable parameters. This approach lends itself better to design intuition and experience than topology optimization, which is purely computational in nature. Kim, Moon, and Kota [128] used two-part beams and ellipsoid sections as constitutive elements. The settable parameters were the angle between the segments of the beam, and their individual lengths, and the radius and subtended angle of the ellipsoidal segment. Martell and Schultz [165] used series spring combinations where the settable parameters were the individual spring constants. The approach in this book is more of the library synthesis type, but where each element is rhomboidal in nature. Settable parameters are the number of segments, the thickness, overall dimensions, and the angle.

2.1.2.3 Flexural rhomboidal mechanisms

Rhomboidal mechanisms are an intuitive topology and lend themselves well to interconnections in series, parallel, and multi-layer nested configurations and can thus be used to construct muscle-like robotic actuators. Another good rationale for using rhomboidal mechanisms is that automatically generated topology optimizations have converged to structures that are rhomboidal in shape [25].

A rhomboidal strain amplifying mechanism also relates four quantities of interest: the force along the major diagonal (hereafter referred to as the input), the displacement along the input, the force along the minor diagonal (hereafter referred to as the output), and the displacement along the output. The key advantage of the two-port network is this: it can capture the two-way interaction of the rhomboid if the immittance matrix is identified. Furthermore, the interconnection laws for two-port networks can simply and elegantly describe an entire multi-stage compliant device, and if the source impedance (electrical) and terminal impedance (mechanical) are known, the entire response can be predicted. This section focuses on determining what the immittance matrix elements are; multi-stage models will be addressed in Section 2.3.

2.1.2.4 Nested amplification mechanisms

Most of the discussion of nested mechanisms is covered in Section 1.2, but a short description is given here to motivate the remaining discussion in this section. A nested actuator begins with single-layer amplified PZT stacks, concatenates a small number of these in series, and amplifies their displacement multiple times using compliant rhomboidal mechanisms. Each layer of amplification can be expressed as a two-port network.

In essence, Ueda, Secord, and Asada [267] solve the "forward" problem, namely, given a geometry, determine its performance characteristics. The "reverse" problem – given a set of performance specifications, determine the geometry characteristics and PZT arrangement that will meet a set of force–displacement specifications – is not truly solved in their work nor developed in previous chapters. Section 1.3 relies on Finite Element Methods, and although a few evaluations of rhomboid displacement can lead to some valid conclusions about design choices, it is not a formal method for choosing the geometry of the part.

Optimization routines generally must solve the forward problem multiple times in order to find an optimal point. It is possible to incorporate finite element methods into an optimization approach, but it is computationally expensive, and great care needs to be taken with the mesh at interfaces, thin sections, and changes in cross-section. If the geometry is suitably parameterized, one can iterate the design geometry toward optimal with finite element simulation in the loop [40,23], but this requires specialized knowledge of advanced functionality of the finite element software and may not work for all geometries. This is where the two-port models come in. With an analytical expression for each of the elements of the immittance matrix in terms of real, physical quantities such as length, width, elastic modulus, rhombus angle, and so on, the forward problem is reduced to a function evaluation, and the optimization quantities such as search direction, Jacobian, Hessian can be well defined in classical ways. It would be one thing to iteratively compute a finite element model if the nodal information was used at the end, but if all that is being predicted is the input–output behavior, this calculation hardly merits the massive amounts of computation involved.

The analytical model also gives the designer physical insight, and a better sense of what effect changing a geometric parameter will have. Furthermore, the two-port model provides a hierarchical abstraction that gives a sense of how each layer contributes to the overall objectives of the nested rhomboidal strain amplification mechanism. As will be seen in Section 2.3, the rhomboidal strain amplifying layers of a multi-layer nested mechanism cannot be designed independently, but the hierarchical abstraction helps give some order to a complicated optimization problem.

2.1.3 Finding expressions for the immittance parameters using Castigliano's theorem

Since the rhomboidal strain amplification mechanism will be modeled as a two-port network, this work is less concerned with the internal stress and strain fields than the input–output behavior of the mechanism. For this reason, it is natural to use Castigliano's theorem [124], since it provides an input–output relationship between the loads on a structure and a displacement at a given point.

Castigliano's theorem (for a thin structure) states:

$$\delta_j = \frac{\partial}{\partial F_j} \int_0^l U(F_1, F_2, \ldots, F_j, \ldots, F_n; z)\, dz, \tag{2.3}$$

where δ_j is the displacement of a structure at the point of application of the jth load F_j, in the direction of that load, l is the length of the member, U is the strain energy for a differential element of length dz, as a function of loads $F_1, \ldots, F_n$, and coordinate z.

Castigliano's theorem has been used in several previous works to determine the stiffness of flexure hinges [155,252]. Lobontiu and Garcia [152] characterize an entire hinged mechanism in terms of three parameters, the input stiffness, the output stiffness, and the displacement amplification, and optimize the filleted flexure hinges to achieve a balance between amplification and stiffness; however, unlike a two-port model, it does not describe the function of the mechanism across varying loads and control inputs and was not formulated with large-stroke robotic applications in mind. Although each and every layer in a nested mechanism may not move throughout its entire range (form its undeformed to free displacement length), it is important to have a concise model that is valid for wide ranges of motion. The immittance matrix model captures the effects of input displacement on the output force and displacement, but is still based on small angle approximations and retains the simplicity of a linear model.

This analysis begins along the lines of the "chain method" described by Howell [105] for a simply connected body. Because the only quantities of interest are the input and output displacements and forces, the method presented here is much simpler. In particular, the boundary conditions for a multiply connected body are resolved analytically, resulting in an analytical input–output model. In order to determine the two-port model of a general rhomboidal compliant mechanism, certain simplifying assumptions are made that limit its applicability, but which make the analysis tractable. First, it assumes that the mechanism is symmetric about two orthogonal axes, aligned with the input and output directions. This allows derivation of the input–output relationships by analyzing a quarter mechanism, reducing the amount of bookkeeping required. Second, it assumes that forces are applied to a given layer only at the input and the output. Third, it assumes that the compliant mechanism is composed of finitely many thin straight segments rigidly connected in series and fillet radii are small relative to the segment length and have negligible effect.

The method to determine the expressions for the immittances is summarized as follows:

1. It resolves the static indeterminacy in the doubly connected rhomboid, eliminating the internal moment reaction in favor of the input and output forces. The fact that the rhomboid is doubly symmetric about the input and output axis is instrumental in the analysis.
2. Castigliano's theorem is applied in both the input and output directions to describe the force–displacement relationship.
3. Terms are collected so that the two relations found in the second step are expressed in terms of the two-port model immittance matrix relationship.

2.1.3.1 Forces and displacements applied to an elastic quarter-polygon

The model will be formulated from a connected chain of straight segments, each of which may differ in thickness and are angled from one another. Deformation in the "corners", where one segment joins to another, can reasonably be neglected. This chain forms one quarter of the polygon; the remainder of the polygon is this same chain reflected about the center of the polygon and the two axes of symmetry. The end of the chain opposite the point of application of the input force will be fixed; the end where the input is applied will be movable. Consider a general single straight segment i rigidly connected to the preceding segment $i-1$ in this chain. The free-body diagram of this segment is shown in Fig. 2.3. Looking at the free-body diagram, the internal forces in a given segment can be expressed in terms of its geometric parameters and the forces in the preceding segment. N_i is the internal axial load, V_i is the internal shear force, and M_i is the internal moment for segment i. z_i is the distance to an arbitrary point on the segment, which ranges from 0 to z_i^f, the total length of that segment. $d_\theta(z)$ is the angular deflection at the point at distance z_i. θ_i is the orientation of the segment in the undeformed configuration. Beginning with the reactions at the fixed flange (section 0), one can propagate the internal forces and moment outward from segment to segment through the quarter mechanism until the point of application of the input force is reached.

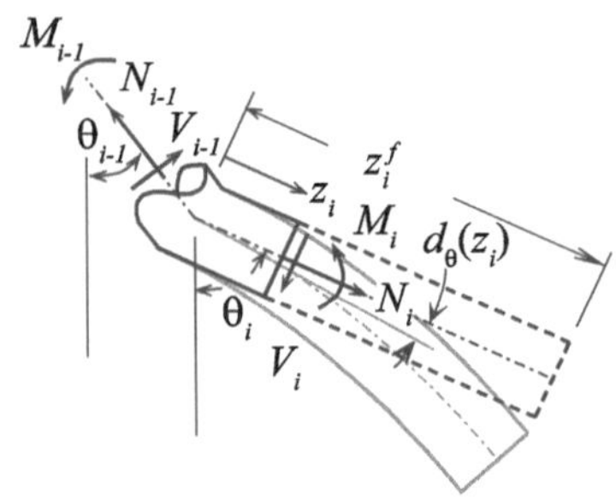

Figure 2.3 Free-body diagram of a general flexible segment with no loads applied to its interior. © 2013 IEEE, reprinted with permission

Applying the equations of static equilibrium to section i, in matrix–vector form, the internal forces and moments in segment i at a distance along the segment z_i can be expressed in terms of those in the preceding segment as

$$\begin{bmatrix} N_i \\ V_i \\ M_i \end{bmatrix} = {}^{i}P_{i-1} \begin{bmatrix} N_{i-1} \\ V_{i-1} \\ M_{i-1} \end{bmatrix} \tag{2.4}$$

where

$${}^{i}P_{i-1} = \begin{bmatrix} \cos(\theta_i - \theta_{i-1}) & -\sin(\theta_i - \theta_{i-1}) & 0 \\ \sin(\theta_i - \theta_{i-1}) & \cos(\theta_i - \theta_{i-1}) & 0 \\ z_i \sin(\theta_i - \theta_{i-1}) & z_i \cos(\theta_i - \theta_{i-1}) & 1 \end{bmatrix}. \tag{2.5}$$

The matrix P_{i-1} projects the forces and moments from the coordinate system associated with segment $i-1$ to that of segment i. To propagate this to the next segment, evaluate (2.4) at $z_i = z_i^f$, which is a geometric parameter set by the design of the rhomboid.

2.1.3.2 Resolving symmetry and boundary conditions

The rhomboidal amplification layer is a doubly connected, statically indeterminate structure. Compliant mechanisms of this type will follow the geometry shown schematically in Fig. 2.4. In order to solve a doubly connected structure, one needs additional compatibility conditions expressed in terms of the displacement, which are developed here. N_k and M_k, where $k \in \{\text{P, Q, R}\}$, are the equivalent moment and axial reactions from the removed section. The circled numbers $\{1, \ldots, N_{seg}\}$ denote the index of each segment. A fixed end condition is applied at point R; it is considered to be a fixed rigidly mounted flange. The mobile end, point P, moves along the center line A–A. R_x, R_y, and M_R are the reactions at the fixed end condition. As these internal reactions at point P are not quantities of interest in an input–output model, they will be eliminated in favor of input–output quantities.

To resolve the static indeterminacy, proceed as follows: imagine that the structure is cut in half along A–A, discarding the right hand (light gray) half. At point P, the

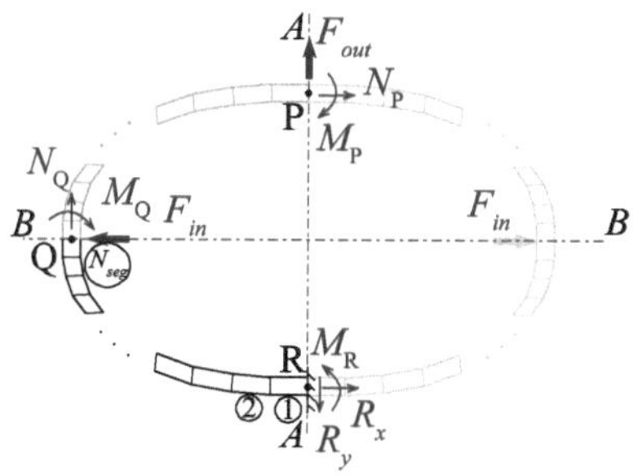

Figure 2.4 Doubly symmetric actuator composed of straight segments. Because of the symmetry, the entire behavior can be determined by analyzing the black structure only. © 2013 IEEE, reprinted with permission

internal stresses imposed by the material of the right half on the left will be replaced by its resultant reactions. Due to symmetry about A–A, each half of the structure will carry half of the applied load F_{out}, and deform by the same amount in the direction of F_{out}. Because of the symmetry about A–A, the right half will impose no shear reaction on the left half. It is as if the two halves were deformed independently by the same amount, then joined lightly together. However, there will be a normal reaction force, N_{P}, imposed by the right half on the left half to ensure that point P remains on line A–A. The right half will also apply a moment reaction M_{P} on the left half, ensuring that the deflection angle of the final segment is continuous at P. These two compatibility conditions allow the determination of these unknown internal reactions.

Because the structure is also symmetric about B–B, the upper and lower quarters will each carry 1/2 of the load F_{in}, and the deformation of each quarter will each contribute half of the displacement of point P (i.e., point Q moves exactly half as far as point P). Furthermore, the deformed shapes of the upper and lower quarter will be symmetric about B–B. This allows the compatibility conditions to be expressed at point Q, simplifying the calculations. Similarly to how the light gray half was discarded and replaced by resultant reactions, the upper, dark gray quarter can be discarded and replaced by its internal reactions N_{Q} and M_{Q}. Because of symmetry about B–B, the internal reaction $N_{\mathrm{Q}} = F_{out}/2$. In addition, the surface of the quarter actuator produced by the cutting plane aligned with B–B in the undeformed configuration will remain parallel to B–B as it deforms. Alternatively expressed, the tangent at Q must remain unchanged from its undeformed configuration, or

$$d_\theta(z_i^f) = \theta_N, \tag{2.6}$$

where d_θ is the function for the deflection angle of segment i. The moment M_{Q} that enforces (2.6) will be eliminated first. In [124,152], expressions for internal reactions are found using Castigliano's theorem, but in this case, since it will require only one integration, it is simpler to proceed by direct integration of the moment and then solve for M_{Q}. Assuming each segment is linearly elastic, has negligible shear deformation, and undergoes small deflections, applying the well-known slope–moment relationship

[124] gives

$$d_\theta(z_i) = \frac{1}{EI_i} \int M_i(z_i)\, dz_i, \tag{2.7}$$

where E is the Young's modulus of the material, and I_i is the moment of inertia of the segment. Integrating, augmenting (2.4) with the deflection equation, and evaluating at z_i^f, the internal forces and deflection angles of each segment can be propagated forward according to

$$\left[\begin{array}{c} N_i \\ V_i \\ M_i \\ \hline d_{\theta i}(z_i^f) \end{array}\right] = {}^{i}P_{i-1}^{aug} \left[\begin{array}{c} N_{i-1} \\ V_{i-1} \\ M_{i-1} \\ \hline d_{\theta i}(z_{i-1}^f) \end{array}\right], \tag{2.8}$$

$${}^{i}P_{i-1}^{aug} = \left[\begin{array}{ccc|c} \cos(\theta_i - \theta_{i-1}) & -\sin(\theta_i - \theta_{i-1}) & 0 & 0 \\ \sin(\theta_i - \theta_{i-1}) & \cos(\theta_i - \theta_{i-1}) & 0 & 0 \\ z_i^f \sin(\theta_i - \theta_{i-1}) & z_i^f \cos(\theta_i - \theta_{i-1}) & 1 & 0 \\ \hline \frac{z_i^{f\,2}}{2EI} \sin(\theta_i - \theta_{i-1}) & \frac{z_i^{f\,2}}{2EI} \cos(\theta_i - \theta_{i-1}) & \frac{z_i^f}{EI} & 1 \end{array}\right]. \tag{2.9}$$

The compatibility conditions state that

$$\left[\begin{array}{c} N_N \\ V_N \\ M_N \\ \hline d_{\theta N}(z_N^f) \end{array}\right] = \left[\begin{array}{c} N_Q \\ V_Q \\ M_Q \\ \hline 0 \end{array}\right]. \tag{2.10}$$

The first three equations are trivial, but the fourth equation can be solved to find the expression for the unknown internal moment M_Q in terms of F_{in} and F_{out}. Once this expression is determined, it can be substituted it into the expression for M_R and propagated outward using Eq. (2.4). Now that the static indeterminacy has been resolved, the expressions for the input displacement (displacement at Q along the direction of F_{in}), and the output displacement (displacement at R along the direction of F_{out}) can be found using Castigliano's theorem.

2.1.3.3 Applying Castigliano's theorem to find displacement expressions

Because the integration limits in this application of Castigliano's theorem are finite constants, and for a linearly elastic structure, the strain energy/per unit length will be continuous, the derivative can be brought inside the integral [101]. The expression for displacement then becomes

$$\delta_j = \sum_{i=1}^{N} \int_0^{z_i^f} \left(\frac{M_i}{EI_i} \frac{\partial M_i}{\partial F_j} + \frac{2(\nu+1)V_i}{EA_i} \frac{\partial V_i}{\partial F_j} + \frac{N_i}{EA_i} \frac{\partial N_i}{\partial F_j} \right) dz_i, \tag{2.11}$$

where ν is the Poisson's ratio of the material, and A_i is the cross-sectional area of the segment. Since each of the internal forces and moment is linear in the loads F_j, one can use Eq. (2.4) to propagate the partial derivatives forward segment by segment, with the internal forces and moment replaced by their partial derivatives. Assuming each summand is calculated iteratively, $[\ N_{i-1}\ V_{i-1}\ M_{i-1}\ ]$ and their partial derivatives will not depend on z_i, and thus are pulled out of the integral. The output displacement can then be expressed as

$$\delta_j = \sum_{i=1}^{N} \begin{bmatrix} N_{i-1} \\ V_{i-1} \\ M_{i-1} \end{bmatrix}^T Q_i \begin{bmatrix} \frac{\partial N_{i-1}}{\partial F_j} \\ \frac{\partial V_{i-1}}{\partial F_j} \\ \frac{\partial M_{i-1}}{\partial F_j} \end{bmatrix}, \tag{2.12}$$

where

$$Q_i = \int_0^{z_i^f} {}^iP_{i-1}^T \begin{bmatrix} \frac{1}{EA_i} & 0 & 0 \\ 0 & \frac{2(\nu+1)}{EA_i} & 0 \\ 0 & 0 & \frac{1}{EI_i} \end{bmatrix} {}^iP_{i-1}\, dz_i. \tag{2.13}$$

Performing the integration with respect to z_i and assuming each segment has a rectangular cross-section of width t_i results in

$$\begin{aligned}
Q_i &= \frac{1}{24EI_i} \begin{bmatrix} q_{11} & q_{12} & q_{13} \\ q_{21} & q_{22} & q_{23} \\ q_{31} & q_{32} & q_{33} \end{bmatrix}, \\
q_{11} &= 2(z_i^f t_i^2 \cos^2(\theta_{i+1} - \theta_i), \\
&+ \ ((2\nu + 1) z_i^f t_i^2 + 4 z_i^{f\,3}) \sin^2(\theta_{i+1} - \theta_i)), \\
q_{12} = q_{21} &= ((2\nu + 1) z_i^f t_i^2 + 4 z_i^{f\,3}) \sin(2(\theta_{i+1} - \theta_i)), \\
q_{13} = q_{31} &= 12 z_i^{f\,2} \sin(\theta_{i+1} - \theta_i), \\
q_{22} &= 2(z_i^f t_i^2 \sin^2(\theta_{i+1} - \theta_i) \\
&+ \ ((2\nu + 1) z_i^f t_i^2 + 4 z_i^{f\,3}) \cos^2(\theta_{i+1} - \theta_i)), \\
q_{23} = q_{32} &= 12 z_i^{f\,2} \cos(\theta_{i+1} - \theta_i), \\
q_{33} &= 24 z_i^f.
\end{aligned} \tag{2.14}$$

Applying (2.12) requires a change of variable, $F = F_{in}/2$. This is necessary because the quarter actuator carries a half of the input force, and taking the partial derivative with respect to the full input force F_{in} is a misapplication of Castigliano's theorem and will yield an erroneous result. To find the input displacement δ_{in}, apply (2.12) with $F_j = F$. This quantity will be multiplied by 2, since each half of the actuator undergoes this displacement. Applying (2.12) with $F_j = N_{\mathrm{Q}}$ yields the displacement of point Q. Because of symmetry about B–B, the displacement at point Q can simply be multiplied by 2 to get the displacement of point P, or δ_{out}.

Because each segment is a linearly elastic, Euler–Bernoulli beam with point loads, one would anticipate that the displacement expression will be linear in the loads multiplied by a multivariate cubic polynomial in dimensions with units of length (width of the amplifier, height, and so on). In a manner of speaking, this will be seen to indeed be the case, but the expression is complicated due to substituting in the expressions for the internal reactions along section A–A and B–B. This analysis solved (2.8) for the unknown moment reaction M_Q, and back substitution puts the coefficient of the M_Q term in (2.8) in the denominator, which will be first order in these quantities. However, looking at the *relative order* of the displacement expression (order of the numerator – order of the denominator), it is 3, the same order as a simple beam. If the structure is not rectangular, there are also trigonometric factors as part of these expressions. So even with only a few geometric parameters, the analytical expression can quickly become long and complicated.

Collecting terms in the expressions for δ_{in} and δ_{out} with respect to the loads F_{in} and F_{out}, the denominator D_M will be common, up to a multiplicative constant. Therefore, these equations can be written as

$$\begin{bmatrix} \delta_{in} \\ \delta_{out} \end{bmatrix} = \frac{1}{D_M} \begin{bmatrix} C_{11} & C_{12} \\ C_{21} & C_{22} \end{bmatrix} \begin{bmatrix} F_{in} \\ F_{out} \end{bmatrix}, \tag{2.15}$$

where C_{ij} are quartic factors of the geometric parameters. The compliance matrix in this equation is precisely an immittance matrix of a two-port network. (Choosing displacement to be analogous to voltage makes it an admittance matrix.) Its inverse is precisely the stiffness (impedance) matrix introduced in [267]. According to Maxwell's reciprocity theorem [4], at DC, the compliant mechanism will be a reciprocal two-port network, i.e., it has a symmetric stiffness matrix. Therefore, following the notation in Section 1.12 [267], the stiffness form is written as

$$\begin{bmatrix} F_{in} \\ F_{out} \end{bmatrix} = \begin{bmatrix} s_1 & s_3 \\ s_3 & s_2 \end{bmatrix} \begin{bmatrix} \delta_{in} \\ \delta_{out} \end{bmatrix} = \boldsymbol{S} \begin{bmatrix} \delta_{in} \\ \delta_{out} \end{bmatrix}, \tag{2.16}$$

where s_1 corresponds to the stiffness in the input direction when the output is blocked, s_2 corresponds to the stiffness in the output direction when the previous amplification layer is completely rigid, and s_3 describes the cross-coupling from input to output, e.g., how much force is generated at a fixed output due to a given input displacement. s_3 also has units of stiffness. This matrix is positive definite; physically speaking, the combined effect of the stiffnesses in the input and output load's own direction will always be greater than the cross-coupling effect, due to the storage of strain energy in the compliance of the mechanism.

2.1.4 Connecting strain amplifiers and amplified stacks together

The goal of this book is to expound on how amplified piezoelectric stacks can be used as artificial muscles. Like muscle fibers, their true advantage is realized when they are used together, so that the aggregate of each of their individual efforts effects a force or displacement of the environment. In order to realistically predict how the entire "muscle" will behave under activation, we must understand what happens in a series combination when some units are active and others are inactive. Inactive units don't simply "move over", they deform, and this deformation must be accounted for in light of the two-port model. This section will develop a more reasonable model for this behavior. Secondly, one would reason that it is not possible to simply amplify displacement using more and more amplifying layers *ad infinitum*. There is an important relationship between the impedance of the entity being amplified and the strain amplification layer itself. If these do not match well, the strain amplifying layer will attenuate, rather than amplify the displacement.

2.1.4.1 Series combinations of internal subunits

The examples of prior sections have all involved series connections of component subunits before strain amplification. "Component subunits" could refer to individual Cédrat amplified piezoelectric stacks, or "higher level" subunits consisting of 2-layer structures: a series combination of Cédrat units with a strain amplifier surrounding them. Taken to its furthest logical conclusion, a subunit could mean any number of piezoelectric stacks and strain amplifying layers, provided it is amplified further at some subsequent stage to produce the complete actuator. Series combination within a strain amplifying structure increases the displacement of the resulting actuator in two ways: first, the displacement of the subunits are additive, resulting in a larger input displacement to be amplified when all units are active, and second, the increased length of this series combination makes the major diagonal of the rhombus longer relative to the minor diagonal, producing more displacement. Even though the two-port model dispenses with this idea of an amplification factor for cases where the impedance of the amplifier is significant relative to that of the subunit driving it, this principle still qualitatively holds.

To quantitatively predict what happens when some subunits are active and others inactive, however, there is a need to estimate how much of an effect adding an additional driving subunit to a given series combination will have on the displacement and force capability. Simply multiplying individual displacements by the number of units in the series combination does not adequately predict how the series combination will behave. This is because the strain amplifying layer constrains the endpoints of the series combination somewhat and contraction of a single unit in the series chain will deform each of the others in addition to transmitting displacement at the output. When analyzed using the two-port model, however, a much more accurate picture is given.

Consider the following example: Fig. 2.5A shows a strain amplifying mechanism driven by 4 subunits in series. Fig. 2.5B shows a strain amplifying mechanism with an identical stiffness matrix, but with a single driving subunit and rigid spacers to transmit its action to the input of the mechanism. Will the actuator in Fig. 2.5A produce four times the displacement of the one in Fig. 2.5B? One might initially be inclined to say yes, but in reality the situation is more nuanced than that. It depends on the two-port properties of the strain amplifier (gray) compared to the driving subunit impedance. What happens as the chain gets longer and longer? The impedance of the series chain

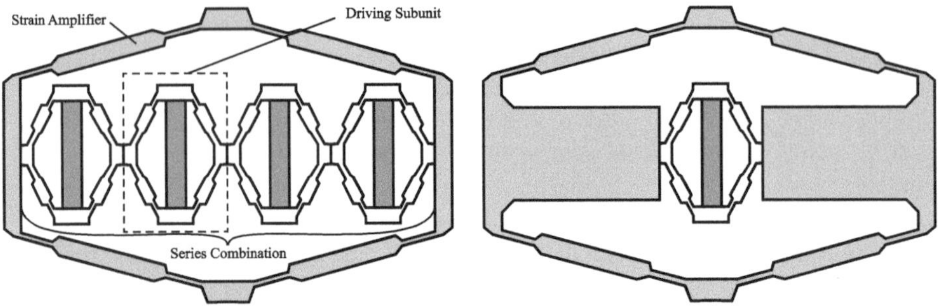

(A) Four subunits in series (B) Single subunit with rigid spacer

Figure 2.5 Comparison of an actuator with the same outer layer and different numbers of internal subunits.

of driving subunits gets smaller and smaller. This means that the longer the chain gets, the more the behavior deviates from the idealized "amplification factor" model. This provides some guidance as to how long a series combination should be.

The strain amplifying layer shown in the figure will have a two-port model with parameters s_1, s_2, and s_3. Each of the driving subunits will have a lumped stiffness $\tilde{k}$. No matter how many individual strain amplifying layers are present, each will feel to the environment like a linear spring that is the combination of all strain amplifying layers along with the piezoelectric stack at its core. A method for calculating this lumped stiffness will be presented in Section 2.3. Thus if one or several driving subunits in the series combination are activated, the remaining subunits in the chain will appear as passive springs of stiffness $\tilde{k}$.

Consider a chain of N subunits, N^{ON} of which are active. If we are interested in the output of the actuator only, and not in the individual displacements of the driving subunits, the order in which the active and inactive units are placed does not matter. The equations of equilibrium and compatibility, when combined to compute the output displacement and force will yield the same result [215]. So the series chain of driving subunits can be represented as in Fig. 2.6, with the active units on the left and the inactive on the right. This model will return the correct input force and displacement to the strain amplifying layer even though their units may not physically be arranged in this way. Upon activation the active units undergo a (fictitious, since they are not necessarily adjacent) displacement x and impart a force F to the inactive units. That is to say, the active subunits are modeled as a force generator, but in order to move, they must also overcome their own stiffness. The end of the series combination displaces by an amount δ^{in}. Because the series chain is bonded to the input (major axis of the rhomboidal strain amplification mechanism), this displacement of the chain is the same as the input displacement to the amplifying mechanism (2.16).

Using this model, we can calculate the force produced by a series chain including the effects of each subunit's impedance. When a subunit is activated, its resting length changes from its tabulated dimensions, to δ_s^{free} (a datasheet value), the free displacement of that individual subunit. The force F shown in the figure can then be expressed

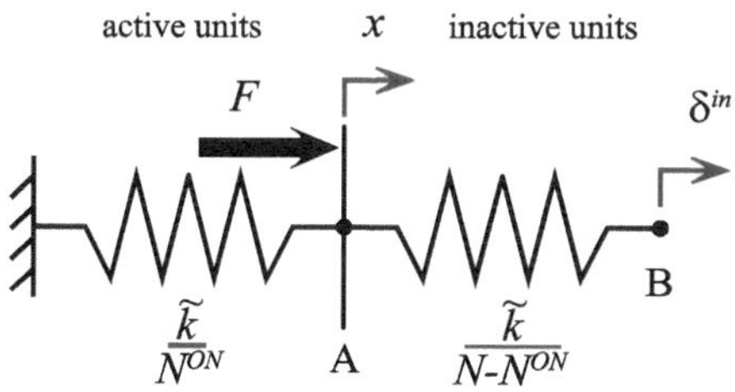

Figure 2.6 Active and inactive subunits represented as springs.

as

$$F = \frac{\tilde{k}}{N^{ON}}(N^{ON}\delta_s^{free} - x). \tag{2.17}$$

Applying equilibrium at point A and solving for x results in

$$x = \frac{N^{ON}}{N}\left((N - N^{ON})\delta_s^{free} + \delta^{in}\right). \tag{2.18}$$

Substituting this into (2.17) yields

$$F = \frac{N^{ON}}{N}\tilde{k}\delta_s^{free} - \tilde{k}\frac{\delta^{in}}{N} = \frac{N^{ON}}{N}F_s^{block} - k\frac{\delta^{in}}{N}, \tag{2.19}$$

where F_s^{block} is the blocked force of a single subunit (also a datasheet value). Substituting into (2.16) and solving in the blocked and free cases gives

$$\delta^{free} = -\frac{N^{ON}}{N}\frac{s_3 F_s^{block}}{s_3^2 - s_2(s_1 + k/N)}, \tag{2.20}$$

$$F^{block} = \frac{N^{ON}}{N}\frac{s_3 F_s^{block}}{s_1 + k/N}. \tag{2.21}$$

Using the equations above, we can compare the free displacement δ^{free} and blocked force F^{block} of the entire actuator for the case shown in Fig. 2.5A with the one shown in Fig. 2.5B that has only a single driving subunit. In fact, we can do this for any number of driving subunits N (provided they fit). We simply consider the ratio of the absolute value of expression (2.21) with an unspecified N to the absolute value of the same expression with $N = 1$, and likewise for (2.20). This results in the following expressions:

$$R_\delta = \frac{N^{ON}(\Delta_S + s_2\tilde{k})}{N\Delta_S + s_2\tilde{k}}, \tag{2.22}$$

$$R_F = \frac{N^{ON}(s_1 + \tilde{k})}{Ns_1 + \tilde{k}}, \tag{2.23}$$

where $\Delta_S = s_1 s_2 - s_3^2$ is the determinant of the stiffness matrix of the strain amplification layer (shown in gray in the figures). One useful metric is to find the minimum N^{ON} for which both R_F and R_δ exceed 2. This is the number of units that need to be activated in order for the mechanism in Fig. 2.5A to double the performance of that in Fig. 2.5B.

Activating each and every subunit in the chain means $N^{ON} = N$. Looking at (2.23), when $\tilde{k} \gg s_1$ (the amplifying structure is extremely compliant in the input direction when compared to a subunit), R_F approaches N^{ON}, meaning that activating one driving subunit produces the same blocked force as the single-driving-subunit actuator, and the force produced is multiplied by the number of units active N^{ON}. In this case the 4-unit chain produces up to 4 times the blocked force of its counterpart. This is only an upper bound; as s_1 becomes significantly close to $\tilde{k}$, the 4-unit chain produces less than 4 times the blocked force. It also underscores the importance of the relative stiffnesses of various layers in the hierarchy. Looking at (2.22), it is clear that R_δ approaches N as Δ_S approaches 0. A mechanism with a singular stiffness matrix is exactly the idealized or hinged mechanism described in Section 1.3. This means that the series chains should in general be kept short, and the strain amplifying mechanism should have a stiffness matrix with a small determinant, i.e., close to the idealized mechanism.

As N becomes large, R_F approaches an upper bound of $1+\tilde{k}/s_1$. Likewise, R_δ has an upper bound of $1 + s_2\tilde{k}/\Delta_S$. Evaluating these simple metrics can be helpful to the designer when choosing how many subunits to put in series at a given layer in the hierarchy. Figs. 2.7 and 2.8 show the progression of R_δ and R_F, respectively, with increasing N at maximum activation. The example shown has a strain amplifier with a stiffness matrix

$$S = \begin{bmatrix} 80e3 & 5770 \\ 5770 & 660 \end{bmatrix} \tag{2.24}$$

and a stiffness of the driving subunits of $\tilde{k} = 384e3$. Note that a chain of 20 driving subunits with all of them active produces a free displacement of less than 5 times that of a single unit with rigid spacers and the blocked force falls off as well as the number of units increase. In this case, where $s_1 < \tilde{k}$ but they differ by less than order of magnitude, we see substantial diminishing returns from making the series chain longer and longer.

2.1.5 Effectiveness of multiple layers and figures of merit

Nested mechanisms can produce more aggressive force–displacement tradeoffs than a single compliant mechanism, reaching larger stroke lengths without drastically increasing the resting length of the actuator. Each time a layer is added, some strain energy is stored in the compliant mechanism, so unlike the idealized mechanism, the amplification effect is not simply multiplicative. The two-port network formalism allows one to analyze the effect of adding successive layers of strain amplification, determine if adding an additional layer is really a good idea, and estimate the point of diminishing returns.

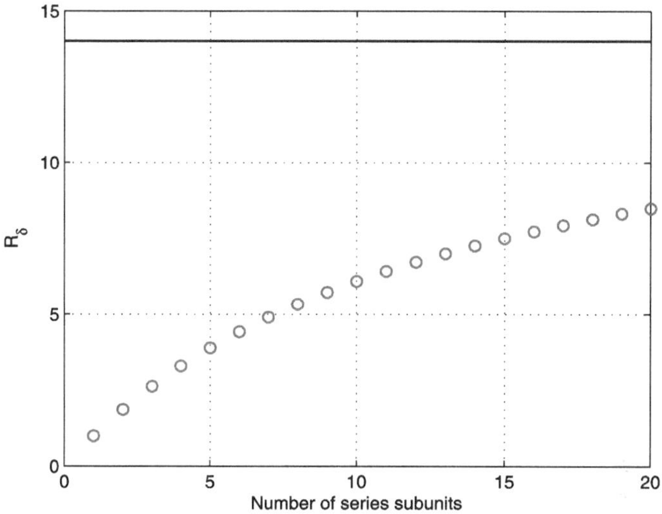

Figure 2.7 Free displacement ratio for series combinations of second layer subunits within the third layer amplifying mechanism. Black line is the theoretical upper limit.

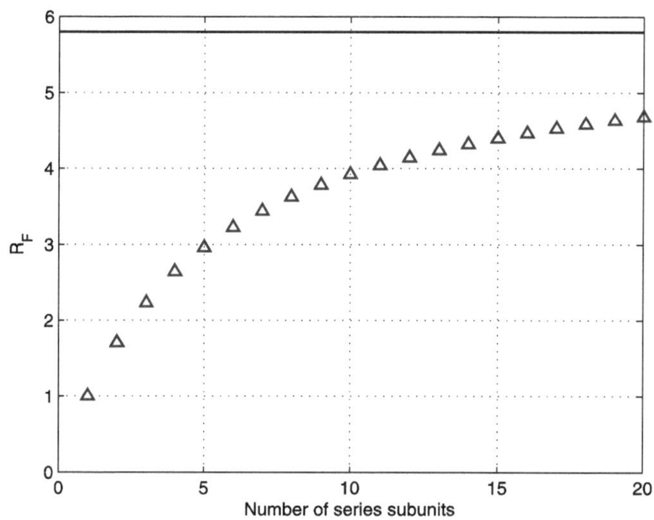

Figure 2.8 Blocked force ratio for series combinations of second layer subunits within the third layer amplifying mechanism. Black line is the theoretical upper limit.

2.1.6 Amplifying still further with additional strain amplifying mechanisms

Suppose that you have an amplified piezoelectric stack from a supplier such as Cédrat, Dynamics Structures and Materials, or Piezo Drive. Many of these suppliers have amplification factors of 20 or more, but the strain rate of these are still too small for most robotics applications. More displacement can be achieved by concatenating these in series, but if the series combination is too long relative to the stroke, it is not usable. For many applications such as signaling, camera positioning, and simple grasping, the force provided is more than is needed.

With the displacement too low, and the force too high, this is begging for a different force–displacement tradeoff. Of course, one could adjust the characteristics of the strain amplification mechanism to get different s_1, s_2, and s_3 in theory, but in practice this is more difficult. We saw in earlier sections that s_1, s_2, and s_3 are functions of the very same geometric parameters, and that in general thin sections and small rhomboid angles lead to higher free displacement when the piezoelectric stack is active. The commercial offerings by Cédrat and others have pretty aggressive tradeoffs already. Manufacturing constraints may prevent any further increase in strain at the expense of force. We also saw in the previous section that making the series chain longer and longer in the input direction (transverse to operation) in hope of getting more displacement in the output direction runs into diminishing returns with even modest numbers of driving subunits in series.

As has already been mentioned before, the solution is nesting, or amplifying the amplified stack (or more likely, a series combination of amplified stacks). The driving subunit of the previous section is then an amplified piezoelectric stack rather than a stack itself. This brings about another tradeoff for increased displacement at the expense of force and brings us closer to the realm of macroscopic displacements. This would result in a much more modest force, but in many applications this is acceptable, such as positioning stages, signaling, and electronic pick and place.

Nesting is not without its pitfalls, and approaching nesting carelessly can even result in an actuator with *smaller* displacement that the series chain of driving subunits by itself. Increasing displacement is not as simple as taking on more and more strain amplifiers until a large enough stroke is achieved. The deformable behavior of the strain amplifiers needs to be considered and this means looking at the two-port models of each strain amplifier included and how they interact.

2.1.6.1 Example: reusing a strain amplifier you have already designed

The following, somewhat naïve example illustrates what is meant in the preceding discussion. Take, for instance, the amplified piezoelectric stack shown in Fig. 2.9A. It has an angle of less than 45° and appears quite thin, so it should amplify the displacement of the piezoelectric stack, similarly to any of the commercial products available. But based on the displacement of those models, we would not expect this displacement to be at the level required in robotics applications. Since he wants more amplification, the designer suddenly hatches an ingenious plan: he can take another amplification mechanism, just like the one he made for the piezoelectric stack, turn it around, and make the "three-dimensional" mechanism shown in Fig. 2.9B! Of course, given the strain energy is stored in not one, but two strain amplification layers, he would not expect the increase in displacement to be multiplicative, but it should move further than the unit in Fig. 2.9A, right? Looking at the two-port model of each strain amplifying layer and how they interact shows us that this is not such a great idea after all.

(A) two dimensional mechanism (B) three dimensional mechanism

Figure 2.9 Multi-layer nested geometry with reuse of a compliant mechanism. © 2013 IEEE, reprinted with permission

Using the procedure described in the previous sections, the designer can find the s_1, s_2, and s_3 of the strain amplifier used in building the "three-dimensional" multi-stage strain amplification mechanism. In order to represent the cascaded connection of two strain amplification mechanisms, it is more enlightening to use a different immittance matrix, the *forward transmission* matrix. This represents same two-port network for the same compliant strain amplification matrix, but instead of choosing displacements at the ports as the input and forces at the ports as the output, the "input" to the network is a mixed-unit vector of force and displacement along the major axis of the rhomboid, and the output is the force and displacement along the minor axis. The forward transmission matrix can be calculated directly following the procedure in [42] and is expressed as follows:

$$\begin{bmatrix} -\delta_{out} \\ F_{out} \end{bmatrix} = \frac{1}{s_3} \begin{bmatrix} s_1 & -1 \\ -\Delta_S & s_2 \end{bmatrix} \begin{bmatrix} \delta_{in} \\ F_{in} \end{bmatrix}, \tag{2.25}$$

where Δ_S represents the determinant of the stiffness matrix S in (2.16). δ_{out} is negated to represent the change of direction of displacement with each added layer in the hierarchy. Some readers might consider the forward transmission matrix to be more fundamental and more intuitive – after all, isn't the major diagonal the "input" to the strain amplification layer and the minor diagonal the "output"? As we will see in the next section, it is relatively easy to experimentally measure and tabulate the stiffness matrix. Designing an experiment to characterize the forward transmission matrix would be more difficult, in part because unlike the stiffness matrix, the forward transfer matrix does not have consistent units among all its elements.

To find a forward transmission matrix for a set of two nested strain amplification mechanisms, as in Fig. 2.9B, we simply take the product of the two forward transmission matrices. In our example, they will be the same, but to derive a more general relationship to be used later, we will need to distinguish between the "inner" and "outer" strain amplification mechanisms. So the s_1, s_2, and s_3 for each mechanism will be given a leading superscript, beginning with the innermost. For example, ${}^1s_3$ refers to the off-diagonal term of the inner mechanism (visible in both Figs. 2.9A and 2.9B) and ${}^2s_1$ is the input stiffness of the outer mechanism with the output blocked. After adding the superscripts and taking the product to compute the combined forward transfer characteristic, we get

$$\begin{bmatrix} \delta_{out} \\ F_{out} \end{bmatrix} = \frac{1}{{}^1s_3^2 s_3} * \begin{bmatrix} {}^1s_1\,{}^2s_1 + {}^1\Delta_S & -{}^1s_2 - {}^2s_1 \\ -{}^2\Delta_S\,{}^1s_1 - {}^1\Delta_S\,{}^2s_2 & {}^2\Delta_S + {}^1s_2\,{}^2s_2 \end{bmatrix} \begin{bmatrix} \delta_{in} \\ F_{in} \end{bmatrix}. \tag{2.26}$$

(The negative sign cancels out due to the direction being reversed twice.) Returning to our example, which uses the same mechanism twice (${}^1s_j = {}^2s_j$), the leading superscripts become unnecessary. In this case (2.26) becomes

$$\begin{bmatrix} \delta_{out} \\ F_{out} \end{bmatrix} = \frac{1}{s_3^2} \begin{bmatrix} s_1^2 + \Delta_S & -s_1 - s_2 \\ -\Delta_S(s_1 + s_2) & \Delta_S + s_2^2 \end{bmatrix} \begin{bmatrix} \delta_{in} \\ F_{in} \end{bmatrix}. \tag{2.27}$$

2.1.6.2 Displacement amplification figure of merit

So is the "three-dimensional" mechanism really better than the "two-dimensional" amplified piezoelectric stack? Naturally, it is difficult to evaluate this across a wide range of operating conditions, but it is simple enough to check the *free displacement*, or the displacement of the output flange when the piezoelectric stack is activated and no load is applied. A good dimensionless figure of merit that will tell whether the additional strain amplification mechanism added is a good idea is the ratio of the free displacement of the "three dimensional mechanism" to the free displacement of the amplified piezoelectric stack we started with. This can be generalized to complex strain amplification structures with many layers of amplification, although it is most useful for comparing the effects of adding one layer at a time. For our example with identical strain amplification layers, after a little algebra, this figure of merit works out to be

$$FOM = \frac{|\ \delta_{3D}^{free}\ |}{|\ \delta_{2D}^{free}\ |} = \frac{s_3}{s_1 + s_2}. \tag{2.28}$$

If this figure of merit is greater than one, adding the additional strain amplifying mechanism will increase the displacement still further. However, since the stiffness matrix of an individual strain amplification mechanism must be positive definite, $s_3 < \sqrt{s_1 s_2}$, and this figure is upper bounded by $\frac{\sqrt{s_1 s_2}}{s_1 + s_2}$. Therefore, the figure of merit must lie below the surface shown in Fig. 2.10. Unless one has a really firm grasp on what the immittance parameters mean, this can be difficult to understand in terms of design implications. Instead, using the analytical expressions for the two-port model of the mechanism developed in earlier sections, one can plot the figure of merit as it varies with *geometric* parameters. The figure of merit with changes in the thickness and rhombus angle of the strain amplifying mechanism are plotted in Fig. 2.11. The maximum value on this plot is around 0.5! This shows that the designer's plan in our example is not such a good one after all; the free displacement of the "three-dimensional" mechanism is actually lower!

This is not to suggest a figure of merit greater than one cannot be achieved when the two layers are allowed to vary independently, in fact, previous work [267,222,

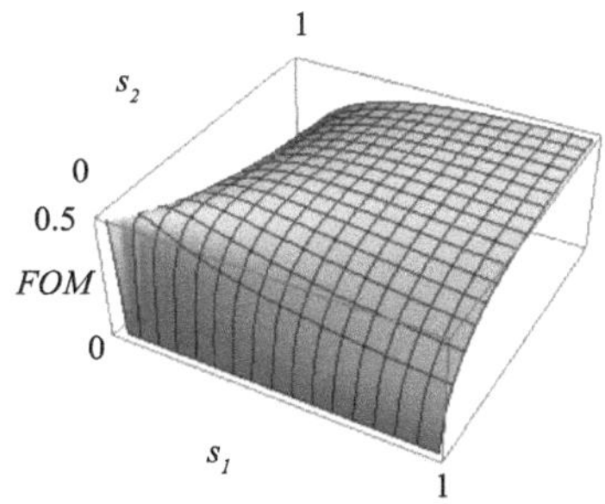

Figure 2.10 Bounding surface on figure of merit over s_1, s_2. © 2013 IEEE, reprinted with permission

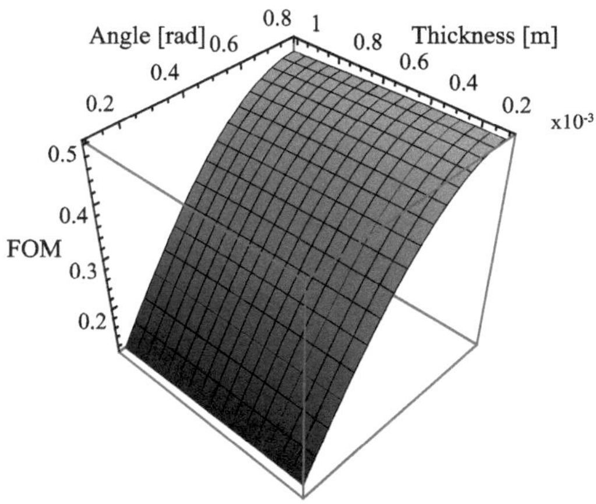

Figure 2.11 Variation in figure of merit with angle and thickness. © 2013 IEEE, reprinted with permission

246] indicates that this is possible. In each of these cases, the geometry of the outer layer is drastically different from the inner layer. The process for optimizing multilayer mechanisms is quite involved and is addressed in Section 2.3. This example shows how analytical expressions for the two-port immittance parameters from Section 2.3 combined with the results in this section allow for quick evaluation of nested topologies. In this way, one can determine whether there is a good range of geometric parameters for a given performance characteristic, something that is not easily done with purely numerical tools, such as finite element methods.

2.1.6.3 Matching the actuator to the driving subunit for greater displacement

The previous discussion teaches us that we need to use caution when using a rhomboidal strain amplifying mechanism to make sure that it will indeed amplify the displacement of the input units driving it. If nothing else, it makes the point that there is more involved in increasing free displacement than adding more and more rhomboids until it works. In fact, engineering intuition should suspect that something like this might happen: if this were nothing more than a multiplicative tradeoff, one would

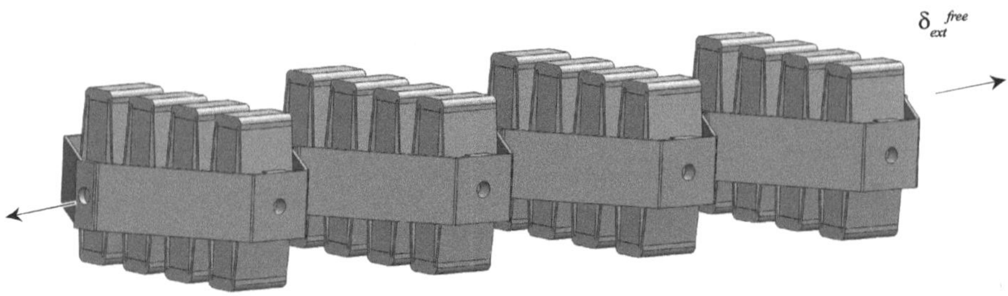

Figure 2.12 Existing actuator for which a higher free displacement is desired.

conclude that in theory unlimited displacement could be produced from a single piezoelectric stack just by including enough stages of amplification. So is multi-stage strain amplification something destined to fail? Far from it – after all, if it were, what would be the point of this book? Prior research [246,222,267] has found combinations of rhomboids that produce greater displacement than the single-stage amplifier included within it. What is clear about each of these examples is that the outermost rhomboid is substantially more flexible than the inner ones.

When designing an actuator, it might seem natural to proceed as follows:

1. Design a strain amplification mechanism for the piezoelectric stack to make the FOM as large as possible within manufacturing constraints.
2. Check the free displacement: did it produce enough amplification?
3. If not, make a series combination of the amplified stacks and nest them inside another strain amplifier, making the FOM for this new strain amplifier as large as possible.
4. Check the amplification again, and if it is still not enough, continue making series combinations and nesting until the desired amplification is reached.

While this seems appealing at first, further analysis will show that this does not work well. This is because an amplified piezoelectric stack (with one or several layers of amplification) does not necessarily make an effective driving subunit for yet another strain amplifying mechanism.

Suppose you have a cellular actuator consisting of a certain number of piezoelectric stacks with a single or multi-layer strain amplification mechanism. An example is shown in Fig. 2.12. The one shown in the figure has two strain amplification layers, but the analysis to follow will work with any number. Perhaps this is a legacy actuator that was used in some other application, and you want to adapt its performance to a higher displacement, lower force application. Alternatively, this could have been an actuator designed using the iterative procedure above, and it has fallen short of the desired displacement. We will call this the "existing" actuator, with a free displacement δ_{ext}^{free}. We are considering adding *one more* strain amplifying rhomboid, lightly sketched in Fig. 2.13 to give the idea. If we get the immittances of this rhomboid just right, we hope that the displacement of this "augmented" actuator, which we will call δ_{aug}^{free} will be larger than δ_{ext}^{free}.

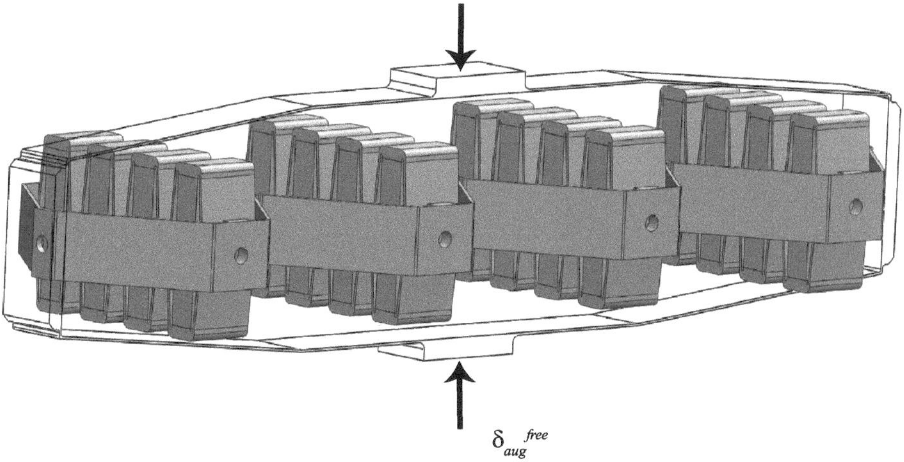

Figure 2.13 Adding an additional strain amplifying mechanism to amplify the displacement of the actuator in Fig. 2.12 still further.

To test out the effect of various design choices on the augmenting strain amplifier, we can evaluate the same FOM we used in the last section, but this time we use the displacements of the actuators shown in these two figures. Rewriting (2.28) with this new nomenclature, it becomes:

$$FOM = \frac{|\delta_{aug}^{free}|}{|\delta_{ext}^{free}|}. \tag{2.29}$$

In order for the augmenting layer to have some benefit, FOM must be greater than 1.

The two-port description of the actuator's strain amplification mechanisms are as follows:

$$\begin{bmatrix} \delta_{ext}^{out} \\ F_{ext}^{out} \end{bmatrix} = \frac{1}{\prod_{k=1}^{n} {}^{k}s_3} \quad S_n \begin{bmatrix} \delta^{in} \\ F^{in} \end{bmatrix}, \tag{2.30}$$

$$S_n = \begin{bmatrix} s_{11} & s_{12} \\ s_{21} & s_{22} \end{bmatrix}, \tag{2.31}$$

$$\begin{bmatrix} \delta_{aug}^{out} \\ F_{aug}^{out} \end{bmatrix} = \begin{bmatrix} {}^{n+1}s_1 & -1 \\ -{}^{n+1}\Delta_S & {}^{n+1}s_2 \end{bmatrix} \begin{bmatrix} \delta_{ext}^{out} \\ F_{ext}^{out} \end{bmatrix}, \tag{2.32}$$

where F^{in} and δ^{in} are the input force and displacement from the piezoelectric stack, F_{ext}^{out} and F_{aug}^{out} are the forces at the output of the existing and augmented mechanisms, respectively, and δ_{ext}^{out} and δ_{aug}^{out} are the displacements of the existing and augmented mechanisms. It should be noted that δ_{ext}^{out}, the displacement of the existing actuator when it is nested within the augmenting strain amplification layer, is not the same as δ_{ext}^{free}, the amount it moves when nothing is constraining its motion. S_n is the product of all the forward transmission matrices [42] (with $1/{}^{k}s_3$ pulled out front) of layers 1 through n in the existing actuator. The matrix in (2.32) is the forward transmission matrix of the augmenting strain amplification mechanism. ${}^{n+1}\Delta_S$ is the determinant of the stiffness matrix for the additional layer. Thus the first two equations describe the existing actuator (consisting of n layers), and the third equation describes the augmenting strain amplification layer ($(n+1)$th layer).

The free displacement expressions for the augmented actuator can be found by setting $F_{aug}^{out} = 0$ and $F_{ext}^{out} = 0$ in (2.30) and (2.32) and solving for δ_{ext}^{out} and δ_{aug}^{out}. Taking the ratio of the absolute values and simplifying, the FOM becomes

$$FOM = \frac{{}^{n+1}s_3}{{}^{n+1}s_2} - {}^{n+1}\Delta_S \frac{s_{11}}{s_{21}}. \tag{2.33}$$

If the existing actuator has just one layer of strain amplification, $n = 1$, and the augmenting layer becomes layer 2, then the FOM becomes

$$FOM = \frac{{}^{2}s_3}{{}^{2}s_2 + {}^{1}s_2 \frac{{}^{2}\Delta_S}{{}^{1}\Delta_S}}. \tag{2.34}$$

Once again the leading superscript is being used to denote to which layer the immittance parameters s_1, s_2, and s_3 refer. Any valid geometry for a strain amplification mechanism will have a positive definite stiffness matrix, meaning for any layer k ${}^{k}s_3$ is upper bounded by $\sqrt{{}^{k}s_1\,{}^{k}s_2}$. Substituting this into (2.34) reveals that a necessary condition for FOM to be greater than 1 is

$$\sqrt{\frac{{}^{2}s_2}{{}^{2}s_1}} + \frac{{}^{2}\Delta_S}{{}^{1}\Delta_S}\sqrt{\frac{{}^{1}s_1^2}{{}^{2}s_1\,{}^{2}s_2}} < 1. \tag{2.35}$$

While this expression does not tell us the optimal design procedure for a multilayer strain amplification mechanism, it does give us some important insight. Both determinants ${}^{k}\Delta_S$ are positive, and only the positive square root has physical significance. This means that each of the two terms in the expression must individually be less than one. Several conclusions can be drawn from this. The first term involves only the parameters of the augmenting layer. In order for the adding of another layer to be effective, it must have a much higher stiffness in the input direction than the output direction. The second term involves both layers. It will be small if ${}^{2}\Delta_S \ll {}^{1}\Delta_S$. For an idealized (hinged) mechanism, ${}^{k}\Delta_S = 0$. This means that the outermost layer should be as close as possible to an idealized layer, whereas the inner layer should have some numerically significant stiffness behavior. The second factor in the second term can be troublesome, since a large ${}^{1}s_1$ reduces the figure of merit. Since ${}^{1}s_2$ does not appear in the second factor of the second term, it to some degree can be used to increase the determinant. This is not a silver bullet, because we have already seen that s_1 and s_2 depend on the same geometric parameters and cannot be set independently.

The forgoing analysis shows that if our existing actuator already has a good FOM, it will already have a compliant outer layer with a small determinant and stiffer inner layers. This makes it difficult to augment the actuator with another strain amplifying mechanism to increase the displacement, because when this outer layer strain amplifying mechanism becomes an internal layer, the same properties that made it a good outer layer make it a bad internal layer!

Looking at the FOM shows us that trying to improve upon an existing actuator with an additional strain amplifying mechanism is not a very effective strategy. This, by extension, means that the strategy outlined above, designing each strain amplifier in the hierarchy for as much free displacement as possible and adding layers once you have maxed out the free displacement will usually not work. Two and three-layer structures can perform very well, but we are forced to start from scratch and design all layers concurrently. This approach will be demonstrated in Section 2.3 for three-layer cellular actuator for a camera positioning mechanism. Now that we understand the role that the immittances of each strain amplification mechanism plays in the overall displacement, we have a better picture of the advantages a multi-layer strain amplification mechanism has over one with fewer layers: it is not that we are "amplifying the displacement more", but rather that more layers give us more design parameters that we can tune to improve the strain amplifying behavior when we run into real-world constraints. If we reach a limit on the angle of a rhomboid, for instance, with multiple rhomboids, there might be able to address it by changing the angle of a rhomboid elsewhere in the hierarchy. One of the most obvious constraints is the envelope of the actuator. With three or more layers, the actuator will grow in all 3 directions of space, rather than just one or two. This may result in a better form factor than one and two-layer actuators.

2.1.7 Discussion

In this section, a method was derived using Castigliano's theorem that presents a iterative method of determining an analytical two-port model of a rhomboidal strain amplifying mechanism. The input–output behavior is described neatly by the two-port model across the entire working range by multiplying forces and displacements by a 2×2 stiffness matrix. The iterative method produces each element of the stiffness matrix as a function of the geometric parameters in closed form. This is the much-needed connection between the physical design of the rhomboid and its behavior in a nested device.

This method is specific to the types of strain amplifiers envisioned to be used in nested piezoelectric cellular actuators – they are assumed to be doubly symmetric, planar, and composed of a sequence of straight segments joined at the end. Thus, it applies to a wide variety of polygonal shapes, and the thickness of each straight segment can vary from one to another. The iterative process also assumes that forces are applied at only two places: the input and the output. These two locations correspond to the two axes of symmetry.

This formulation is made with distributed compliance in mind, but could be applied to hinged structures by including certain segments that are much thinner than those before and after, as long as the hinge is a straight segment. Rhomboids made with circular flexure hinges could be derived by a similar procedure but the closed-form expressions would be much more complicated. An alternative formulation for mechanisms with flexure hinges was presented by Lobontiu [151]. Spatial (three-dimensional) closed topology mechanisms for transmitting motion with a monolithically (or at least mostly monolithically) manufactured part are gaining interest. A functioning mechanism of this type was produced by Tanık and Parlaktaş [250] that converted rotary motion into reciprocating motion in a different plane.

2.2 Calibration of two-port network models

2.2.1 Model validation by finite element methods

The procedure in Section 2.1 produces an analytical two-port model for a wide class of geometries [220]. In order to implement this method, the designer must choose a useful set of geometric parameters that will be varied. The immittance matrix will be a function of these parameters. To demonstrate how this procedure works, this section considers an octagonal mechanism of constant thickness ($t_{ai} = t_a \ \forall\, i$), varies its parameters, and compares the analytical model's predictions to that returned by a Finite Element Method (FEM) simulation under the same loading and displacement conditions. The numerical values of the geometric parameters were varied so as to show the applicability of the model and assess its limitations.

2.2.1.1 Geometric parameterization

The geometric parameters that describe the mechanism are shown in Fig. 2.14, along with the deformed shape that this elastic body will tend to assume under a tensile load at the output. Each quarter of the octagon consists of three segments, one that begins at the input and is perpendicular to the input direction, and angled section, and one that reaches to the output and is perpendicular to the output direction. The width of the entire rhomboid is w, the length of the section perpendicular to the input is h, and the height of the angled section is d. The entire rhomboid that fits in a box is $w \times h + d$. The angled section makes an angle θ with the input direction. The depth of the rhomboid (direction mutually perpendicular to the input and output directions) is b, which, along with the thickness t_a, will be constant for all segments in this example. For the purpose of validation of the analytical model $w/2 = 1$ m and $b = 2$ mm were chosen as general characteristic dimensions, while d, h, and θ were varied. As long as the size permits the device to be analyzed as a continuum, the mesh size will scale with the geometry of a given part, meaning this is a sufficiently general procedure to give confidence in the analytical modeling approach.

The procedure in Section 2.1 for this set of parameters gives the compliance matrix elements, C_{11}, C_{21}, and C_{22}. The expressions are shown in Appendix A.4. The denominator common to all elements in the compliance matrix is

$$D_M = 6144\, E\, I\, \left(h + w - 2d(\cot\theta - \csc\theta)\right). \tag{2.36}$$

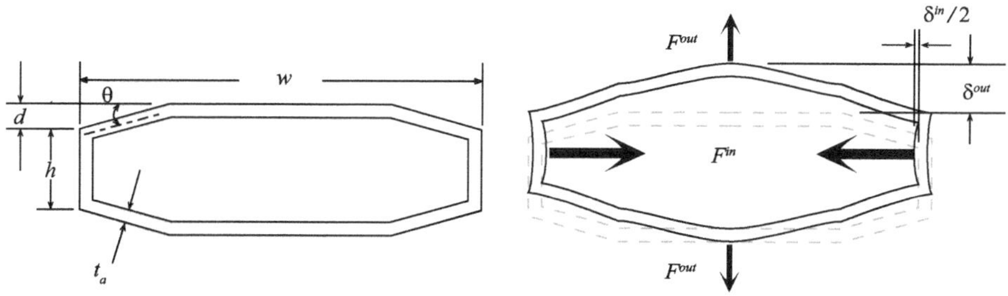

Figure 2.14 General octagonal rhomboidal shape showing the geometric parameters. The depth into the page of the shape will be denoted b, i.e., the final implementation will be machined from a plate of thickness b. © 2013 IEEE

These were evaluated in an iterative fashion described using Wolfram Mathematica with the collection of terms performed using the `Collect[]` function. E is the elastic modulus of the material used (High Silicon Bronze). Because b and t_a are constant, the denominator is greatly simplified; note the presence of a constant I (the area moment of inertia) in the denominator. If b and t_a were to vary from segment to segment, the moment of inertia would also vary from segment to segment and D_M would be much more complex. The immittances in stiffness matrix form (which will be measured experimentally for one implementation in Section 2.2.2) are obtained by Cramer's rule as

$$s_1 = \frac{C_{11} D_M}{C_{11} C_{22} - C_{21}^2}, \tag{2.37}$$

$$s_2 = \frac{C_{22} D_M}{C_{11} C_{22} - C_{21}^2}, \tag{2.38}$$

$$s_3 = \frac{-C_{21} D_M}{C_{11} C_{22} - C_{21}^2}. \tag{2.39}$$

This mechanism is comparatively simple, but the expressions describing it are still rather complicated, even with just a few parameters. This should be borne in mind when parameterizing the geometry.

2.2.1.2 Comparison for various geometries

The test matrix to evaluate the fidelity of model was as follows: d, θ, and t_a were varied one at a time with the sum $h + 2d$ held equal to w. θ was held at 45° when it was not being varied, d was held at 500 mm when not being varied, and t_a was held at 2 mm when not being varied. Loads were chosen for each geometry such that the mechanism would have a small, but visually discernible displacement at true scale. Values that did not result in a simple convex polygon were not used.

The expressions in (A.48)–(A.50) and (2.37) were evaluated numerically using Wolfram Mathematica according to the test matrix. One half of the compliant mechanism was simulated for each set of parameters using Dassault Systèmes' SolidWorks

Table 2.1 Mechanism displacements when varying d, h, $F_{in} = 0.002$ N, $F_{out} = 0$ N.

d [mm]	δ_{in}			δ_{out}		
	FEM	Analytical Model	% difference	FEM	Proposed Method	% difference
2	1.481	1.488	0.47	−1.780	−1.787	0.39
5	1.486	1.487	0.06	−1.782	−1.788	0.34
10	1.479	1.485	0.41	−1.785	−1.791	0.34
50	1.469	1.473	0.27	−1.809	−1.816	0.39
200	1.404	1.409	0.35	−1.928	−1.816	6.16
500	1.225	1.230	0.41	−2.140	−2.127	0.61
900	0.932	0.937	0.54	−1.863	−1.870	0.37
990	0.847	0.852	0.59	−1.696	−1.700	0.23

Table 2.2 Mechanism displacements when varying θ, $F_{in} = 0.002$ N, $F_{out} = 0$ N.

θ [°]	δ_{in}			δ_{out}		
	FEM	Analytical Model	% difference	FEM	Proposed Method	% difference
85	1.462	1.468	0.49	−1.830	−1.836	0.33
80	1.442	1.448	0.42	−1.880	−1.884	0.21
70	1.398	1.404	0.43	−1.966	−1.971	0.26
60	1.344	1.350	0.44	−2.043	−2.048	0.26
30	0.967	0.971	0.41	−2.011	−2.018	0.36

Simulation finite element modeling software, along with a sliding end constraint so that the mechanism would behave as the full mechanism would. The reason it was simulated this way was to avoid any difficulties with the finite element solver due to the doubly connected structure. The results for several different geometries are shown in Tables 2.1–2.6. h and d were varied under constant load. Variation in t_a (which is the most important quantity in the moment of inertia) greatly affected the total displacement of the output under a given load, hence the load was varied so that the deformation produced for each test would be large enough to be numerically significant, but small enough that gross plastic yielding would not occur in a specimen of that geometry.

The analytical model produced in this example shows good agreement with finite element methods (within 1% in most cases) for thin mechanisms. As the ratio between t_a and w approaches 0.1, the accuracy begins to suffer, but most useful compliant mechanisms will be thin and be within this range of accuracy. So for thin mechanisms, this geometric parameterization has shown good agreement with Finite Element models, but it is much less computationally intensive and is much less laborious, since no CAD model needs to be created to try out a set of parameters. The iterative method

Table 2.3 **Mechanism displacements when varying t_a, F_{in} variable, $F_{out} = 0$ N.**

t_a [mm]	F_{in} [N]	δ_{in}			δ_{out}		
		FEM	Analytical Model	% difference	FEM	Proposed	% diff
2	0.02	1.481	1.488	0.47	−1.780	−1.787	0.39
5	0.2	7.814	7.874	0.76	−13.50	−13.61	0.81
6		4.515	4.557	0.93	−7.804	−7.88	0.97
10	1	4.372	4.922	1.53	−8.377	−8.51	1.59
25	12	3.641	3.782	3.89	−6.285	−6.535	3.98
50	60	2.196	2.370	7.90	−3.779	−4.087	8.15
100	1200	5.131	5.978	16.5	−8.740	10.24	17.2

Table 2.4 **Mechanism displacements when varying d, h, $F_{in} = 0$ N, $F_{out} = 0.002$ N.**

d [mm]	δ_{in}			δ_{out}		
	FEM	Analytical Model	% difference	FEM	Proposed Method	% difference
150	−9.43	−9.46	0.33	28.57	28.66	0.31
250	−9.84	−9.88	0.37	27.56	27.65	0.32
500	−10.57	−10.63	0.33	24.52	24.60	0.35
900	−9.31	−9.35	0.45	18.66	18.74	0.47
990	−8.40	−8.52	1.45	16.96	17.04	0.49

Table 2.5 **Mechanism displacements when varying θ, $F_{in} = 0$ N, $F_{out} = 0.002$ N.**

θ [°]	δ_{in}			δ_{out}		
	FEM	Proposed Method	% difference	FEM	Proposed Method	% difference
45	−10.57	−10.63	0.33	24.52	24.60	0.35
35	−10.47	−10.51	0.36	24.69	24.78	0.35
30	−10.05	−10.09	0.41	25.09	25.18	0.36
27.5	−9.56	−9.69	1.4	25.48	25.5	0.08

will work for other geometric parameterizations as well, but the expressions may be more complicated.

2.2.2 Experimental results

A procedure is developed in this section for measuring the immittance parameters s_1, s_2, and s_3 experimentally for a strain amplification mechanism. It is performed

Table 2.6 **Mechanism displacements when varying t_a, $F_{in} = 0$ N, F_{out} variable.**

t_a [mm]	F_{in} [N]	δ_{in}			δ_{out}		
		FEM	**Proposed**	**% diff**	**FEM**	**Proposed**	**% diff**
2	0.02	−10.57	−10.63	0.33	24.52	24.60	0.35
5	0.02	−0.6753	−0.6806	0.78	1.563	1.575	0.77
	0.2	−6.725	−6.806	1.2	15.45	15.75	1.94
10	2	−8.376	−8.508	1.58	19.39	19.68	1.50
25	24	−6.285	−6.535	3.98	14.56	15.13	3.91
50	60	−3.780	−4.087	8.12	8.778	9.472	7.91
100	1200	−8.748	−10.24	17.06	20.53	23.86	16.2

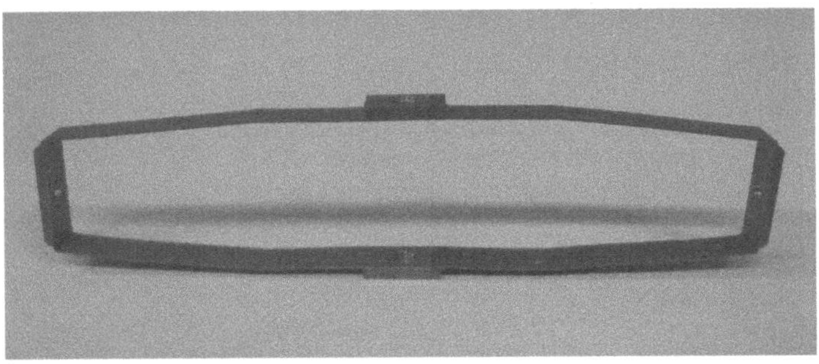

Figure 2.15 Mechanism characteristics used in experiment. Parameters are as in Fig. 2.14. © 2013 IEEE, reprinted with permission

on a strain amplifying mechanism that is used in a three-layer nested amplification mechanism described in Section 2.3. The outermost layer of the nested structure was used because it was the largest of the three, and as such, is the easiest to mount in the fixtures used. It is an example with the same parameterization in Section 2.2.1.1: d, w, h, θ, and t_a. A photograph of the mechanism is shown in Fig. 2.15 and the numerical values of its parameters are found in Table 2.7. How these values were selected will be described in the design of the nested mechanism in Section 2.3. A rigid section was added to the middle to increase the height, because the chosen value of h is not enough to allow for the internal layers of the nested architecture to be placed within this mechanism. It is assumed that any deformation of this rigid section is negligible. The mechanism was machined from a 5 mm thick plate of C655 High Silicon Bronze by wire Electrical Discharge Manufacturing (EDM), by Jack's Machine of Hanson, MA. The C655 alloy has an elastic modulus of 105 GPa and a Poisson's ratio of 0.346.

2.2.2.1 *Experimental procedure*

s_1, s_2, and s_3 can be identified directly from experimental data by a least squares regression, but several experiments are needed so that the calculation will be well

Table 2.7 Parameter values used in experiment.

Parameter	Value	Units
d	2.4	mm
h	1	mm
θ	6	°
t_a	0.15	mm

conditioned. Each corresponds to a different loading/imposed displacement condition. Although any loading–displacement condition is a valid data point, this procedure takes advantage of the ease of measurement of the fixed and free conditions, because these are as valid configurations as any other. Testing in a fixed or free condition gave implicit knowledge that either a force or displacement was zero, meaning that fewer sensors were needed. This strain amplification mechanism is not very stiff in the out of plane direction, and would not perform well in a cantilevered configuration, so both sides of a given mounting flange needed to be connected to the fixture. This meant that imposed and measured displacements were the sum of two quantities, one for each side, rather than marking distance from a fixed flange. The fixture was made from aluminium, which was sufficiently rigid compared to the strain amplification mechanism.

The first experiment (shown in Fig. 2.16) varied the output displacement with the input fixed and measured input and output force. The output displacement could be set precisely using a pair of NAI Aperture micropositioning stages. Forces were read from Futek LSB200 load cells for each output displacement value and the data recorded.

The second experiment (shown in Fig. 2.17) varied the input displacement with the output free and measured input force and output displacement. Once again, input displacement (the independent variable) was set using the micrometer stage and force was measured using the load cells. To measure output displacement (a dependent variable) two Microepsilon OptoNCDT laser position sensors were used. Looking at the figure, it is easy to see why the free condition was so important to this experiment: placing a force sensor to measure output force would block the laser beam, making it impossible to measure the displacement.

2.2.2.2 Results and data analysis

These two experiments constitute an independent set of measurements, making it possible to identify s_1, s_2, and s_3 from the data gathered. Rearranging Section 2.1 with the measured, imposed, or implied displacements on the one side and the forces on the other results in the following form:

$$\boldsymbol{D}\begin{bmatrix} s_1 \\ s_2 \\ s_3 \end{bmatrix} = \boldsymbol{b}, \tag{2.40}$$

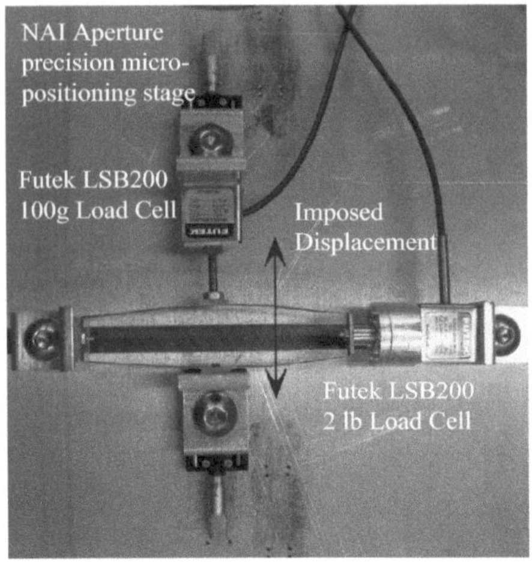

Figure 2.16 First experiment (input fixed). © 2013 IEEE, reprinted with permission

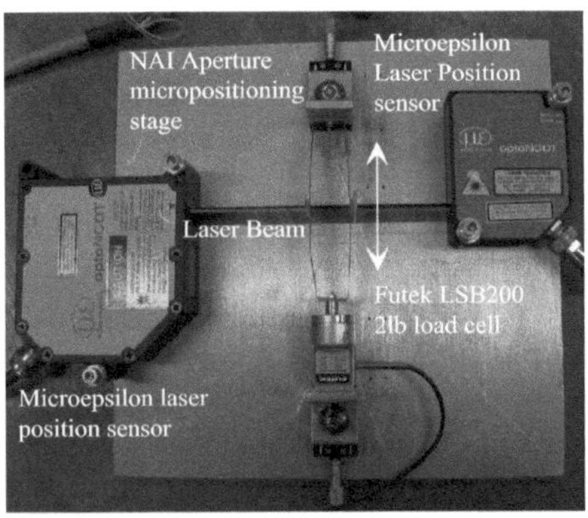

Figure 2.17 Second experiment (output free). © 2013 IEEE, reprinted with permission

where each row is one set of measured data. By plugging in the values of displacement and force, one quickly sees the reason two experiments were needed. If a fixed or free condition is used, with only one of the two experiments described, the matrix $\boldsymbol{D}$ is not of full column rank, and the parameters cannot be found. However, with both experiments, a least squares regression [248] returns the result. It is worth noting an important pedagogical fact that engineering students should understand: there is a connection between the notions of linear independence of the columns of $\boldsymbol{D}$ and independence of experiments (or experiments that measure the behavior of the mechanism in a physically different way).

The results of the experiments are shown in Table 2.8. The analytical model predicts all parameters to within 18% of their measured values. This deviation is most likely caused by nonlinearities from the small angle assumption being violated, mis-

Table 2.8 Measured and modeled immittances.

Parameter	Analytical Model [N/m]	Experiment [N/m]
s_1	83,400	80,000
s_2	561	660
s_3	6560	5770

matches in material properties, and effects of residual stresses from the manufacturing process and transport.

2.2.3 Discussion

The two-port model is a simple model, a first order linear approximation. The key thing to remember is that the two-port model is the simplest model that takes into account the fact that output displacement will have an associated input displacement, which means the PZT crystal will not be operating at its blocked force. It also captures to some degree the complex impedance relationship when multiple nested layers interact. This will be described in the next section. This is a great leap forward in accuracy compared to a scalar "amplification factor," which in most cases drastically overpredicts the blocked force and free displacement achievable. Nonlinear models that iteratively compute the displacements and take into account changes in angle, etc., would be more accurate, but more difficult to implement. The match between experimental results and the analytical model is good enough that the analytical model can be used to make design decisions, with calibration routines or feedback control used in the implementation to fine-tune the mechanism.

The formulation of this model makes the traditional mechanics of materials assumptions of linearity and small deformations. In fact, the errors on each of the immittance parameters (which have units of stiffness) are on the order of errors found in experimental evaluations of flexure hinge stiffness [154,193,297]. One of the strengths of this modeling approach is that it works equally well for distributed compliance as well as localized compliance. The mechanism shown in Fig. 2.15 takes advantage of this. However, the drawback of distributed compliance is that the deformation can vary more widely with manufacturing tolerances. So it is understandable that the errors are slightly larger than for hinged mechanisms.

In addition, the mechanism considered is used to produce an aggressive force–displacement tradeoff, since the strain rate of a PZT stack is so small. At this operating point, the surrounding region of the design space is very sensitive to parameter variations. The surfaces in Fig. 2.18 show the changes in immittance when the geometric parameters are perturbed from the operating point. When θ is small, varying θ while fixing d can be problematic because it can result in degenerate geometries, so to preserve valid geometries, $d = d' \tan\theta / \tan\theta'$ was chosen when generating the surface shown, where d' and θ' are the nominal values. Although a small change in one of the

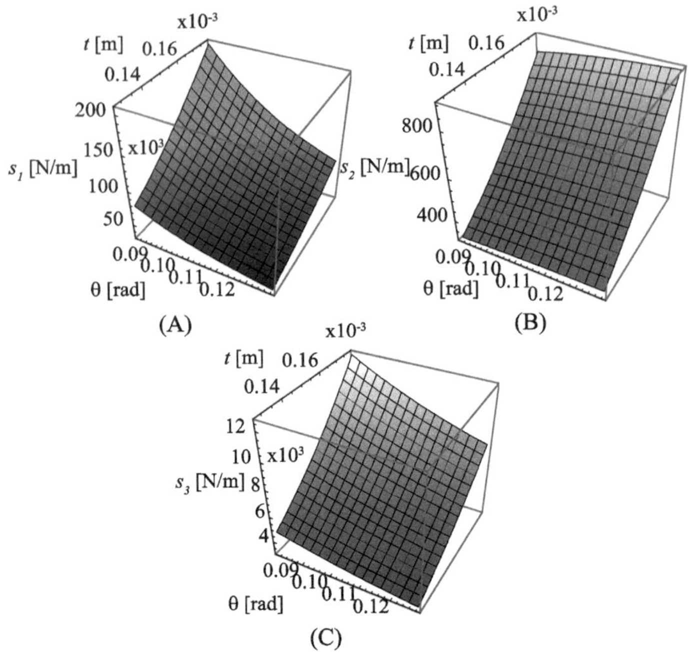

Figure 2.18 Variation in immittance parameters: (A) s_1, (B) s_2, and (C) s_3 with geometry. © 2013 IEEE, reprinted with permission

geometric parameters may affect blocked force and free displacement only slightly, the immittances can vary much more.

Because the mechanism tested was so fragile, some permanent deformation occurred either through the manufacturing process or shipping and handling. It was assumed that the immittances would not be largely changed by this deformation, but the residual stresses may have contributed to some of the error. There was some compliance inherent in the micropositioning stages and the load cells and this may have introduced some error in measurement. There also may have been some variation in the material constants from typical tabulated values. This could be improved by a more repeatable manufacturing process.

This section demonstrated a procedure for verifying that a fabricated strain amplifying rhomboid indeed had the immittance properties predicted by the modeling process of Section 2.1. By and large the model gives a reasonably good prediction, well within 20%. The procedure produced the immittances s_1, s_2, and s_3. Why do the experiments evaluate the stiffness matrix, rather than characterizing the mechanisms' input–output behavior, e.g., the amount of strain amplification? The reason is that a single set of immittances completely characterizes the amplification mechanism's behavior. The stiffness matrix is convenient as all elements of the matrix have the same units. This method uses measurement of force and displacement, which can be gathered using a wide variety of commercially available sensors.

In practice, since the rhomboidal mechanisms are of a prototype nature and are very thin, it is likely that some of the idealized features of the model do not hold. Even though the wire EDM process does not have side loads during machining as in milling or turning, the discharge process introduces residual stresses [68]. The thin sections of the rhomboid can pose some difficulty – permanent deformations are introduced meaning that some elastic deformation occurs when lining up with the fixture. As the manufacturing process improves, with associated fixturing and control of the process, the immittance matrix for a rhomboid should be known more precisely.

The mechanisms characterized in this section were of the distributed compliance type. Flexure hinge types of compliant mechanisms, where compliant sections are interspersed with sections of localized compliance are also popular. This framework is already set up to handle flexure hinges composed of short rectangular sections of constant thickness. Modeling and characterization of other types of flexure hinges have a long history in the precision engineering community, beginning with the work on right circular flexure hinges of Paros and Weisbord [191], who developed the equations for a circular geometry. Later works have analyzed other geometries and include detailed finite element and experimental validation [155,152,175,251,252]. Recent work has also examined flexure hinges made from silicone elastomers [284].

2.3 Modeling of actuator arrays: the nesting theorem: three-layer structure

2.3.1 *Actuator compliance for nested amplified piezoelectric actuators*

Given the discussion in section "Soft robots for unstructured environments", it is natural to think that one would be interested in the impedance properties of nested amplified piezoelectric stack actuators, and whether the impedance perceived by the environment can be "tuned" or enforced by changing the terminating electrical impedance on the piezoelectric stack. The answer to this is given by the nesting theorems [219] of Section 2.3.3. Piezoelectric stacks themselves have high impedance; that of lead zirconate titanate (PZT) stack actuators are on the order of 10^6 N/m [88]. A force applied to the output of an amplified piezoelectric stack actuator will experience much higher deformations than the same force applied to the piezoelectric stack. This is because in an amplified piezoelectric stack actuator very little deformation is borne by the piezoelectric stack itself. Large internal deformations of the amplification structure itself result in a much smaller perceived stiffness from the environment than that of the piezoelectric stack alone.

The stiffness that is perceived from the environment is an aggregate of all the strain amplifying layers present in the actuator. Because the displacement amplification factor of each individual layer has a dual impact on the force transmitted through the layer, each layer does not contribute equally to the stiffness the way series and parallel combinations of springs would. The nested relationship governs the relative contributions of the stiffness of each rhomboidal layer. If a large number of layers are present in the actuator, this may give sufficient design freedom to produce an actuator that not only has a high enough amplification factor to produce satisfactory strain rates at free displacement along with sufficient blocked force, but also to set individual strain amplification mechanism parameters to set the stiffness perceived from the environment to be in the range desired. Put another way, if each strain amplification layer accomplishes a portion of the amplification needed, there will be many ways to get to the amplification needed, and this large space of designs with adequate amplification properties can be searched for a solution with good stiffness properties.

2.3.2 Antagonist pairs of compliant actuators

If one were to step back and think about it, the human skeletal muscular system is quite curious, particularly from an engineering perspective. While engineering systems have no problem with actuators running bidirectionally, or even using ingenious mechanisms that make opposite directions of an actuator drive distinctly different unidirectional functions, the human body does not exhibit such economy. Far from it: each joint is actuated by not one, but *two* (or even more) muscles. The most familiar example is the biceps and triceps muscles that actuate the elbow. Contraction of the biceps muscle flexes the elbow whereas contraction of the triceps extends it. The human spine is also an antagonistic system, albeit much more complex, involving a carefully balanced relationship between the erector spinae, abdominals, trapezius, iliapsoas, and numerous intrinsic (roteres, multifidus, semispinalis) muscles [6]. The reason for this is physiological: to contract, at the sarcomere level, actin and myosin filaments slide together. When deactivated, the bonds no longer hold strongly, but they do not push apart back to the muscle's resting length. Only an external force can disengage them. This is provided by that muscle's *antagonist*, another muscle or set of muscles whose contraction would tend to lengthen the muscle in question.

The fact that muscles are contractile does not strictly require antagonist pairs for each and every joint. It is well known that for a robot with n joints and unidirectional tendon transmissions, a system with $n + 1$ tendons can reach any desired set of joint angles using strictly tensile tendon forces [176]. Pollard and Gilbert [197] compare these arrangements in an anthropomorphic robotic finger and examines their consequences on actuator torques. Be this as it may, many human body systems exhibit antagonist pairs.

This apparent lack of economy in a system that has every incentive to be minimal, from the standpoint of speed, bone density and metabolic function is puzzling. Wouldn't it be more suitable for organisms to be equipped with an actuator that could both pull *and* push? It stands to reason that there must be some functional reason for being architected thus. From a robotics standpoint, there are two possibilities when confronted with an intriguing biological feature. Either it is there for some functional, kinematic, or operational reason, and therefore replicating its fundamental principles could endow similar function in a robotic device, or instead it is there for a reason completely unrelated to motion, arising from self-replication, metabolic concerns, or chemical factors, for example. The latter category can normally be safely disregarded by a roboticist. Antagonistic pairs may initially seem to be in this latter category, but there is indeed a kinematic functional aspect of antagonistic actuation by muscle systems that is of interest to roboticists. Most of these relate to muscle systems' abilities to vary the impedance about a joint. This modulation of impedance is responsible for balance and subtle interactions with the environment. Hogan postulates that co-activation of antagonist joints are used by the central nervous system to offset gravitational destabilization [103].

Is a piezoelectric cellular actuator a contractile-only actuator like human muscle? Not completely – because the strain amplification layers have a stiffness, deactivating the piezoelectric stack will return the actuator to its equilibrium length. So a single piezoelectric cellular actuator could be used in bidirectional applications as it doesn't need anything to pull it back. But further consideration reveals that this is not any different in principle from more traditional unidirectional actuators (e.g., single-chambered pneumatic cylinder) that have a return spring; it is just that the return spring is already present due to the need for strain amplification. So a unidirectional piezoelectric cellular actuator has the same drawbacks as any return spring-based actuator, namely, that the reverse direction is governed primarily by the return spring–load mass dynamics. Although the user can apply control input in the forward direction to slow the system down, there is no way to return more rapidly than this in the reverse direction.

By putting two piezoelectric cellular actuators in an antagonistic configuration, there is no "preferred direction" as in the return spring case, and actuation effort can be applied in either direction without relying on passive return dynamics. In addition, this makes the control effort symmetric in nature. Co-contraction is also a possibility, giving greater flexibility.

A simple example of such an antagonistic setup is illustrated by the single-joint robot in Fig. 2.19. The joint compliance is important in many emerging applications. This has provided motivation for numerous variable impedance actuator implementations [213,108]. Several of these implementations use co-contraction to modulate the impedance of the joint.

So does the piezoelectric cellular actuator fall into the variable impedance category? The modeling process in Section 2.1 clearly uses Hooke's law and the assumption of small displacements, so the spring behavior of each amplification layer is linear. As the literature makes clear, a nonlinear spring profile is necessary for producing variable joint impedance [192]. This becomes evident when looking at Fig. 2.19; if one takes the derivative of the expression for the net force from both actuators, a constant offset in force (from co-contraction) is differentiated out. Although some have suggested looking into nonlinear factors (large displacements, the change in angle of the rhomboids with displacement) not treated in Section 2.1 this would require dispensing with the linearly elastic mechanics of materials foundation on which the analysis is based, and is unlikely to produce a closed-form model. Furthermore, even if such a nonlinear model were to be produced, there is no guarantee that the nonlinearity of the profile would be able to be tuned to produce the variable impedance behavior desired. The most fruitful path for producing variable impedance behavior in piezoelectric cel-

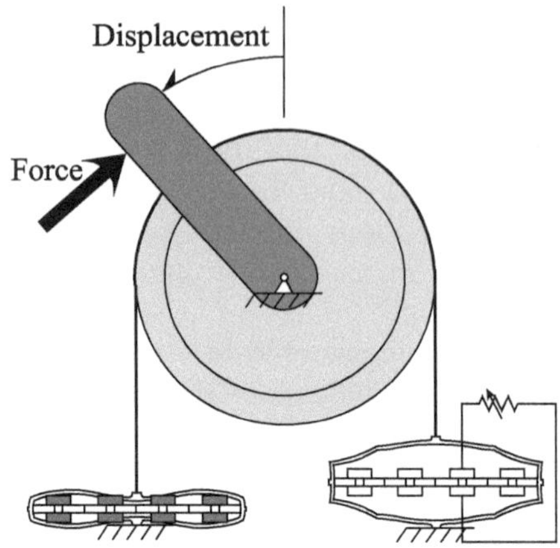

Figure 2.19 Single degree of freedom robotic joint. The agonist (left) is activated and contracts. The antagonist (right) behaves as a passive stiffness. This affects the relationship between applied load and joint angle.

lular actuators is using mechanical stops to produce a nonlinear switching condition [222]. This will be described further in Section 4.1.

Since the piezoelectric cellular actuator cannot be readily exploited as a variable impedance actuator via co-contraction, it is likely to be to activated only in differential mode as shown in Fig. 2.19. Although there is not additional functionality that can be produced by co-contraction, further study of the antagonist is warranted for two reasons. First, when the agonist contracts, however, it will have to overcome the passive spring force of the inactive antagonist. This spring force results from a complex interconnection of all the amplification layers, at whose core lies the piezoelectric material. What will the combined stiffness opposing the motion of all layers together, each of which has a two-port model? Second, because the piezoelectric material transduces energy between the mechanical and electrical domains, it is tremendously tempting to design a system that sets the terminating electrical impedance of each piezoelectric stack in the antagonist, and attempt to vary the impedance felt at the joint that way. In the subsequent sections, we find answers to both questions. We see that when many layers are present, the impedance of the passive antagonist can be computed quickly by ignoring the innermost layers and as more and more layers are added, the strategy of varying the joint impedance by changing the antagonist's terminating impedance quickly becomes ineffective.

2.3.3 The first and second nesting theorem: evaluating the perceived stiffness based on the stiffness of each layer

Assuming linear deformations of each strain amplifying mechanism within the actuator, each can be given a two-port model description as described in Section 2.1. The theorems presented in the following section show how the stiffness perceived by the environment can be calculated or estimated from the knowledge of the two-port models of each individual layer. Furthermore, the theorems show that when designing a multi-layer strain amplification mechanism, the perceived stiffness from the environment is dominated by the outermost mechanisms in the nested hierarchy [219].

2.3.3.1 Two-port models of strain amplifying mechanisms

The amplification mechanism of each layer can be described by a two-port network model, an intellectual device taken from circuit theory that concisely describes input–output behavior while abstracting away internal variables [42]. Abdalla et al. [4] used two-port networks to optimize a planar compliant mechanism for a known load and to describe the electro-mechanical transduction of a piezoelectric stack. The mathematical relationship between the input and output to the mechanism is described by the following:

$$\begin{bmatrix} F^{in} \\ F^{out} \end{bmatrix} = \begin{bmatrix} s_1 & s_3 \\ s_3 & s_2 \end{bmatrix} \begin{bmatrix} \delta^{in} \\ \delta^{out} \end{bmatrix}, \tag{2.41}$$

where F^{in}, F^{out}, δ^{in}, and δ^{out} denote the forces and displacement at the input and output ports, respectively. s_1, s_2, and s_3 have units of stiffness and are functions of the geometry and material properties. Using this modeling technique for rhomboidal strain amplifying mechanisms assumes that the material is linearly elastic and undergoes small changes in angle. This is valid for the piezoelectric cellular actuator because the outermost layers tend to have small angle values, and the innermost undergo smaller motions by virtue of the fact that their displacements are being amplified by subsequent stages. A procedure to determine these quantities for rhomboidal mechanisms is described in [220]. The matrix in (2.41) is the stiffness matrix, one of several *immittance* matrices [42] that is used to describe the two-port network relationship. The strength of the two-port network approach is that it accurately accounts for the effects of both the control force applied by previous layers and environmental or load forces from subsequent layers. Using this framework, it is possible to model systems that do not have the high input impedance/low output impedance characteristics necessary in most branches of circuit analysis. An alternate way of expressing this is to say that the two-port (or multi-port) network succeeds in modeling devices that have significant "back effects."

The innermost, or "zeroth," layer of a nested cellular device will be a piezoelectric stack (or some other active material) of known characteristic. Typically, the manufacturer tabulates the totally blocked, or clamped, force and the mechanical stiffness. This can be represented as a Norton equivalent circuit, where the blocked force is represented by a current source, and the mechanical stiffness is represented by the Norton resistance. The displacement of this device, which is analogous to the voltage across the Norton resistance, will depend on the load impedance. Several of these zeroth layer units can be combined in series, and their Norton circuits combined according to the methods of circuit analysis. The voltage across the Norton equivalent resistance corresponds to the input displacement of the first layer displacement amplification mechanism. The current to the load impedance corresponds to the input force. When the amplification mechanism is represented as a two-port network, this is the voltage and current at the left-hand port. With appropriate mechanical analogies, this can be expressed as

$$F_1^{in} = F_0^{block} - k_0 \delta_1^{in}. \tag{2.42}$$

Because series combinations of springs add compliances, not stiffnesses, the Norton equivalent resistance is set to $1/k_0$. Interconnections between networks are described in terms of matrix operations on the immittance matrices and are well known [42]. Therefore any combination of two-port networks can be readily analyzed. The topological relationships between the various layers in the hierarchy will be represented by the electrical connections of their two-port models.

2.3.3.2 Collapsing of nested two-port models

Since the impedance characteristic at the input to the first amplification layer is known, and the immittances of the amplification mechanism's two-port model are known, the entire connection can be "collapsed" and replaced with the Norton equivalent circuit that represents the characteristic at the output of the first layer. This can be performed repeatedly, up to the outermost layer, which is connected to the load. Fig. 2.20 illustrates collapsing connections of two-port networks. If the input to a given amplifying mechanism is a series combination of subunits, as it is in the camera positioner application, this can be mathematically represented by replacing the series combination with a fictional unit with the same blocked force and stiffness as the series combination. The following derives the equations that describe the collapsing process.

Consider an active material with blocked force F_0^{block} and stiffness k_0. The output of this Norton circuit is applied to the left-hand port of a two-port network representing the first layer amplification mechanism. The displacement of the zeroth layer, δ_1^{in}, is analogous to the voltage across the Norton resistance. Using (2.42) and (2.41), the two-port relationship can be represented by the following equation:

$$\begin{bmatrix} F_0^{block} - k_0 \delta_1^{in} \\ F_1^{out} \end{bmatrix} = \begin{bmatrix} s_1 & s_3 \\ s_3 & s_2 \end{bmatrix} \begin{bmatrix} \delta_1^{in} \\ \delta_1^{out} \end{bmatrix}. \tag{2.43}$$

Solving the upper of the two equations for δ_1^{in}, we get

$$\delta_1^{in} = \frac{F_0^{block} - s_3 \delta_1^{out}}{s_1 + k_0}. \tag{2.44}$$

Substituting this into the lower of the two equations and collecting terms we obtain

$$F_1^{out} = \frac{s_3 F_0^{block}}{s_1 + k_0} + \frac{s_2(s_1 + k_0) - s_3^2}{s_1 + k_0} \delta_1^{out}. \tag{2.45}$$

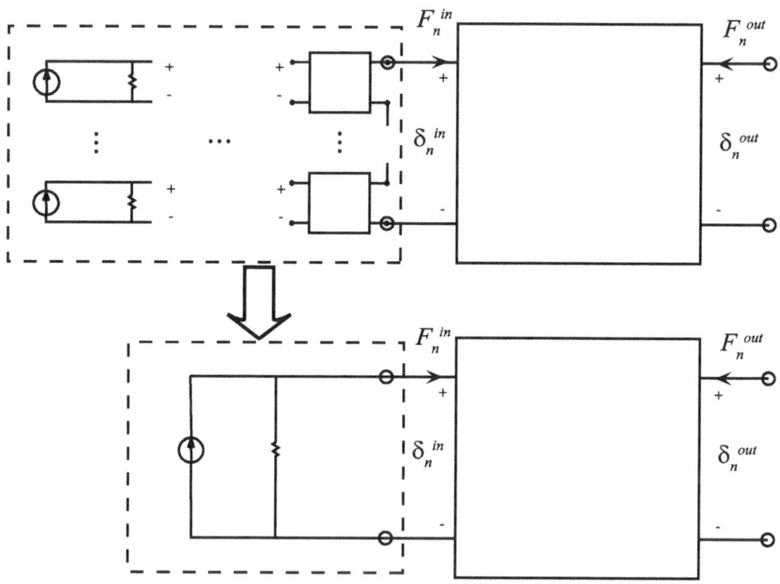

Figure 2.20 Collapsing of two-port networks. Each square box represents a two-port network, with a voltage and current at the right- and left-hand ports. The entire hierarchy within the dashed lines is collapsed and replaced with its Norton circuit. © 2013 IEEE, reprinted with permission

Define the following:

$$F_1^{block} = \frac{s_3 F_0^{block}}{s_1 + k_0}, \tag{2.46}$$

$$k_1 = \frac{s_2(s_1 + k_0) - s_3^2}{s_1 + k_0}, \tag{2.47}$$

Eq. (2.45) can be written as

$$F_1^{out} = F_1^{block} + k_1 \delta_1^{out}. \tag{2.48}$$

This has physical significance: F_1^{block} is the fully blocked force of the displacement amplified active material, and $k_1\delta_1^{out}$ is the stiffness as seen from the output of the combination times the output displacement of the compliant mechanism. Since this is an equivalent stiffness for the nested connection, we refer to it as the *lumped stiffness*. In the Norton equivalent circuit, F_1^{block} corresponds to the current source, and $\frac{1}{k_1}$ corresponds to the equivalent Norton resistance. This procedure can be performed repeatedly for each amplification mechanism in the hierarchy.

2.3.3.3 Displacement of the mechanism against the passive actuator

Modeling the camera platen as a rigid body that rotates about a frictionless pivot, considering motion to be quasistatic, and representing the antagonistic pair as a con-

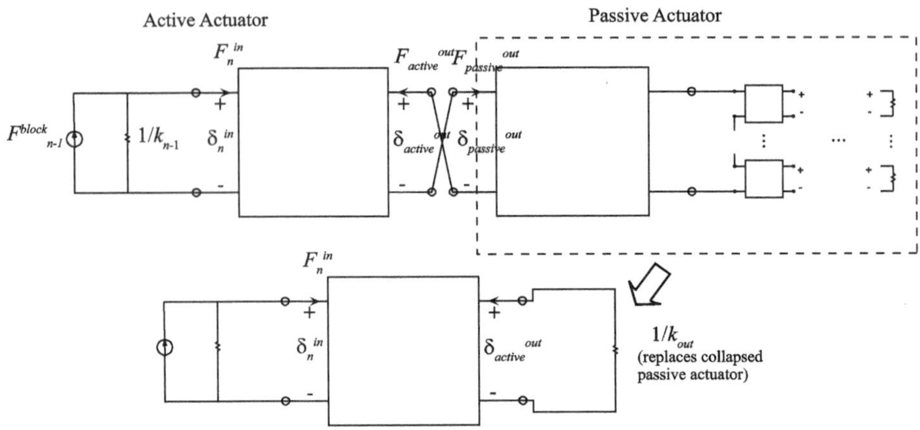

Figure 2.21 Two-port representation of antagonistic pairs. The active actuator is in black and has all but the outermost layer collapsed and represented by its Norton circuit. The passive actuator is in gray and is will be replaced by a simple stiffness. © 2013 IEEE, reprinted with permission

nection of two-port models results in the configuration shown in Fig. 2.21. From this model, we see that the following relations hold:

$$\delta_{active}^{out} = -\delta_{passive}^{out}, \tag{2.49}$$

$$F_{active}^{out} = F_{passive}^{out}, \tag{2.50}$$

where δ_{active}^{out}, $\delta_{passive}^{out}$, F_{active}^{out}, and $F_{passive}^{out}$ are as shown in Fig. 2.21. Let us assume that the antagonistic actuators are identical. Because the passive actuator is OFF, the current source F_0^{block} will be zero. The stiffness k_0 at the zeroth layer of the passive actuator will depend on the electrical terminating impedance on the leads of the active material. It will be shown later that the effect of the terminating impedance on the stiffness is minimal.

Because the passive actuator is not energized, it can be completely collapsed and simply be represented as a terminal resistance $\frac{1}{k_{eq}}$ across the active actuator. This is illustrated in Fig. 2.21. The displacement of the antagonistic mechanism corresponds to the voltage across the right-hand port of the two-port model of the outermost layer of the active actuator.

Using (2.43) and Ohm's law, solving for δ_n^{out} we get

$$\delta_n^{out} = -\frac{s_3 F_{n-1}^{block}}{s_1 s_2 - s_3^2 + k_{eq}(s_1 + k_{n-1}) + s_2 k_{n-1}}. \tag{2.51}$$

The free case has $k_{eq} = 0$, resulting in a displacement of

$$\delta^{free} = -\frac{s_3 F_{n-1}^{block}}{s_1 s_2 - s_3^2 + s_2 k_{n-1}}. \tag{2.52}$$

Taking the ratio of the two gives

$$\frac{\delta_n^{out}}{\delta^{free}} = \left(1 + \frac{k^{out}(s_1 + k_{n-1})}{s_1 s_2 - s_3^2 + s_2 k_{n-1}}\right). \tag{2.53}$$

Here we have to make an assumption regarding the electrical terminating impedance of the piezoelectric stack on the passive antagonist muscle. We will assume that the electrical impedance is chosen such that the mechanical stiffness of the piezoelectric stacks on the antagonist side match those on the agonist side. It will be seen later that this does not have much of an effect on the final stiffness observed from the environment and therefore the specific choice does not matter. Using (2.46), we can write

$$k^{out} = \frac{s_2(s_1 + k_{n-1}) - s_3^2}{s_1 + k_{n-1}}. \tag{2.54}$$

Substituting into (2.53), we get $\delta_n^{out}/\delta^{free} = 1/2$. Therefore, in an antagonistic pair, the actuator must be designed so that its free displacement is double the stroke length corresponding to the angle of travel specification.

2.3.3.4 Nesting of layers

If a passive actuator is in an antagonistic pair arrangement with an active actuator, it will appear to the active actuator as a stiffness. This stiffness will depend on the immittances of the previous layers, as well as the terminal stiffness of the zeroth layer. This subsection will consider an amplified mechanism with n layers, with outermost layer (connected to the load) denoted layer n. The innermost layer (connected to the active material) will be denoted layer 1. Immittances for a given layer will be denoted with a leading superscript. k will be used to count layers outward from the active material, and $j \in \mathbb{N} \mid n - j \in [\ 0,\ k\ ]$ will be used to count layers inward from outermost. The stiffness of a passive actuator with n layers is

$$k_{passive} = {}^{n}s_2 - \cfrac{{}^{n}s_3^2}{{}^{n}s_1 + {}^{n-1}s_2 - \cfrac{{}^{n-1}s_3^2}{{}^{n-1}s_1 + {}^{n-2}s_2 - \cfrac{{}^{n-2}s_3^2}{{}^{n-2}s_1 + {}^{n-3}s_2 - \ddots}}}. \tag{2.55}$$

For the jth layer, denoting

$${}^{j}\gamma = {}^{n-j}s_2 - \frac{{}^{n-j}s_3^2}{{}^{n-j}s_1 + k_{n-j-1}} \tag{2.56}$$

where k_{n-j-1} is the lumped stiffness of the $(j-1)$th layer subunit and subsumes all terms for any nested unit index (numbered from the outside in) $i > j$, we can state that

$$-\frac{{}^{n-j}s_3^2}{{}^{n-j}s_1} < {}^{j-1}\gamma - {}^{n-j}s_2 < 0. \tag{2.57}$$

Therefore, approximating the stiffness of the passive actuator by truncating remaining terms in the denominator will result in a conservative estimate of the displacement for a nested structure.

2.3.3.5 Estimating stiffnesses of actuators with large numbers of layers

Eq. (2.55) has a form of a continued fraction. This results in some desirable properties. The following subsection will show that for an actuator with a large number of layers, the stiffness of the entire actuator can be approximated to a desired degree of accuracy by replacing the continued fraction representing the remaining layers by an arbitrary constant. In the literature [120,208], these approximations are known as *convergents*.

We intend to show that the sequence of convergents P_j is Cauchy, and therefore, all convergents beyond some finite k lie within some interval of size ϵ of P_k on $\mathbb{R}$.

Lemma 1. *All denominators for the continued fraction expansion of the stiffness of a nested linearly elastic mechanism are positive.*

Proof. Each compliant mechanism has a positive definite immittance matrix. Therefore, the determinant for any layer, ${}^k s_1 {}^k s_2 - {}^k s_3^2 > 0$. Because any realizable nested actuator will contain a finite number of compliant mechanisms, the continued fraction expansion terminates. Let us refer to the sequence of continued fraction expansions for the lumped stiffness at the output of each layer, beginning with the innermost, as z_k. The zeroth term of the sequence, z_0, is constant. Because the stiffness is derived from passive elastic material, z_0 will be positive, and all elements of the immittance matrix are positive. We can express the remaining terms of the sequence by the recursive relation:

$$z_k = {}^k s_2 - \frac{{}^k s_3^2}{{}^k s_1 + z_{k-1}}. \tag{2.58}$$

This can be rewritten as

$$z_k = \frac{{}^k s_2 {}^k s_1 - {}^k s_3^2 + {}^k s_2 z_{k-1}}{{}^k s_1 + z_{k-1}}. \tag{2.59}$$

If $z_{k-1} > 0$, then $z_k > 0$. The proof is completed by induction on k. □

When a force is applied at the output of a compliant mechanism, it not only causes a deformation in the output direction, but also a deformation in the input direction as well. This deformation at the input has an effect equivalent to a force transmitted back to the output, even when no load is applied at the input. Essentially, Lemma 1 says that the effective stiffness due to this back, then forward transmission, s_3^2/s_1, will always be less than the input clamped output stiffness, s_2.

Lemma 2. *In the set of all terminating continued fractions of the form:*

$$\left\{ p_k = a_0 - \cfrac{b_1}{a_1 - \cfrac{b2}{a_2 - \cfrac{b3}{\ddots \cfrac{}{a_{k-1} - \cfrac{b_k}{c_k}}}}} \;\middle|\; a_i, b_i > 0\ \forall i,\ c_k \in \left(\frac{b_k}{a_{k-1}},\ c_{max}\right]\right\} \tag{2.60}$$

with a_i, b_i constant, the maximum p_k is achieved by choosing the maximum of c_k, c_{max}.

Proof. For $k = 1$, $p_k = a - \frac{b}{c}$. It is clear that this quantity is maximized by the largest possible c. For the general case, define the quantity $a_{k-1} - \frac{b_k}{c_k} = c_{k-1}$. c_{k-1} is maximized by $c_k = c_{max}$. Therefore, by induction, choosing the maximum c_k maximizes all denominators, and p_k is maximized. □

$$p_n = {}^n s_2 - \cfrac{{}^n s_3^2}{{}^n s_1 + {}^{n-1} s_2 - \cfrac{{}^{n-1} s_3^2}{{}^{n-1} s_1 + {}^{n-2} s_2 + \cfrac{\ddots}{{}^{n-j+2} s_1 + {}^{n-j+1} s_2 - \cfrac{{}^{n-j+1} s_3^2}{{}^{n-j+1} s_1 + z_j}}}} \tag{2.61}$$

Physically speaking, this is equivalent to saying that the terminal impedance condition that results in the stiffest possible actuator as seen by the environment can be realized by clamping the input of the innermost mechanism.

Theorem 1. *The sequence of convergents of the continued fraction expansion of the stiffness of a nested mechanism, with successive denominators for $n - j < k$ replaced by ${}^{k-1} s_1 + {}^k s_2$ is Cauchy.*

Proof. Any continued fraction can be represented by Eq. (2.61). This considers the denominators from the top level n, down to level $n - j$; z_j represents the continued fraction expansion for the remaining layers. It can be approximated by

$$z_j \approx {}^{n-j}s_2 - \frac{{}^{n-j}s_3^2}{{}^{n-j}s_1 + {}^{n-j-1}s_2}. \tag{2.62}$$

Let y_j represent the convergent (2.61) where z_j is replaced by the approximation in (2.62). Physically speaking, this represents the stiffness perceived at the output when the innermost layer considered in the approximation (layer j) has its input fixed. Accordingly, y_j has units of stiffness. y_{j-1} represents the previous convergent in the sequence, replacing the final z_j simply by ${}^{n-j}s_2$. By Lemma 1, the quantity $\frac{{}^{n-j}s_3^2}{{}^{n-j}s_1+z_j}$ will be positive for all j. Also from Lemma 1, it can be determined that this quantity will not exceed ${}^{n-j}s_2$. y_{j-1} and y_j are identical, except for the terminating constant. Therefore, they are both of the form of the continued fraction in Lemma 2. y_{j-1} terminates with the constant $c_k = {}^{n-j+1}s_1 + {}^{n-j}s_2$ and y_j terminates with the constant $c_k = {}^{n-j+1}s_1 + {}^{n-j}s_2 - \frac{{}^{n-j}s_3^2}{{}^{n-j}s_1+{}^{n-j-1}s_2}$. By Lemma 2 and induction on j, we can say the sequence of convergents is monotonically decreasing, or

$$y_{j-1} > y_j \;\forall j. \tag{2.63}$$

Applying Lemma 1 at level n, we can say the sequence is bounded below by 0. Since the sequence is monotonically decreasing and bounded below, it must be Cauchy. □

For any Cauchy sequence, for any arbitrary ϵ, we can find a k such that all terms in the sequence beyond k are within a distance ϵ from one another [299]. Physically speaking, this means that for an actuator consisting of nested compliant mechanisms, we can approximate its stiffness to some desired accuracy by considering merely the outer k layers and considering the lumped stiffness of the $(k-1)$th layer to be infinite. Alternatively, the kth layer can be said to be clamped at its input.

Theorem 1 says that outer layers, regardless of whether they are rigid or compliant, have the greatest effect on the stiffness of the overall actuator. Therefore, the innermost layers do not need to have as strict tolerances as the outermost layers to achieve a uniform stiffness from actuator to actuator. It also means that the electrical terminating impedance of the passive actuator is not critical for actuators with large numbers of layers, justifying the assumption in Section 2.3.3.3. The procedure of Theorem 1 provides an upper bound on the true stiffness.

An alternate choice is to approximate z_j by ${}^{n-j}s_2 - \frac{{}^{n-j}s_3^2}{{}^{n-j}s_1}$. The sequence of these approximants (which are not convergents, strictly speaking, since they do not result from a strict truncation) will be denoted by x_j. Like y_j, this has units of stiffness. This is equivalent to the stiffness perceived at the output when the innermost mechanism considered in the approximation is free to deform in the input direction. It will be shown that y_j and x_j have the same limit.

Theorem 2. *The sequences y_j and x_j converge to the same limit.*

Proof. Any continued fraction of the form in Lemma 2 can be represented by the recursive relation:

$$
\begin{aligned}
B_{j+1} &= a_{j+1}B_k - b_{j+1}B_{j-1}, \\
A_{j+1} &= a_{j+1}A_k - b_{j+1}A_{j-1}, \\
y_{j+1} &= \frac{B_{j+1}}{A_{j+1}}, \\
B_{-1} &= 1, \\
B_0 &= a_0, \\
A_{-1} &= 0, \\
A_0 &= 1.
\end{aligned} \tag{2.64}
$$

This is shown in Rockett and Szüsz [208]. In this case $a_{j+1} = {}^{n-j+1}s_1 + {}^{n-j}s_2$ and $b_{j+1} = {}^{n-j+1}s_3^2$. Using a similar procedure, it is simple to demonstrate that x_{j+1} can be written as

$$
x_{j+1} = \frac{B_{j+1} - k_{j+1}B_j}{A_{j+1} - k_{j+1}A_j} \tag{2.65}
$$

where $k_{j+1} = \frac{{}^{n-j}s_3^2}{{}^{n-j}s_1}$. Note that each term of x_j is written in terms of the B_j and A_j used to construct y_j. We then construct the sequence $y_j - x_j$ and show that it approaches zero in the limit. After algebraic manipulations, the $(j+1)$th term of the sequence can be written as

$$
y_{j+1} - x_{j+1} = \frac{y_j - y_{j+1}}{\frac{A_{j+1}}{k_{j+1}A_j} - 1}. \tag{2.66}
$$

The numerator approaches zero for large j by Theorem 1. To show that the entire sequence goes to zero it is sufficient to show that $\frac{A_{j+1}}{A_j} \neq k_{j+1}$. This can be proved by contradiction.

Assume that for all j greater than some k the sequence $\frac{A_j}{A_{j-1}} = k_j$. Then $\frac{A_{j+1}}{A_j} = k_{j+1}$. This can be rewritten as

$$
A_{j+1} = k_{j+1}A_j = a_{j+1}A_j - b_{j+1}A_{j-1}. \tag{2.67}
$$

Collecting terms and substituting in the immittances, we get

$$
{}^{n-j}s_3^2 A_{j-1} = \left({}^{n-j}s_1 + {}^{n-j-1}s_2 - \frac{{}^{n-j-1}s_3^2}{{}^{n-j-1}s_1} \right) A_j. \tag{2.68}
$$

We can divide this quantity by A_{j-1}. Since j is assumed to be greater than k, the sequence $\frac{A_j}{A_{j-1}} = k_j$. Substituting in for this quantity in terms of the immittances, we

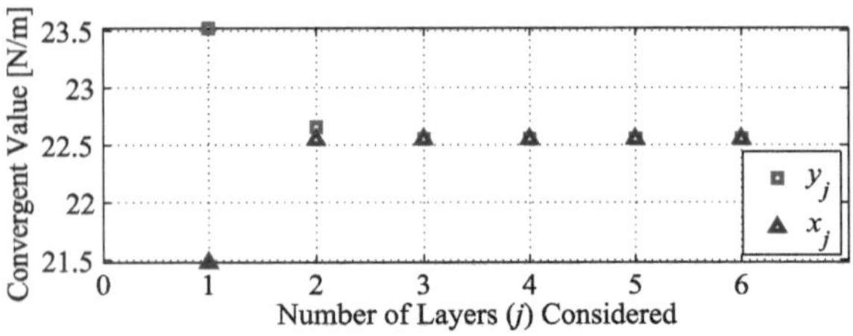

(A) Convergence of input fixed (y_j) and input free (x_j) approximations

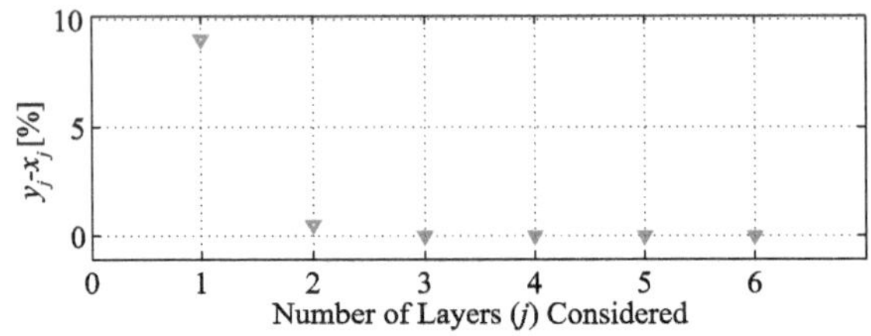

(B) Error convergence

Figure 2.22 Convergence of the various stiffness approximations for a hypothetical many-layered mechanism with randomly chosen immittances. y_k is the input-fixed approximation through layer k, and x_k is the input-free approximation. © 2013 IEEE, reprinted with permission

get

$$ {}^{n-j}s_3^2 = \frac{{}^{n-j}s_3^2}{{}^{n-j}s_1}\left({}^{n-j}s_1 + {}^{n-j-1}s_2 - \frac{{}^{n-j-1}s_3^2}{{}^{n-j-1}s_1}\right). \tag{2.69} $$

Simplifying, we get

$$ \frac{{}^{n-j-1}s_2\,{}^{n-j-1}s_1 - {}^{n-j-1}s_3^2}{{}^{n-j-1}s_1\,{}^{n-j}s_1} = 0. \tag{2.70} $$

All ${}^k s_i$ are finite, and the numerator of (2.70) is simply the determinant of the immittance matrix for layer $n - j - 1$. This must be strictly positive for all j, hence the contradiction and the result is proved. □

A simple numerical example of the convergence is shown in Fig. 2.22. Fig. 2.22A shows the convergence rates of the input-fixed and input-free approximations y_j and x_j, and Fig. 2.22B shows the error bound convergence (the difference between the two approximations). The hypothetical mechanism is said to have a large number of layers, with the parameters of the 6 outermost known. To show that the theorems still hold even if the mechanism is constructed haphazardly, two-port network immittance parameters were chosen randomly for each of the 6 layers with ${}^k s_1 \in (0, 18)$, ${}^k s_2 \in (0, 27)$, and ${}^k s_3 \in (0, \min[\sqrt{{}^k s_1 {}^k s_2},\ 16])$. The restriction on ${}^k s_3$ ensure that the immittance matrix is positive definite and physically consistent. Other trials perform similarly.

The lumped stiffness of the remaining unconsidered layers will be in the interval $[0, \infty)$. x_j represents the stiffness of the actuator when this lumped stiffness for the remaining layers is zero. y_j represents the stiffness of the actuator when this lumped stiffness for the remaining layers is infinite. For a finite nonzero compliance, the stiffness of the overall actuator should be between the values of x_j and y_j. Since these are the same in the limit, the compliance of the innermost layers has a negligible effect on the overall stiffness of the passive actuator for actuators with many layers.

2.3.4 The three-layer structure

2.3.4.1 Beneficial properties of three-layer structures

We have seen in earlier sections that with two layers of strain amplification, strains on the order of human muscle (20% as per Madden et al. [161]) can be achieved. Following the argument of the prior sections, strictly speaking, from a theoretical perspective, only a single layer is needed. Since the amplification factor a varies inversely with the tangent of the angle θ of the rhombus, one could simply decrease this angle until the desired amplification ratio is reached. Reporting the strain rate only, however, ignores something extremely relevant to implementation, namely, the dimensions of the actuator in the direction transverse to motion.

With a single strain amplification mechanism, the actuator needs to be quite wide in one direction transverse to the actuation direction, yet is thin in the other direction transverse to the actuation direction. In fact, the APA1000XL, the largest amplified piezoelectric stack actuator produced by Cédrat Technologies, is nearly four times wider than it is long, over eight times wider than it is thick, with a strain rate of 1.8%! This is not a convenient form factor for most applications because seldom is a wide, thin cavity available for the actuator. This leads us to a key advantage for multi-layer structures: as the rhomboidal shape gains a large major to minor diagonal ratio (flattens) to achieve high levels of amplification, the actuator can grow in *both* directions transverse to the actuation, resulting in a more convenient envelope for the actuator itself.

The development of Section 2.3.3 is valid for any number of amplification layers, so the question begs itself, "how many layers is the right amount"? This is not an easy question to answer definitively, but in general more layers will mean more elastic material that needs to be deformed. This means more energy must be stored within the deformation of the actuator to produce displacement, meaning less of the input energy is transmitted to the load. Of course, this is not a total loss because energy is only stored in the actuator, not lost. It can be recovered at other portions in the cycle. Be that as it may, being told that "you need to store this much energy in order to move this much" can be a nuisance.

For this reason, a three-layer structure is very attractive. It is a relatively small number of layers, but sufficient to make possible a number of meaningful design tradeoffs. A three-layer structure:

- Re-aligns the actuation direction with that of the piezoelectric stacks.
- As an odd number of strain amplifiers, is contractile.
- Is the smallest number which has an "internal" layer that touches neither the piezoelectric stack nor the load.
- Fits within an elliptical envelope where the two minor axes are comparable in length.

As the piezoelectric cellular actuator is a technology inspired by human muscle fibers, it is only natural that they be contractile, not extensile, in order to do justice to their inspiration. Although one could make an argument that in mechatronics, there is no strict reason to be contractile, see the "Reverse" [as in pushes] Pneumatic Artificial Muscle [157], for example, there is one key difference: extensile actuators are vulnerable to buckling whereas contractile actuators are not. The interface of the piezoelectric stack to the first layer, type of adhesive if any, amount of preload, and so on involves closely held know-how and trade secrets, so often it is advantageous to simply purchase a commercially available single-layer amplified piezoelectric stack for the first layer. Subsequent layers can be connected with screws. A three-layer structure then leaves two layers to be designed, meaning the designer can exploit the interplay between the two remaining layers.

2.3.4.2 Designing a multi-layer amplification structure

An amplified piezoelectric stack actuator will be expected to fulfil a number of objectives common to any type of actuator. It must have a certain stroke length, or be able to move the load a certain amount. It must be able to impose a certain amount of force. The force specification may be constant, or it may be different for different ranges along the stroke. Knowing that the force the actuator can apply decreases with extension, it makes sense to put a specification on the *blocked force* or the force when the mounting flange of the actuator does not move. The *free displacement* is the maximum contraction the actuator can maintain under maximum activation when no load is applied. The application of any external load will reduce the displacement. In most applications, there will be some load applied to the actuator at the maximum displacement required by the application. The actuator must be able to exceed this minimum force at this location. This means that the free displacement must be *greater* than the maximum displacement required, in order to supply enough force to balance the load expected. Specifications on the system dynamics are also possible. Grossard provides a framework for a single-layer amplified piezoelectric device based on the controllability and observability Gramian [90]. As described in Section 2.3.4.1, the actuator may not exceed certain dimensions, both in the actuation direction and the directions transverse to it. Manufacturing processes will place limits on the properties of individual layers: the thickness of material left by a machining operation, how small a small angle may be, and so on.

All of these represent constraints on the actuator design that must be satisfied. Lobontiu and Garcia [152] present a design process for a single amplification layer based on these sorts of constraints. It is here that we see the value of adding a layer.

If an attempt to design an amplified piezoelectric stack with two layers does not result in a feasible region in the design parameter space, perhaps adding a third layer, expanding the dimensions of the design space, will generate a feasible region.

So while it is indeed possible to accomplish the necessary force–displacement tradeoff for any strain rate desired using a single-layer amplification mechanism with a large enough amplification ratio, any resulting mechanism will likely violate at least one of the constraints imposed by the application. Recall that with the small free displacement of a PZT stack, extremely aggressive force–displacement tradeoffs are necessary. A three-layer structure allows one to make such aggressive tradeoffs and still satisfy performance, geometry, and manufacturing constraints.

2.3.4.3 A three-layer example

An example of an application where nested cellular actuators will be used is in the positioning of a camera. Because the load consists of little more than the inertia of the camera, the free displacement can be very close to the specified displacement. The process of designing a three-layer actuator and the design choices made for the camera positioner described in [217] will be described in the following section. The actuator is designed to be used in an anatagonistic configuration.

2.3.4.4 Amplified PZT base units

The camera positioner is designed to have an angular excursion of $\pm 25°$ to correspond to the range of linear operation of saccadic motions in the human eye [99]. Taking into account the moment arm of the connecting rod of the camera positioner, this mechanism produces a free displacement of 8 mm for the actuator itself. A less rigid specification is that the camera positioner be able to complete saccadic motions with 50 ms so as to have equivalent capability [207]. A specification on the blocked force can be derived from this with knowledge of the camera inertia and friction parameters, but in the application this was evaluated experimentally after the fact. Rough calculations regarding the amplification factor achievable and the resolution required indicated that satisfactory performance of the camera positioner could be expected when each actuator contains 16 PZT stacks; this was used as a baseline for the design.

In order to function properly, the PZT stack must be under a preload to avoid disengaging from the innermost amplification mechanism when inactive. As mentioned earlier, commercially available stacks from the Cédrat corporation, model APA50XS were used as the first layer assembly, so that this step would already be completed. This meant that the first layer in the nested amplification structure was pre-determined, and the remaining layers would need to be designed in light of this choice. A PZT stack with a single layer of amplification is shown in Fig. 2.23. According to the data sheet for the APA50XS, 16 of these units in series have a free displacement of 1.25 mm, which is well short of 8 mm. Since simply placing the single-layer-amplified PZT stacks in series does not produce sufficient free displacement, another layer of amplification is necessary.

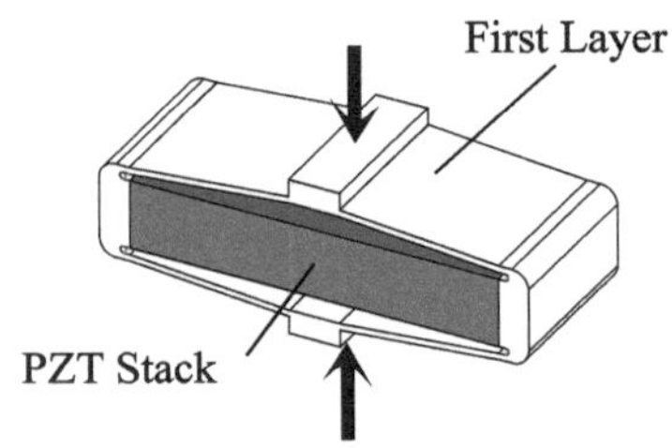

Figure 2.23 PZT stack with a single layer of amplification. © 2012 IEEE, reprinted with permission

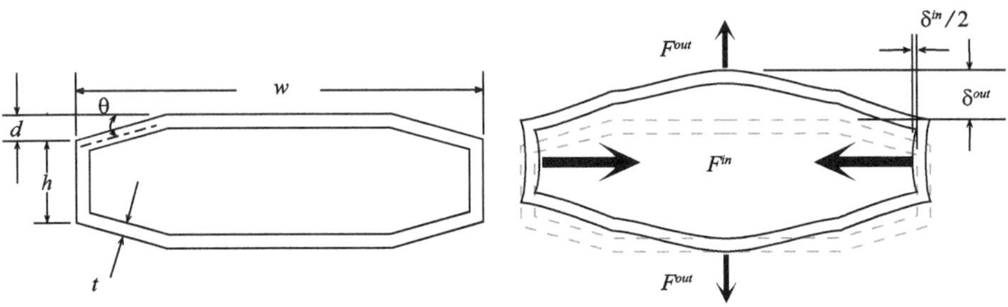

Figure 2.24 Parameterization of a rhomboidal mechanism. © 2012 IEEE, reprinted with permission

2.3.4.5 *The second layer of amplification*

To further increase the strain rate, the Cédrat first layer assemblies can be placed inside another rhomboidal mechanism, as described in previous sections. The method in Section 2.2 is valid for any simply connected doubly symmetric convex polygon, but a polygon with large numbers of segments is not only onerous to analyze, it is also difficult to manufacture reliably. To simplify matters, the design choice was limited to an octagonal shape with the variable parameters shown in Fig. 2.24, where the start point of the angular section and angle of the diagonal portion can vary. To further simplify matters, the mechanism was kept a constant thickness t. The 16 first layer units also need to be broken up into a certain number of two-layer assemblies that will be placed in series. Only regular combinations were examined, e.g., 8 assemblies, each containing 2 first-layer assemblies, 4 assemblies of 4, and so on. Of all these choices, 4 two-layer assemblies, each consisting of 4 first-layer assemblies in series with a custom rhomboidal strain amplifier was selected as it appeared to be able to produce a better overall displacement and had the best aspect ratio for the application in all three dimensions. The two-layer assembly that resulted is shown in Fig. 2.25. As there is an even number of amplification layers, it will extend when the PZT stack is energized.

Since this is a mechanism with multiple layers, each of which has its own set of immittance and geometric parameters, each geometric parameter $(d, h, \theta, t_a, w, b)$ will be given a subscript to denote the layer to which it applies. The layer for the immittance parameters are given a leading superscript, as in Section 2.3.3.4. Those corresponding to the Cédrat-supplied amplification mechanism are given subscript 1,

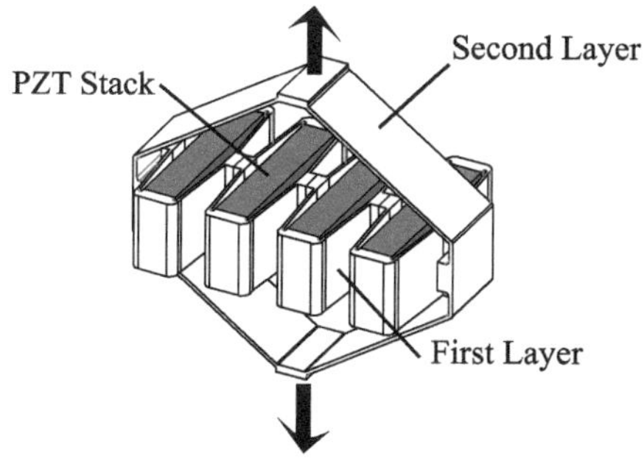

Figure 2.25 PZT stack with two layers of amplification. © 2012 IEEE, reprinted with permission

the interior amplification layer – subscript 2, and the outermost layer – subscript 3. Section 2.2 shows how to obtain the immittance matrix elements ${}^j s_1$, ${}^j s_2$, and ${}^j s_3$ as expressions of the geometric parameters. The free displacement then comes from the continued fraction expansion

$$ {}^2\delta^{free} = -\frac{{}^2 s_3 \, {}^1 F^{block}}{{}^2 s_1 \, {}^2 s_2 - {}^2 s_3^2 + {}^2 s_2 k_1}, \tag{2.71} $$

$$ {}^2 F^{block} = \frac{{}^2 s_3 \, {}^1 F^{block}}{{}^2 s_1 + k_1}, \tag{2.72} $$

where ${}^1 F^{block}$ is the blocked force and k_1 is the stiffness of the Cédrat APA50XS units.

So a chain of Cédrat APA50XS units will not meet the displacement specification. What about the four two-layer assemblies in series? If we get the amplification mechanism just right, will they make it? An attempt to find the right geometric parameters that will achieve this using Wolfram Mathematica's `NMaximize[]` routine fails to find a feasible region given natural geometric constraints, for instance, the first-layer units must be able to physically fit inside the second-layer unit, and so on. To give a sense of how aggressive the amplification tradeoff is, we reduce the displacement specification by half and get the small feasible region shown in Fig. 2.26. Maximizing blocked force and free displacement seem to be competing goals, and free displacement falls off quickly with thickness.

So, given real-world constraints, a series chain of 4 two-layer assemblies, each of which amplifies 4 first-layer units in series will not meet the specification either, although it gets substantially closer than the chain of 16 Cédrat units. Of course, there is nothing strictly saying that the units have to be arranged in powers of two, so some other two-layer configuration may meet the specification. It turns out that arranging all 16 Cédrat units in a series chain and then amplifying the series chain with an appropriately chosen amplifying rhomboidal structure can meet the 8 mm displacement specification, this has a number of drawbacks. First, since there is an even number of amplification layers, this will be extensile, not contractile, as mentioned earlier. This can be mitigated by using the "bow tie" shaped structure as in Secord and Asada

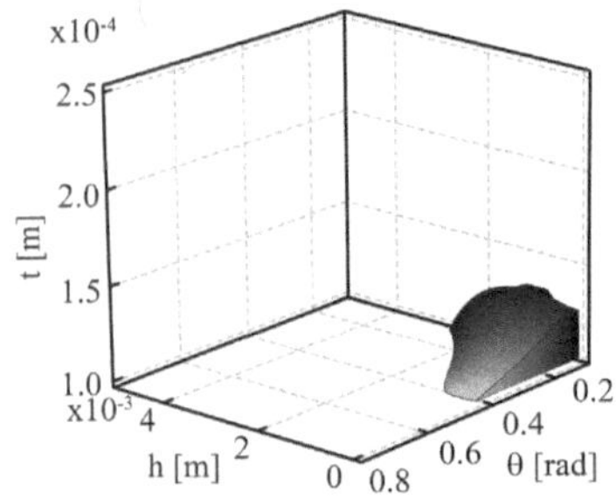

Figure 2.26 Region in θ, h, t space where both good force and displacement properties can be achieved. Plot shown for 4 two-layer units in series with 4 Cédrat APA50XS units each. d is fixed at 1 mm. Displacement threshold is 4 mm total, force threshold is 0.5 N. Notice that the displacement threshold is only half of that required; the specification is not met by this configuration. © 2012 IEEE, reprinted with permission

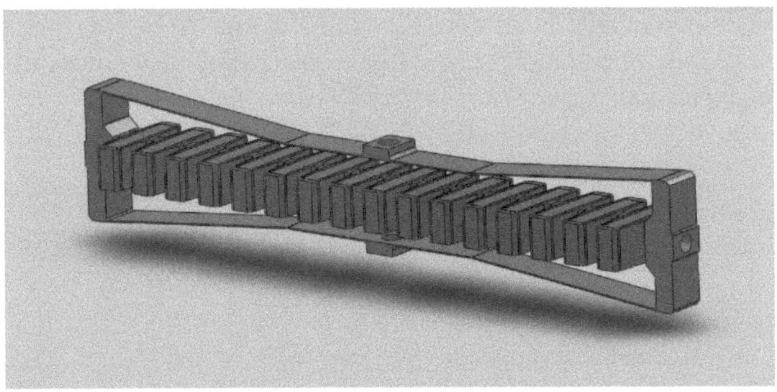

Figure 2.27 A two-layer actuator design that meets the 8 mm displacement specification. This design was rejected in favor of a three-layer mechanism because the "bow-tie" shape was not as good of a use of space and posed manufacturing difficulties.

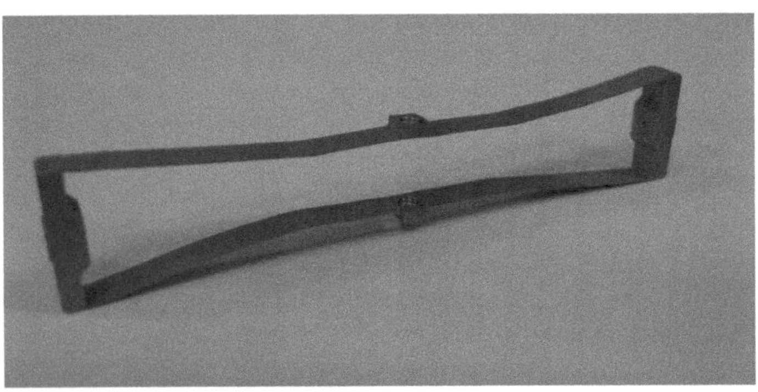

Figure 2.28 Custom strain amplifier for the two-layer actuator that amplifies a series chain of 16 Cédrat APA50XS amplified piezoelectric stacks.

[222] (Fig. 1.37) or that shown in Figs. 2.27 and 2.28, but this puts the mounting flange nowhere near the outside of the actuator, meaning a long connecting rod is needed, and

the envelope is not particularly nice. Second, each APA50XS unit has a tapped hole for an M1 screw, so there is no way to fasten one to another in a series chain and we had to resort to affixing them with epoxy adhesive. Making a chain of 4 APA50XS units with proper alignment in this fashion is difficult enough; making a chain of 16 would be more difficult still! Misalignments would likely cause undesirable operation. Given these drawbacks, it is worth seeing if a three-layer mechanism can give a better result.

2.3.4.6 Three-layer mechanism

A three-layer mechanism shown in Fig. 2.29 gives us the chance to design two custom layers, each with its own immittance matrix and set of geometric parameters. Both the second and the third layer will have the general octagonal shape shown in Fig. 2.14. Although most of the geometric parameters are independent of one another, there are some constraints between the two; the width of the third layer must be the same as that of four second layer assemblies in series, and the second layer must be able to fit within the third layer.

Because the two-port model is a convenient abstraction to the problem of designing multiple amplification layers, it would be nice to proceed as follows:

1. Choose the immittance parameters ${}^js_1$, ${}^js_2$ and ${}^js_3$ for each layer j that results in the desired behavior.
2. Tune the geometric parameters so that each rhomboidal strain amplification mechanism's immittance matrix matches the desired one.

This would decouple the design of each layer, but unfortunately, this strategy does not work; the immittances ${}^js_1$, ${}^js_2$, and ${}^js_3$ are coupled, being analytical expressions of the same geometric parameters. For instance, changing θ to make the value of s_1 better may adversely affect s_3. Trying to optimize a weighted cost function of blocked force and free displacement to get a good balance between the two also performs poorly. This gives another motivation for a three-layer structure; one can exploit the interplay between various layers to find an optimum value, since the immittance matrices of each layer are largely decoupled from one another. This increase in the size of the design space brings with it challenges (mainly with regard to managing more geometric parameters), but it also gives the designer a greater opportunity to find a good tradeoff between performance objectives.

The idea of each layer as an "amplifier" brings to mind the idea that each layer accomplishes a portion of the amplification. If 16 of the Cédrat units have a free displacement of 1.25 mm, and 8 mm is needed, then if the second and third layer combine to amplify by a factor of 6.4, that should work, right? Recalling the argument from the previous section reveals that this line of reasoning is incorrect. In order to get 8 mm of displacement at the output of the third layer, an input force to the third layer needs to be applied by the second layer that is large enough to deform the third layer. If the second layer is at its free displacement, it cannot apply any force, and if the input force to the third layer is zero, there will be no displacement, because the compliance matrix is nonsingular. So the second layer must be at some displacement short of its free displacement when the actuator is fully active. Similarly, the first layer cannot be

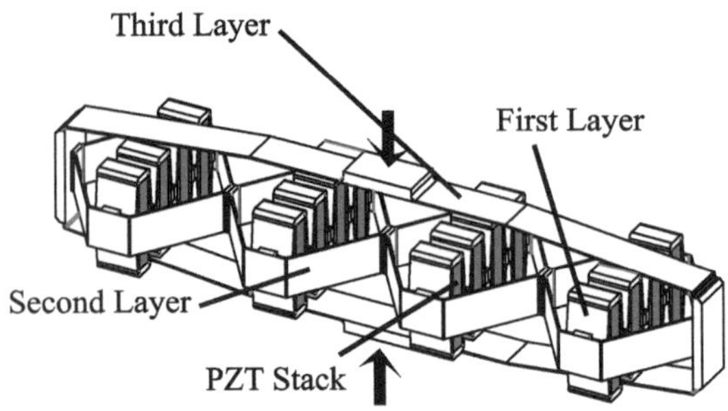

Figure 2.29 PZT stack with three layers of amplification.

at its free displacement, but must be at some displacement short of that, because it must be able to deform not only a free second layer, but the second layer constrained by the stiffness of the third layer. So in general this means that the layers get progressively more compliant the further out they are, and the innermost layers are operating closer and closer to their blocked configuration. The two-port network models, however, do allow the designer to calculate the blocked force and free displacement using the continued fraction expansions.

The force and displacement behavior of a given strain amplification layer depends on its interaction with the other layers, so it is not possible to optimize each layer separately. The examples in Section 2.1 explain why this is so. However, it is not a bad idea to use a displacement-optimized two-layer structure as an *initial condition* for a global optimization over the three-layer structure. For this example, the optimization problem for the three-layer actuator must be conducted over 9 parameters with transcendental functions and has no guarantee of convexity. This requires some simplification to evaluate the most significant parameters and search among those for the best design.

The approach taken in this example was to find some parameter ranges where the free displacement is exceeded, then to search the level surface corresponding to the desired free displacement for the maximum blocked force on the surface. Because searching a design space with 9 parameters at once was too complex, the search was broken into steps, changing a few parameters at a time. While this method was effective in finding a design that met the criteria, it is not guaranteed to be globally optimal.

As a initial condition for the search routine, the lumped stiffness, or stiffness viewed from the output, of the two-layer unit by itself was maximized. Since there was a smaller number of parameters, Wolfram Mathematica's `NMaximize[]` subroutine was able to solve this constrained optimization, even though there were trigonometric functions involved in the constraints. Holding this second-layer unit geometry fixed, a second optimization was performed that varied the third layer parameters to maximize free displacement. Searching in this way allows us to see if the free displacement specified is even achievable. This step resulted in a design with free displacement of 12.9 mm, exceeding the specification, and also gave geometric parameters in the neighborhood of the final design, which helps with getting stuck in local minima.

Since the blocked force is related to the maximum bandwidth the camera positioner can achieve, one way to choose the design is to look at all designs that just satisfy the displacement specifications and choose the one with the maximum blocked force. While a constrained optimization on the free displacement is appealing, with 9 geometric solutions, the optimization routine did not run effectively. Instead, the following search procedure was used: beginning from the set of geometric parameters that produced the 12.9 mm displacement, the search procedure slowly backs away from this maximum displacement toward the 8 mm specification varying two or three of the most significant parameters at a time. This traced out the level surface, which could be searched for maximum displacement.

2.3.4.7 Design constraints

The geometric parameters in this example were subject to several constraints which arise from manufacturing considerations and basic geometry, such as maintaining the convexity of the rhomboid.

- Each rhomboid will be machined by wire EDM techniques, for which the minimum thickness t_a is 0.1 mm.
- The length of the angled section along the actuation direction d must exceed the desired displacement of the amplification stage. Otherwise the rhomboid will fold up into a rectangle and the two-port model no longer applies. In this case the minimum d_3 was 2.4 mm.
- Although it is theoretically possible to have very small angles, there must still be a discernible angle when manufacturing tolerances, assembly misalignments, deformations during motion, and preloads are applied. In this case the angle of all stages was limited to $6° < \theta < 45°$.
- The length of each layer in the output direction must match the length of the subsequent layer in the input direction. This was relaxed slightly between the second and third layers, allowing for a rigid "spacer" no longer than 5 mm on each side. Thus, w_3 was allowed to vary in a 10 mm range.
- Each rhomboid must have its major axis along the input direction, i.e., $2d + h < w$.
- d, θ, and w must be chosen so that the geometry results in a simple convex polygon; anything else is considered a degenerate geometry and is not included.

In addition, the actuator must not impinge on itself when activated. For this reason, a rigid section was added to the middle of the third layer to make room for the first layer units.

2.3.4.8 The final design

The geometric parameter with the most drastic effect on the force–displacement tradeoff was the outer layer thickness t_{a3}. High displacement actuators tend to have low t_{a3}, θ_3, and d_3 values, with high w_3. A plot of blocked force vs. t_{a3} and t_{a2} is shown in Fig. 2.30. The surface is truncated where the free displacement drops below the specified value of 8 mm. Notice that the maximum is on the edge of the surface, indicating the maximum blocked force occurs where the displacement spec is just barely met.

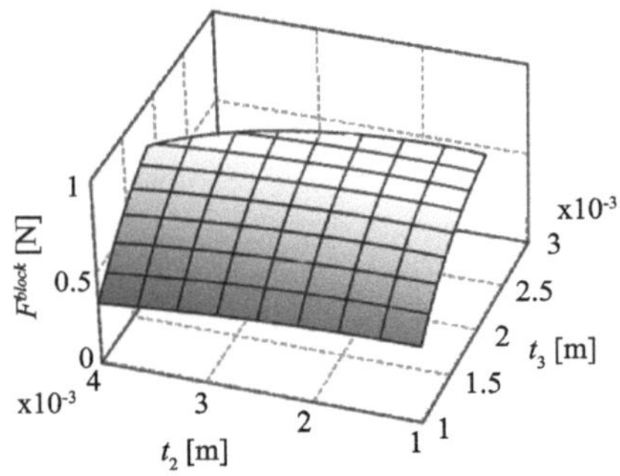

Figure 2.30 Variation in (simulated) blocked force with t_{a3}, t_{a2}. Portions of the surface not satisfying the minimum free displacement have been removed. © 2012 IEEE, reprinted with permission

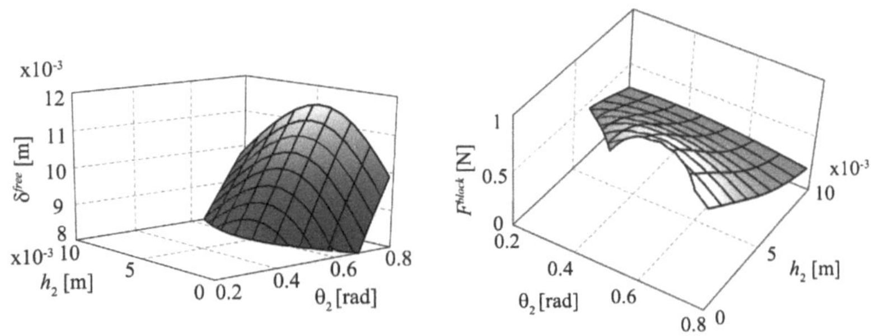

Figure 2.31 Variation in (simulated) performance with h_2 and θ_2. © 2012 IEEE, reprinted with permission

The effect of parameter selection on the second layer is more subtle, but it is important to the performance of the design. The second layer thickness, t_{a2}, affects the shape of the level surface corresponding to the free displacement specification, and the other second layer parameters do likewise. As can be seen in Fig. 2.31A, the free displacement level surface is curved in h_2–θ_2 space. Traveling along this level surface (curve) in Fig. 2.31B, the maximum blocked force with 8 mm free displacement is seen to occur somewhere around 0.5 radians.

Iterating this approach over t_{a2} several times eventually results in a design for the second layer strain amplifier that produces satisfactory blocked force at the output of the third layer. This means that the 16 Cédrat units are sufficient for this application. If a satisfactory solution could not be found, a more involved technique or more iterations may find a design with better performance, but it is likely that a few more Cédrat units would be needed in the design. The geometric parameters are summarized in Table 2.9. The resulting three layer actuator design has a free displacement of 8.1 mm and a blocked force of 0.907 N. By way of comparison, the Cédrat APA1000XL, the largest in their amplified piezoelectric stack actuator series, has a free displacement of 1.05 mm and has a footprint nearly 7.5 times larger than the three-layer actuator presented here.

Table 2.9 Final design of three-layer actuator.

Quantity	Value	Units
d_3	2.40	mm
h_3	1.00	mm
θ_3	6	°
t_{a3}	0.170	mm
w_3	68.8	mm
d_2	6.60	mm
h_2	3.00	mm
θ_2	35.1	°
t_{a2}	0.250	mm
${}^3F^{block}$	0.907	N
${}^3\delta^{free}$	8.10	mm

2.3.5 Discussion

This section describes the process for designing and understanding the behavior of cellular actuators with three or more layers. This can quickly pose a large challenge as the number of strain amplification layers increases and as the number of parameters quickly becomes large. As established earlier, the individual strain amplification layers cannot be designed separately, as the performance of the actuator depends heavily on how the individual layers interact.

The procedure conducted in this section used a few heuristics to start off the optimization process. Given a certain number of piezoelectric stacks to start with, their properties, and the desired specifications (blocked force, free displacement, etc.) determining whether there is a feasible region or not is non-trivial. The method employed here maximizes free displacement at each stage to try and find an acceptable initial condition for optimization. Once the free displacement specification is exceeded, the optimization allows parameter changes that decrease the free displacement. The boundary corresponding to the free displacement meeting the specification is searched for the maximum blocked force.

If these heuristics fail to find a set of parameters that meet the required displacement, this may be a sign that more or longer piezoelectric stacks are required. There are many ways that this procedure can be improved or made more systematic in nature.

The actuators presented in this section are meant to be used in an antagonistic configuration. For soft robotics applications the stiffness of the actuators can be just as important as its stroke length and force. Looking at the input–output nature of nesting and the two-port models of each layer, we see that the stiffness of the passive antagonist takes on the form of a continued fraction. Using properties of continued fractions, we see that the stiffness is dominated by the outermost layers in the hierarchy. This means that making changes at the innermost levels of the hierarchy can be less effective when there are large numbers of amplifying layers.

This brings us to an important question: How many layers *should* there be in a piezoelectric cellular actuator? Given the complexity of designing an effective three-layer actuator (where the first layer strain amplifier was pre-determined) it is unlikely that actuators with more than three layers will be popular or more than an intellectual curiosity. With only three layers, it is possible that changing the terminating electrical impedance or mechanical stops on the first layer units will still have a noticeable effect. In this case, the piezoelectric cellular actuator could be used in variable impedance applications.

Several more general multi-stage compliant amplification structures do exist. Su and Yang [249] analyze a multi-stage lever mechanism for force amplification. Interestingly enough, they come to a conclusion that is dual to the one presented in this section, namely that the best force amplification is achieved when subsequent stages are much stiffer than prior ones. Examples also exist of using the amplification feature in reverse (where the outermost layer is the input and the piezoelectric crystal is the output) for energy harvesting [292]. Malosio and Gao [163] present a dual-stage rhomboidal amplifier where both amplification stages are optimized concurrently using Box–Behnken methods and Finite Element simulation. Kim, Kim, and Kwak [129] present a "three-dimensional" bridge mechanism that amplifies a prior design whose displacement was found to be insufficient. While it is not "nesting" in the sense of this book (it is really more of a series combination), the "telescoping" piezoelectric actuator of Alexander and Brei [9] is related in spirit to the work presented here. It represents an innovative method of increasing displacement while placing successive units within one another to decrease the overall length of the actuator required for a given stroke.

2.4 Representation and characterization of complex actuator arrays

2.4.1 Graph-theoretic modeling

Methods for graphically representing complex multi-body systems and obtaining governing equations include bond graphs and graph-theoretic modeling (GTM), or linear graphs [211]. McPhee has presented a series of publications on the applications of linear graph theory to flexible multi-body systems [169,211,212,231]. The key concept is to introduce a matrix, named incidence matrix, to represent a complex topology of a multi-body system. This graph-theoretic approach enables automatic generation of dynamic equations [55,82]. While these methods do generate the equations of motion for the dynamic systems of interest to the current work, they are highly general and generate redundant equations requiring careful manual selection of state variables, or cut-sets, to have physical meaning to the reduced echelon form required to systematically generate the equations of motion. This manual selection is not conducive to the automatic generation required to analyze many different actuator topologies and differing internal cell structures. Additionally, the incidence matrix in the graphical techniques consists of a list of connections between all elements in the system while the fingerprint method presented below uses the self-contained cell structures to separate the internal cell dynamics from the topology dynamics, thus greatly simplifying and shrinking the incidence matrix. [38] presents a similar simplifying method called Newton–Raphson Mixed Nodal Tableau by separating internal dynamics of photovoltaic cells and then "stamping" these repeated dynamics into a larger system as a single element and using GTM to generate the final equations of motion. This allowed for a simpler process for generating the dynamic equations of motion and allowed for the nonlinear dynamics of the photovoltaic system while treating the rest of the system as linear. [38] focused on electrical systems, specifically photovoltaic power systems, resulting in an application specific algorithm with fixed equations that cannot be easily applied to mechanical systems.

The fingerprint method presented in this section is to represent complex cellular actuator array topologies and develop a deterministic method for generating and exploring possible topologies given a number of cells. The fingerprint method helps to generate the dynamic equations of motion for varied cellular actuator array topologies with a method that accommodates any valid linear internal cell dynamics, well-structured internal cell dynamics, and outside dynamic structure.

2.4.2 Cell

A cell consists of a modular set of dynamic elements including a flexible linear actuator, or a non-flexible linear actuator connected in series with a flexible element(s), and any elements needed to properly model the modular component of the array. Cell flexibility is needed to mitigate the differences in length between cells that are active and those that are not. The actuator can be based on any linear actuation technology so long as (i) the actuator can be represented as a spring, stiff or flexible, with a pure force preloading the spring; (ii) the actuator's pure force follows a known force versus time function, when activated; and (iii) this function is minimally dependent on external cell load and therefore the interaction is negligible. The first criterion is generally true of most actuators, whether they are mostly stiff (like a motor) or flexible (like shape memory alloy). For stiff actuators, a large force acts on a stiff spring to represent the actuator's displacement. The second criterion is generally easy to achieve through measurement of the force generated in the actuator's flexible element when the actuator is fully blocked. For stiff actuators, estimating the stiffness of the actuator and measuring the displacement is usually sufficient to generate the force function. The final criterion implies that the external effects on the cell, whether displacement or force, should have little effect on the actuator's ability to reach the commanded position. This can be met through design, for example, by either limiting outside force or choosing a low stiffness flexible element to combine in series with an actuator, ensuring the actuator can move through its full displacement. Actuator technologies such as piezoelectrics, ultrasonic motors, linear stepper motors, hydraulics, pneumatics, and shape memory alloy (SMA) can all meet the three criteria.

While the theory presented in this work holds for any internal cell model meeting the above criteria, this work focuses on the Hill-type cell model from studies in

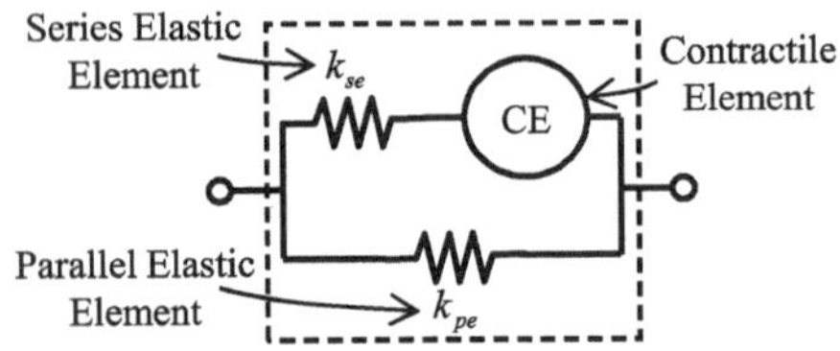

Figure 2.32 Hill-type model.

physiology. This is done to work towards the goal of natural motion and to provide a consistent example for explanation purposes. Different cellular actuator array applications may have goals requiring vastly different cell dynamics than is provided by this model. The Hill-type cell model consists of a parallel elastic element connected in parallel with the combination of a series elastic element and a contractile element, which themselves are connected in series. This is shown in Fig. 2.32. In most accepted Hill-type model literature, the parallel elastic and series elastic elements are modeled as springs while the contractile element does not have consistently defined dynamics. In many cases the contractile element is simply treated as a function of force versus time. In physiology, this is done due to lack of knowledge of the dynamics of the contractile element, an open area of research; however, the current work uses it as a method of model simplification. In this work the actuator, or contractile element in the Hill-type model, is treated as a spring acted on by a pure force preload. The result is a shortening of the neutral length of the actuator's "spring" by the amount of the displacement of the actuator.

The series elastic element represents the connective material between actuators and can be anything from stiff metal to springs, silicone, or even cloth. With fragile actuators the series elastic element should have lower stiffness to protect the actuator; however, this weakens the overall force output of the system. The Hill-type cell model can be modified to add an additional damper in parallel with the series elastic element or the contractile element to represent the damping effects of many actuator or material choices. This allows for additional design options. The parallel elastic element represents spring-like forces carried across the cell but not seen by the series elastic and contractile elements. These are generally kept small to avoid internal compressive forces but are maintained through the theory to allow for greater design flexibility. The methods presented in the following theory are generalized to accept any internal dynamics as long as the above criteria are met. They do not require the internal structure of the cell to follow a Hill-type cell model.

While also not required, this method focuses on operating each cell in a fully-ON or fully-OFF fashion. This allows actuator hysteresis to be avoided, can simplify modeling, control, and physical wiring and electrical requirements. With a large number of cells, proper discretization of the overall array output force/displacement is still maintained.

In many cases, such as with PZT and SMA actuators, the binary control scheme also allows nonlinear effects to be decoupled from linear dynamics. The nonlinear effects can then be modeled using a forward-loop input shaping instead of a dynamic effect, simplifying modeling and simulation. Specifically in the case of SMA actuators, the actuator can be modeled as a spring with a pure force input acting across it while heating and cooling are accounted for by adding a nonlinear delay to the input force signal.

2.4.3 Connecting structures

Cell array actuators are collections of cells connected in various arrangements, or topologies, to provide the large-scale motion required of a muscle system. The topology of the array is critical to determining the array's properties. Having more cells in series tends to give an array more displacement, more cells in parallel gives more force and higher robustness, and having a non-uniform structure can give higher force discretization and more fine-tuned control.

If the above cell criteria are met, internal dynamics of the cell can be decoupled from the dynamics of the array. This is done by treating the cell as an input–output function, taking in the positions and velocities of the cell endpoints (right side and left side) and outputting the force at those endpoints. The array itself consists of the cells and the connecting structures between the cells. The connecting structures are assumed to have mass and are called masses. All cells connect directly to masses, and all masses connect directly to cells or the endpoints. No mass is connected directly to any other mass and likewise for cells.

2.4.4 Incidence matrices

The methods presented in this work build on the concepts of graph-theoretic modeling (GTM) presented in Section 2.4.1. In GTM, incidence matrices are used to represent the connections between cells (topology) and then mathematical operations are carried out on these incidence matrices to develop the dynamic equations of motion. This work follows the same concept, but defines the incidence matrices differently to better suit the application to actuator arrays.

The driving simplification behind modeling actuator arrays comes from the ability to separate the dynamics about each cell, essentially treating cells as input–output functions relating endpoint motion and inputs to output forces. For this reason, the incidence matrices for the current work are separated into incoming connections (connections to the right of a cell; incoming relative to masses) and outgoing connections (connections to the left of a cell; outgoing relative to masses) as shown in Fig. 2.33. The incoming connections matrix is $\boldsymbol{H}$ and the outgoing connections matrix is $\boldsymbol{G}$. $\boldsymbol{H}$ is an $N \times M$ matrix and $\boldsymbol{G}$ is an $M \times N$ matrix where N is the number of cells in the array and M is the number of masses. The elements of both $\boldsymbol{H}$ and $\boldsymbol{G}$ are either 1, representing a connection between the associated cell (row in $\boldsymbol{G}$, column in $\boldsymbol{H}$) and mass (column in $\boldsymbol{G}$, row in $\boldsymbol{H}$), or 0, representing no connection. This is shown in

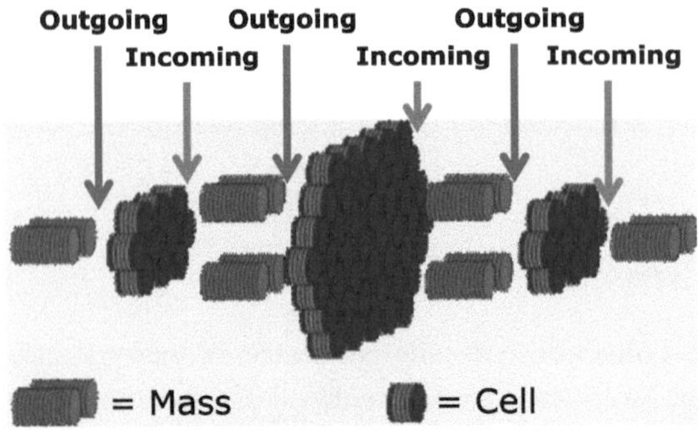

Figure 2.33 Explanation of incidence matrix components for a layer based actuator array topology: Outgoing connections are represented by $\boldsymbol{G}$ and incoming connections are represented by $\boldsymbol{H}$.

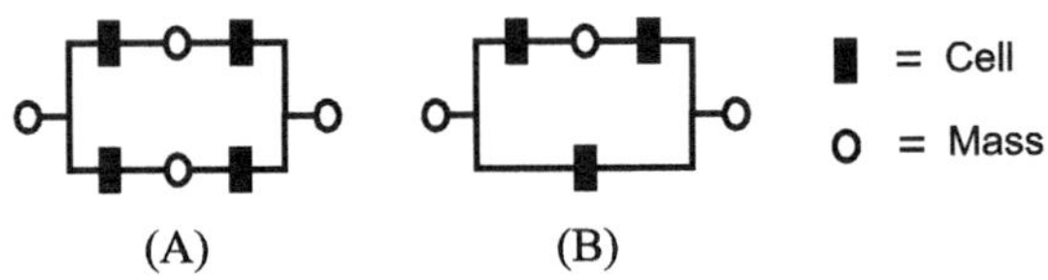

Figure 2.34 Examples of (A) a layer based actuator array structure and (B) a non-layer based actuator structure. The layer based array has two cells on each path between array endpoints while the non-layer based array has one path with one and one with two. With identical cells, the non-layer based array would likely generate internal compressive forces.

Eq. (2.73):

$$\begin{aligned}\boldsymbol{G}_{m,n} &= \begin{cases}0, & \text{if mass } m \text{ is not connected to the left of cell } n,\\ 1, & \text{if mass } m \text{ is connected to the left of cell } n,\end{cases}\\ \boldsymbol{H}_{n,m} &= \begin{cases}0, & \text{if mass } m \text{ is not connected to the right of cell } n,\\ 1, & \text{if mass } m \text{ is connected to the right of cell } n.\end{cases}\end{aligned} \tag{2.73}$$

2.4.5 *Fingerprint method basics*

The layer based structure shown in Fig. 2.33, where components are aligned in the orthogonal direction of the direction of linear actuation, is convenient for describing physical connection from layer to layer. Fig. 2.34 shows examples of (A) a layer based structure and (B) a non-layer based structure. In the layer-based approach, all cells are assumed to have identical relaxed lengths, identical unloaded actuated lengths, and are arranged such that all topological paths between array endpoints contain the same number of cells (and masses).

When an actuator array is a layer based structure, the connection of the actuator units, or topology, can be represented by using a "fingerprint," a layer based set of two row matrices encoding the connection information of the array. The fingerprint transcription consists of segmenting the array into layers just after the cells and before

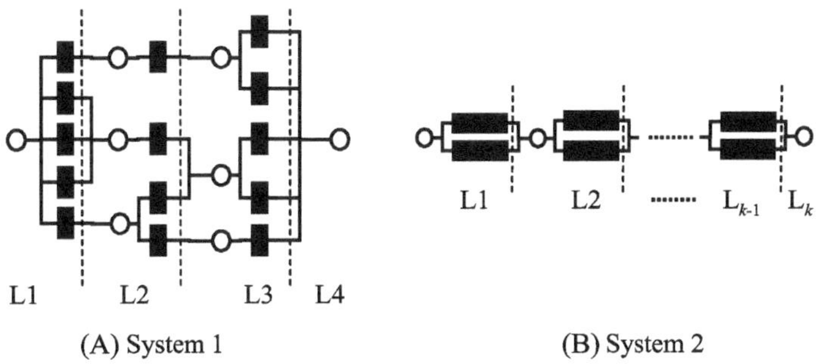

Figure 2.35 Example array topologies. © 2011 SAGE Publications, reprinted with permission

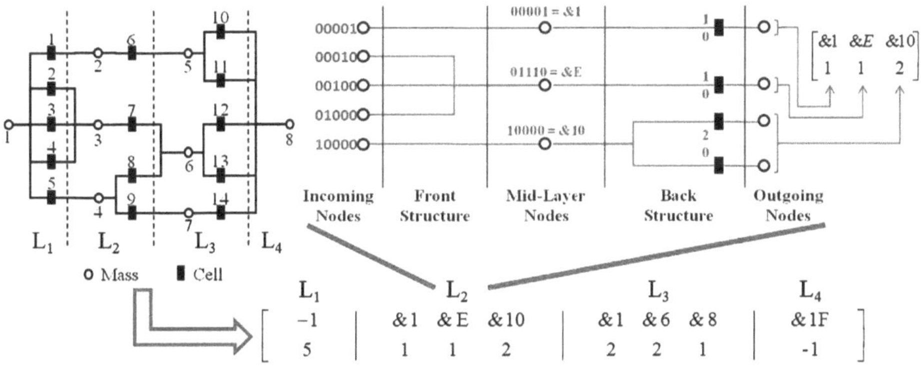

Figure 2.36 Example of building a fingerprint from an actuator array topology.

the masses those cells connect to. The layer can then be represented as an incoming structure and an outgoing structure relative to the layer's masses. For each mass, a hexadecimal number represents a binary encoding of the cells from the previous layer that the masses connect to.

Consider two actuator topologies shown in Fig. 2.35. The array in Fig. 2.35A has the following fingerprint:

$$\left[\begin{array}{c|ccc|ccc|c} -1 & \&1 & \&E & \&10 & \&1 & \&6 & \&8 & \&1F \\ 5 & 1 & 1 & 2 & 2 & 2 & 1 & -1 \end{array}\right]. \tag{2.74}$$

Fig. 2.36 shows how the array topology shown in Fig. 2.35A is described by a fingerprint. The second layer (L2)'s incoming connection is represented by [&1 &E &10]. In L2, &1 corresponds to 00001, which means that the first mid-layer node connects only to the first incoming connection point. &E corresponds to 01110, meaning the second mid-layer node connects to incoming connection points 2, 3, and 4. &10 corresponds to 10000, meaning the third mid-layer node connects to incoming connection point 5. The outgoing connection can be represented by 2 numbers per mid-layer node. These

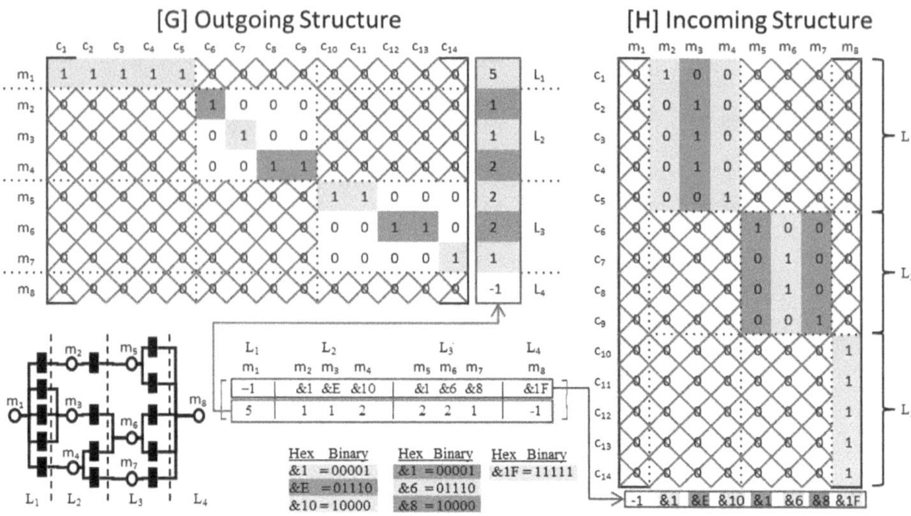

Figure 2.37 Incidence matrix representation of a fingerprint.

two numbers represent the number of cells and the number of links connected to that mid-layer node. The back structure for each mass is represented by the number of cells connected to that mass. For this array topology, L1's outgoing connection is represented by [5], L2 by [1, 1, 2], L3 by [2, 2, 1], and L4 by [−1] where −1 is used to signify the end of the actuator array. Also, −1 is used to signify the beginning of the array as there is no previous layer to provide an incoming structure. For the topology shown in Fig. 2.35B, in which the same intermediate structure repeats, the fingerprint is given by

$$\left[\begin{array}{c|c|c|c|c} -1 & \&3 & \cdots & \&3 & \&3 \\ 2 & 2 & & 2 & -1 \end{array}\right]. \tag{2.75}$$

2.4.6 Fingerprint-to-incidence matrix relationship

Matrix ***H*** can be populated automatically from the first row of the fingerprint. Column one is skipped as there are no left-connecting cells to the leftmost mass. Columns two through the end are populated as shown in Fig. 2.37. Each subsequent layer begins one row down from the previous layer's lowest entry. Also, ***G*** can be populated automatically from the second row of the fingerprint. The last row is skipped as there are no right-connecting cells to the rightmost mass. Rows one through the second last row are populated as shown in Fig. 2.37. Each subsequent layer begins one column right of the previous layer's rightmost entry and contains a number of 1's corresponding to the number of outgoing cells for each mass of the layer, or the elements of the second row of the fingerprint.

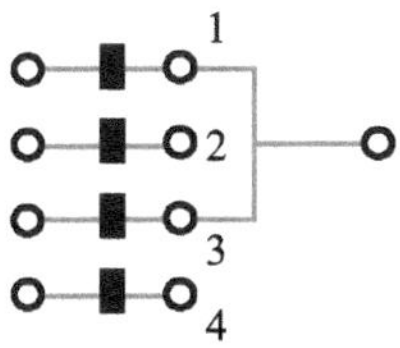

Figure 2.38 Front section autogeneration example. © 2011 SAGE, reprinted with permission

2.4.7 Automatic generation of actuator array topologies

In order to perform an in-depth survey of actuator array properties, a method of generating all possible actuator array topologies given a fixed number of cells is needed. The auto-generation problem can be broken into a layer-by-layer iterative approach. Each layer is then further broken down into its front section, which contains the information for bringing the incoming nodes down to a certain number of mid-layer nodes, and its back section, which contains the information for expanding each mid-layer node to a certain number of cells and outgoing nodes.

For each front section, all of the possible methods for connecting the incoming nodes are explored using a recursive function. This function takes a binary input showing which incoming nodes are not yet connected for the current test case. For example, in Fig. 2.38, incoming nodes 2 and 4 are not connected, thus the input would be 1010. The function iterates upward in binary from 1 to $2^4 - 1$ where 4 is the length of 1010. Each binary number has 0's concatenated to the left side until it is of the same length as 1010 and is then subtracted from 1010 element-wise. If there are any resulting -1's then that binary number is discarded, otherwise that binary number is stored in hex in array $\boldsymbol{c}$ in order. This becomes front structure number as part of a fingerprint layer. In the example illustrated in Fig. 2.38, the first entry is &2 representing a connection to only incoming node 2. The next layer of recursion then takes in an input of $1010 - 0010 = 1000$ and returns &8. When the function finishes, the result is a structure of all the frontal connection possibilities, each of which can have one or more mid-layer nodes. Since each mid-layer node must have at least one cell attached to it, those frontal connections which have more mid-layer nodes than remaining cells are discarded. Another recursive function is run to generate the back structure for each front structure. This function takes in the number of mid-layer nodes and the number of cells left to be placed, and returns all of the back connection possibilities. Fig. 2.39 shows a process tree for how fingerprints were generated for arrays with 4 cells. All arrays begin with a single incoming node and a number of cells. The front end for layer 1 is always -1, and the back end uses between 1 and N cells where N is the total number of cells. The second and subsequent layers take each of the previous layer's configurations and generate all of the possible configurations for that layer. When the function is called with no remaining cells, the last layer is always filled in to connect all mid-layer nodes to a single output node and put -1's for the number of cells.

Fig. 2.40 shows the 23 topologies for arrays with 5 cells, and Fig. 2.41 shows the number of connection possibilities for arrays up to 10 cells and the associated computational effort to generate the possible array structures. The number of topologies for

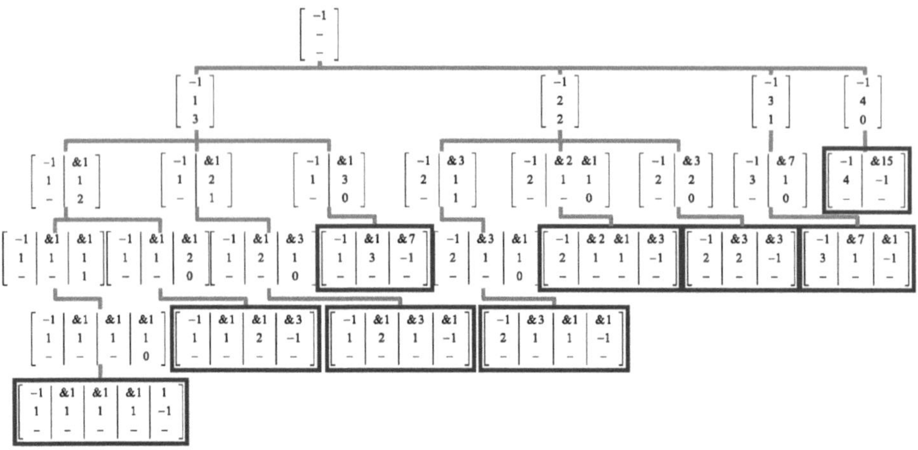

Figure 2.39 Autogeneration process tree for generating fingerprints for arrays with 4 cells. The third row in each representation shows unallocated cells remaining.

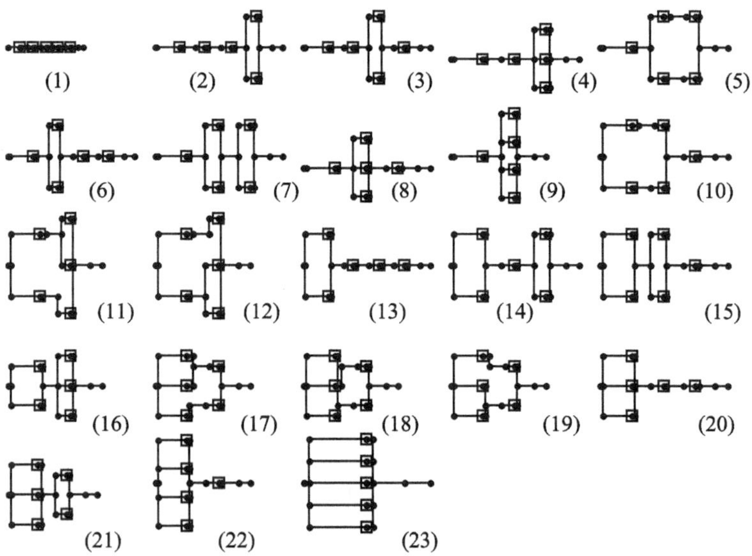

Figure 2.40 Automatically generated 23 topologies for 5 cells. © 2011 SAGE, reprinted with permission

2–8 cells are: 2 topologies for 2 cells, 4 topologies for 3 cells, 9 topologies for 4 cells, 23 topologies for 5 cells, 65 topologies for 6 cells, 199 topologies for 7 cells, 653 topologies for 8 cells, 2283 topologies for 9 cells, and 8467 topologies for 10 cells. For example, the computational time for 10 cells was 324 seconds by MATLAB running on a QuadCore 2.83 GHz processor. The number of topologies and the computation time for generating all of the topologies increases exponentially with the number of cells.

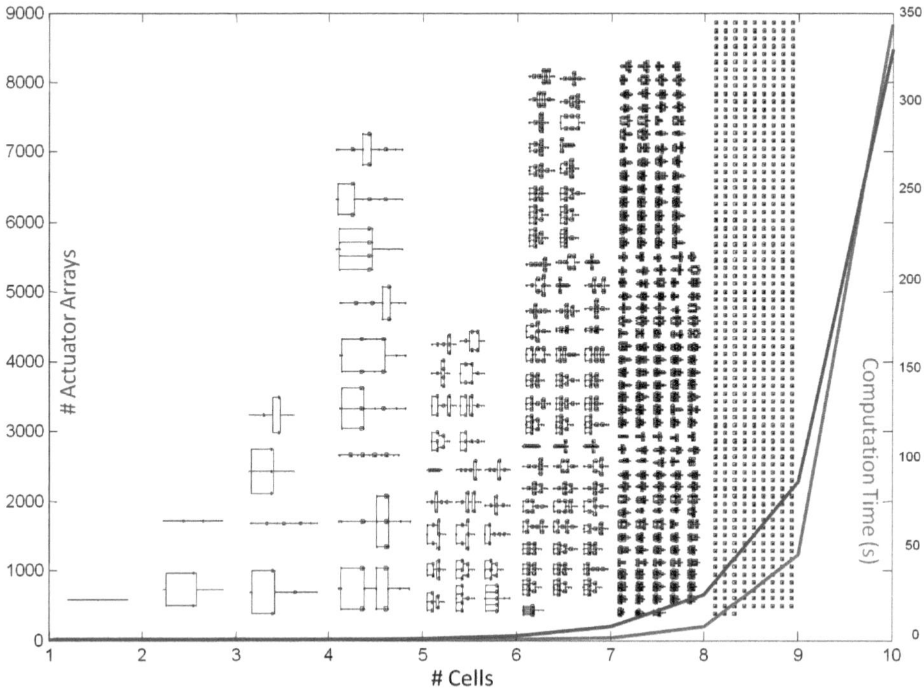

Figure 2.41 Automatically generated topologies and computational effort for 1–9 cells.

A similar process can be used to autogenerate topologies based on the incidence matrices that is not restricted to the layer based arrays; however, that method produces many more topologies. The layer based array structure gives good coverage of possible structures while limiting results to arrays that generally avoid compressive internal forces. For this reason, the incidence matrix autogeneration process was not specifically covered in this section.

2.4.8 Incidence matrix identity and similarity transforms

When comparing two actuator array topologies, whether automatically generated or otherwise, it can be helpful to have a criterion to determine if two topologies are identical. If two topologies are not identical, it can be likewise helpful to determine how similar they are from a topological perspective. Identical topologies are defined as those with the same topological configuration but different numerical labels for the masses and cells. While a mirrored actuator array will contain the same static properties as the original, the forces at the endpoints can be different dynamically, thus the arrays are not considered identical. The same is true for mirrored sub-sections of an array. While two identical arrays may share the same graphical layout, the incidence matrices are different, which complicates the identification. This section gives a method for identifying identical arrays.

This section also develops several transformations which can be used to judge the similarity of two array topologies. Unfortunately, due to the discontinuous nature of the solution space for properties of arrays, these criteria do not always show similarity of the properties of the arrays. This is a limitation of the current methods and could be addressed as future work to develop methods of judging similarity of properties from the incidence matrices directly.

2.4.8.1 Incidence matrix identity transforms

Two identical arrays will always contain the same number of cells and masses, thus their $\boldsymbol{G}$ (and likewise $\boldsymbol{H}$) incidence matrices must have identical dimensions. If this criterion is not met, it is not possible to transform one into the other. Assuming the criterion is met, two transforms are available which when applied to any $\boldsymbol{G}$ and $\boldsymbol{H}$ incidence matrix pair will yield an identical topology.

Transformation 1. This transformation switches mass a and mass b by switching rows a and b in $\boldsymbol{G}$ and columns a and b in $\boldsymbol{H}$ as shown in (2.76). This works because $\boldsymbol{G}$ maps from masses to cells and $\boldsymbol{H}$ maps from cells to masses. The product of $\boldsymbol{G}$ and $\boldsymbol{H}$, $\boldsymbol{GH}$, has both its rows and columns swapped. This is shown in Fig. 2.42 as the transformation between frame (a) and frame (b):

$$\begin{aligned} &\boldsymbol{G}_{(a,:)} \Rightarrow \boldsymbol{G}^{new}_{(b,:)}, && \boldsymbol{H}_{(:,a)} \Rightarrow \boldsymbol{H}^{new}_{(:,b)}, \\ &\boldsymbol{G}_{(b,:)} \Rightarrow \boldsymbol{G}^{new}_{(a,:)}, && \boldsymbol{H}_{(:,b)} \Rightarrow \boldsymbol{H}^{new}_{(:,a)}, \\ &\boldsymbol{G}_{(i,:)} \Rightarrow \boldsymbol{G}^{new}_{(i,:)},\ \forall i \neq \{a,b\}, && \boldsymbol{H}_{(:,i)} \Rightarrow \boldsymbol{H}^{new}_{(:,i)},\ \forall i \neq \{a,b\}. \end{aligned} \tag{2.76}$$

Transformation 2. This transformation switches cell a and cell b by switching columns a and b in $\boldsymbol{G}$ and rows a and b in $\boldsymbol{H}$ as shown in (2.77). This works by a similar reasoning as for Transformation 1. The product of $\boldsymbol{G}$ and $\boldsymbol{H}$, $\boldsymbol{GH}$, is unchanged as it maps from masses to masses. This is shown in Fig. 2.42 as the transformation between frame (b) and frame (c):

$$\begin{aligned} &\boldsymbol{G}_{(:,a)} \Rightarrow \boldsymbol{G}^{new}_{(:,b)}, && \boldsymbol{H}_{(a,:)} \Rightarrow \boldsymbol{H}^{new}_{(b,:)}, \\ &\boldsymbol{G}_{(:,b)} \Rightarrow \boldsymbol{G}^{new}_{(:,a)}, && \boldsymbol{H}_{(b,:)} \Rightarrow \boldsymbol{H}^{new}_{(a,:)}, \\ &\boldsymbol{G}_{(:,i)} \Rightarrow \boldsymbol{G}^{new}_{(:,i)},\ \forall i \neq \{a,b\}, && \boldsymbol{H}_{(i,:)} \Rightarrow \boldsymbol{H}^{new}_{(i,:)},\ \forall i \neq \{a,b\}. \end{aligned} \tag{2.77}$$

Two array topologies are identical if a sequence of Transformations 1 and 2 can be found to move from one topology to another.

2.4.8.2 Incidence matrix similarity transforms

Mirror Transform. The first similarity transform employs the idea of mirroring an array. Since $\boldsymbol{G}$ is a mapping from masses to cells and $\boldsymbol{H}$ is a mapping from cells to masses, the two can be reversed to provide the mirrored mapping. The transform first mirrors $\boldsymbol{G}$ and $\boldsymbol{H}$ about the vertical axis and then takes the transpose of the result.

The same process can be done for sub-arrays so long as the numbers of masses connecting directly to the sub-array on either side of the sub-array are consistent. This

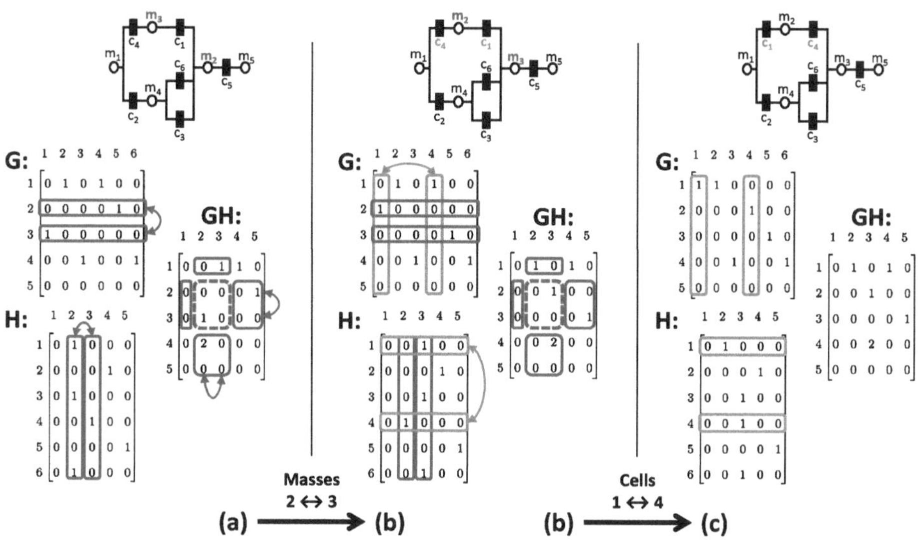

Figure 2.42 Example transitions between identical topologies.

is helpful to show relationships between array topologies but does not imply static property similarity, unless only a single mass connects to either side of the sub-array. Mirror transforms do not imply dynamic property similarity regardless of the number of masses on either side. The exception to this is when an array, or sub-array, is symmetric, though this is not covered directly in the current work.

The sub-array mirror transform begins by using Transforms 1 and 2 to group all of the elements in the sub-array together in both $\boldsymbol{G}$ and $\boldsymbol{H}$. Once grouped, the subsection $\boldsymbol{G}_{([a:b],[c:d])}$ will contain rows a through b and columns c through d of $\boldsymbol{G}$. Likewise, the subsection $\boldsymbol{H}_{([c:d],[a:b])}$ will contain rows c through d and columns a through b of $\boldsymbol{H}$. The two subsections will contain the sub-array and can be transformed by mirroring across the vertical axis and then taking the transpose. This is shown in (2.78):

$$\begin{aligned}
\left[\boldsymbol{G}_{([a:b],[d:c])}\right]^{\top} &\Rightarrow \boldsymbol{H}^{new}_{([c:d],[a:b])},\\
\boldsymbol{G}_{(i,j)} \Rightarrow \boldsymbol{G}^{new}_{(i,j)}&, \forall[i < a], [i > b], [j < c], [j > d],\\
\left[\boldsymbol{H}_{([c:d],[b:a])}\right]^{\top} &\Rightarrow \boldsymbol{G}^{new}_{([a:b],[c:d])},\\
\boldsymbol{H}_{(i,j)} \Rightarrow \boldsymbol{H}^{new}_{(i,j)}&, \forall[i < c], [i > d], [j < a], [j > b].
\end{aligned} \tag{2.78}$$

Cell/Mass Addition Transform. Adding cells and masses to $\boldsymbol{G}$ and $\boldsymbol{H}$ incidence matrices allows for judging the difference between a sub-array and a more complex full array. On its own this provides limited utility; however, similar array topologies generally share common sub-array roots. When a common root is found, the minimum number of steps (cell/mass addition and mirror transforms) required to move from the root to each full array provides a numerical similarity comparison between the two arrays.

Adding elements from a common root provides a better measure than subtracting elements since subtracting an element can make entire sections of an array useless and create a premature root between two topologies. Future work could develop subtraction methods to determine robustness similarity between arrays, though that was not handled as part of this section.

Since cells connect to masses directly and any number of cells can connect to a single mass, adding a cell simply requires selecting the masses to connect to either side of the new cell. Once selected, a new column vector is appended to $\boldsymbol{G}$ with a 1 in the row corresponding to the mass connected on the left and 0 for all other elements. Similarly, a new row vector is added to $\boldsymbol{H}$ with a 1 in the column corresponding to the mass connected on the right and 0 for all other elements. This is shown in (2.79):

$$\begin{gathered} g_m = \begin{cases} 0, & \text{if mass } m \text{ is not connected to the new cell on the left,} \\ 1, & \text{if mass } m \text{ is connected to the new cell on the left,} \end{cases} \\ \begin{bmatrix} \boldsymbol{G} & \boldsymbol{g} \end{bmatrix} \Rightarrow \boldsymbol{G}^{new}, \\ \boldsymbol{h}_m = \begin{cases} 0, & \text{if mass } m \text{ is connected to the new cell on the right,} \\ 1, & \text{if mass } m \text{ is connected to the new cell on the right,} \end{cases} \\ \begin{bmatrix} \boldsymbol{H} \\ \boldsymbol{h} \end{bmatrix} \Rightarrow \boldsymbol{H}^{new}. \end{gathered} \tag{2.79}$$

Since masses connect to cells on either side and only a single mass can connect to any given side of a cell, the process for adding masses also requires adding additional cells. This is treated as a two-step process. First, an empty row vector is appended to the top of $\boldsymbol{G}$ and an empty row vector is appended to the left side of $\boldsymbol{H}$. This creates the additional mass in the system which takes the place of the first mass in the array. The new mass is not yet connected to the rest of the array. The second step utilizes the cell addition transformation to add two cells to the array, one which has a right-side connection to the new mass and the other that has a left-side connection to the new mass.

2.4.9 Robustness analysis

The fingerprint method can be used to analyze robustness of an array topology. The minimum cell loss to controllability failure defines, in the worst case scenario, how many cells would have to break in order to have a complete loss of controllability. A complete loss of controllability is defined as the lack of ability to create a total force in the muscle due to activating cells. This is a measure of the robustness of actuator arrays. Fig. 2.43 shows the result for the array network shown in Fig. 2.35A, in which the worst case scenario to completely lose the force capacity of the array is to lose 3 of the critical units indicated in the figure. Note that the loss of units changes the force capacity and the variability of the array. Therefore, the optimal movements change accordingly due to the failure of cellular units. This robustness measure is useful when designing robotic systems for use in hazardous environments or when

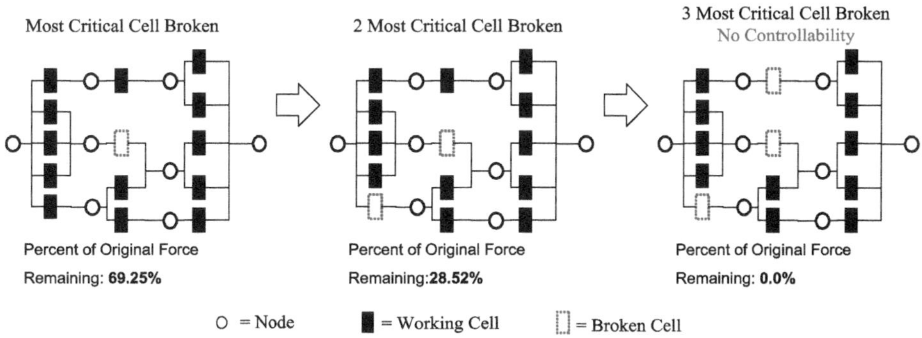

Figure 2.43 Robustness measure: "minimum cell loss to uncontrollability" = 3.

a muscle failure is critical to the application of the robot. The minimum cell loss to controllability failure determines how robust the arm is against a complete failure that is a point which should be greatly avoided. In general, parallel structures increase the number of redundant force paths in a manipulator and this increases the robustness of an array. Serial structures are generally worse in terms of robustness.

2.4.10 Discussion

This section has introduced the fingerprint method to represent complex actuator array topologies. The method contributes to incidence matrix representation for topologies to be more efficiently generated than with traditional graph-theoretic modeling. This compact mathematical representation of topologies enables automatic generation of possible array topologies. The association between the minimum cell loss to controllability failure and the worst failure remaining force has been analyzed. Mathematical processes have been presented for direct generation of corresponding incidence matrices from a fingerprint, and vice versa. The mathematical proof of the identity and similarity transforms is left to future work, though no counter example has been found.

Optimization of an array topology from direct manipulation of the fingerprint may be research of interest. Currently, no direct relationship between changes to the incidence matrix (or fingerprint) and changes to the properties of an array. Having such a relationship would allow efficient search algorithms to be developed that could intelligently explore the possible topologies and determine an optimal solution to a design challenge with less computational effort and time.

The fingerprint method has been successfully extended to dynamic array modeling [158]. The extended fingerprint method aids automation of the discovery of the equations of motion, allows for fast recalculation for different cell array topologies, and provides an intuitive base for future design on cellular actuator arrays. While the dynamics representing a given cellular actuator array could be generated using other means, such as SimMechanics by MathWorks, the expanded fingerprint method allows the dynamics to be calculated with less human effort, less computational effort, and with greater speed, especially when comparing different topologies and internal

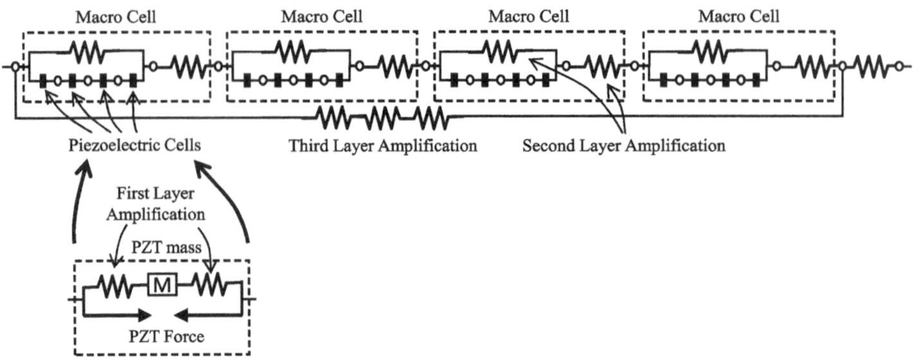

Figure 2.44 Dynamic modeling of a three-layer hierarchical actuator array.

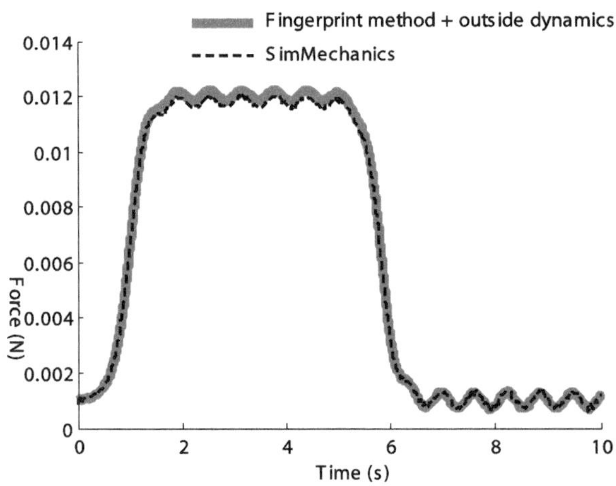

Figure 2.45 Force at the end of the piezoelectric based camera positioner actuator array under isometric contraction.

cell dynamics. In Appendix A.5, this method is used to analyze the dynamic properties of SMA actuator arrays.

Since the fingerprint method is primarily to represent layer-based actuator topologies, it is not directly applicable to the representation of multi-layer mechanisms or *nesting* discussed in Sections 1.2 and 2.3. In nesting, an outer amplification layer encloses inner connections, which does not necessarily follow the layer-based representation. It is still possible to obtain dynamic equations of motion of a multi-layered dynamic cellular actuator by explicitly treating the outer amplification layer as an outside dynamic element [160]. Outside dynamics is defined as the system with which an actuator array (or multiple actuator arrays) interacts. The combination of the actuator array dynamics and the outside dynamics forms the full system dynamic equation. Not every set of outside dynamics will complete a system, and in some cases multiple hierarchical layers of outside dynamics must be added before the system is complete. Fig. 2.44 shows the three-layer hierarchical PZT actuator discussed in Sections 1.3 and 2.3 used for a camera positioning mechanism discussed in Section 4.4. The hierarchical actuator example shown in Fig. 2.44 is numerically evaluated against SimMechanics. Fig. 2.45 shows that the forces at the endpoint of the array as calculated by both methods agree well.

Control of cellular actuators

3

3.1 Minimum switching discrete switching vibration suppression

3.1.1 Control strategies for flexible mechatronic systems

The PZT stacks are not extremely heavy, but they do weigh substantially more than the strain amplifying layers. This mass combined with the compliance of the amplifiers forms a mass–spring system with lightly damped poles, since the only damping effect is material damping. One or even several modes may be significant. The ON–OFF inputs are impulsive in nature, and, given lightly damped system dynamics, one can expect oscillatory behavior. In this way, these actuators bear similarity to human muscle, where efforts of ON–OFF muscle fibers are linked by compliant tissue.

Because the PZT stacks respond quickly to an electrical signal compared to the system dynamics of most mechanical systems, when a command is issued, it is possible to slightly spread out the recruitment of individual PZT stacks. By introducing these impulsive inputs at strategic times, rather than all at once, substantially shorter settling times can be realized. This section presents a control method whereby the activation timings of these discrete-valued inputs can be carefully coordinated to suppress multiple modes of vibration. This method is demonstrated on a replica of the actuator used in [267]. Experimental evaluation of the frequency response of the actuator reveals two significant lightly damped modes of vibration. Commands designed according to this method to suppress these frequencies reduce oscillation considerably [219].

The control of robotic manipulators is an extensive subject. For many applications, linear control of each joint with dynamic coupling treated as an unknown disturbance performs just fine, and can in fact be shown to be Lyapunov stable for manipulators whose links obey the rigid body assumption [48,12]. More sophisticated control methods that cancel the centripetal, Coriolis, and gravitational terms in a feedforward manner abound. Flexibility can cause poor tracking performance or even instability in manipulators. In the collocated case, the distal end of the link does not match the measurement because the bending of the link puts the distal end of the link at an unknown location relative to the proximal end. In the non-collocated case, measurement may be improved, but the potential for multiple poorly known gain crossovers means stability of simple linear joint controllers may lack robustness [77].

Numerous innovative strategies have been proposed to treat flexibility in manipulators from a control perspective. It is not possible to exhaustively describe them all here. Book [28] enumerates many of the challenges common to control of flexible bodies and summarizes several approaches to modeling and mitigating the flexibility. Some examples of these strategies include sliding mode control [41] and dynamic feedback linearization involving the first eigenfunction of an Euler–Bernoulli beam [156]. Nenchev et al. [181] use structural properties of the robot; they exploit kinematic redundancies in macro–micro manipulators so that the transmitted forces lie in the *reaction null space* (i.e., they lie in the null space of the coupling matrix in the

Cellular Actuators, DOI: 10.1016/B978-0-12-803687-7.00003-X
Copyright © 2017 Elsevier Inc. All rights reserved.

dynamic equations) between short links and long flexible links, so the vibration of the longer links is not excited. Ueda and Yoshikawa discuss the stability of task-space feedback control applied to a rigid manipulator with a flexible base. Based on a passivity analysis of the nonlinear closed-loop system, it is reported that the robustness of the end-point control is dependent on posture; a high-gain feedback controller *without* modeling the base flexibility can be used in certain postures [275,274].

This section is not too concerned with the spatial aspect of the flexible link problem, so most of those resources have been omitted. Included are those that have been most influential in the development of the method described in this section. In particular, of most interest are those that can be implemented by discrete pulses because of the discrete nature of the hardware.

3.1.1.1 Command shaping with a continuously variable input

Command shaping refers to a family of techniques that prevents oscillation from being excited in the first place. It only prevents oscillation from being generated by the command, and does not prevent oscillation from being excited by external disturbances. When the most significant natural frequencies of vibration are known, the reference command can be modified outside of the feedback loop, ensuring that the command does not excite any of the known modes of the system. Input shaping [235] is an approach where the reference signal is convolved with a series of impulses, whose spacing is determined by the resonant frequencies of the system. It produces a vibration-free point-to-point motion at the expense of slowing the response. Since an input shaper's transfer function is outside the feedback loop, it cannot produce instability. Input shaping is predicated on the idea that no crucial operations (e.g., manipulation of objects) are being conducted during the transient; it is merely travel time. This is because the commands may produce wild transients before the destination is reached.

There have been several enhancements built on input shaping techniques over the years to increase their performance in a variety of scenarios. In particular, progress has been made for input shaping multi-degree-of-freedom systems. Pao [190] extended the input shaping concept to systems with multiple actuators. She showed that by considering all control inputs simultaneously, one can solve for a shaper that has a faster response than shaping each input separately. Lim, Stevens, and How [150] provide a convex optimization formalism shaping for multiple actuators that incorporates additional effects or performance specifications, including reducing transient oscillation. Díaz, Perira, Feliu, and Cela provide a method to concurrently determine a shaper and add mass at strategic locations that minimizes the additional delay incurred by the shaper [54].

3.1.1.2 Command shaping with discrete inputs

Input shaping convolves a reference command continuously varying with impulses, resulting in a continuously variable command. The input shaper block can then be envisioned as a transfer function block in the Laplace domain containing the Laplace

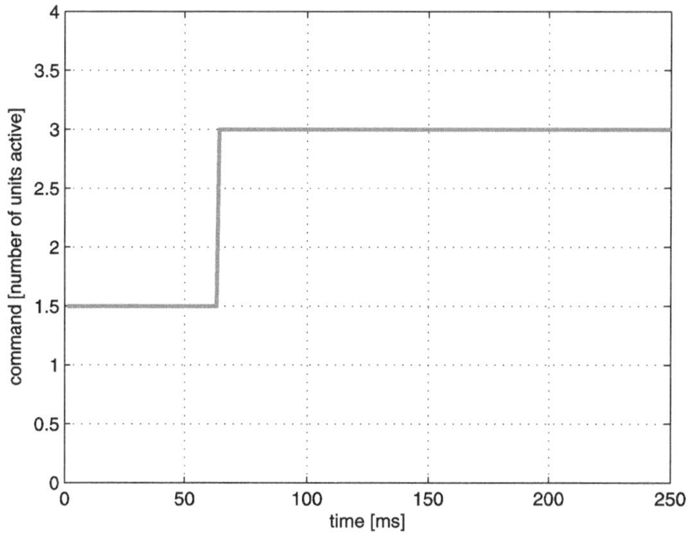

Figure 3.1 Reference command going from 0 to 3 units active after being put through a ZV shaper with a 63 ms delay.

transform of two impulses, or

$$1 + e^{t_s s}, \tag{3.1}$$

where t_s is the delay between the impulses. In the case of a cellular actuator with multiple discrete inputs, even if the reference command is carefully constructed to be implementable by discrete transitions of the individual units, when input shaping this signal the output of the convolution function will most often produce a command with values "in-between" those physically valid values for some portion of the command. So input shaping the cellular actuator command would put us right back at variable voltage levels to each piezoelectric stack. This is illustrated in Fig. 3.1. The actuator is given a command to transit from 0 to 3 active units. This is certainly a legitimate command that can be implemented by the cellular actuator system with discrete PZT stacks. However, after being run through a ZV command shaper (the simplest kind of command shaper, having two impulses), the command that results spends 63 ms at a value of 1.5. This is a fractional number of units to activate, so the ZV shaper command cannot be implemented by a discrete system, demonstrating that input shaping the reference command may result in commands that are inconsistent with the hardware and drive system. This will be the case for any odd-numbered move. Trying to suppress multiple frequencies only compounds the situation. It will be shown later that it is possible to design commands that suppress vibration equally well and that use only integer-valued (allowable) inputs at each instant in time.

Since there are many situations where the physical manifestation of the mechatronic system allows only switching, or where switching provides advantages with regard to simplicity of the control implementation or power savings, switching control laws have been developed for positioning systems with a continuous output. Fiene and Niemeyer [74] develop a direct switching controller for a servo motor current control loop that compares favorably with PWM and linear methods. This method can be extended to the full control loop, resulting in a form equivalent to sliding mode control. Barth and Goldfarb [20] develop a switching control law for pneumatic actuators that is also based on sliding mode control. Shen et al. [229] demonstrate this sliding mode averaging technique for tracking control of a pneumatic cylinder–cart system using only ON–OFF valves. This allows a dramatic simplification in the implementation due to possible use of solenoid valves rather than proportional valves for continuously positioning a pneumatic servo system.

In response to spacecraft applications, in which switching is preferred due to difficulties with continuously varying thrust, Singhose, Seering, and Singer [238] developed the ON–OFF vector diagram approach. They showed that an ON–OFF actuator can provide a vibration-free command in the same manner as input shaping, relying on the same fundamental principle, namely, that time delay produces an array of zeros in the complex plane, which limits the residues of lightly damped poles. Each time the actuator changes from one state to another (forward, OFF, reverse) its effort is associated with an impulse.

If the vector sum of these impulses is zero, there will be no residual oscillation at frequency ω_n after the command is completed. Later works [237,236,227] modify the number and timing of pulses to account for and minimize fuel usage, which is particularly important for flexible spacecraft. In an alternative approach, Song, Buck, and Agrawal [240] combine input shaping with pulse width modulation, rather than using a vector diagram.

Schultz and Ueda [218] employ the vector diagram approach to the cellular actuators described in the previous sections, where each PZT stack can be turned on (charged at max voltage) or off (completely discharged), but cannot be supplied with any intermediate voltage. In this way each channel can be driven with simple ON–OFF MOSFETs, with no need for complex current regulation. In addition, the ON–OFF nature obviates any need for hysteresis compensation. The idea is that the impulsive nature of charging or discharging the stack and the high redundancy provide a means to cancel vibration in a manner that follows from the hardware itself. The algorithm extends the ON–OFF input shaping methods because instead of inputs being restricted to 1, -1 or 0, the command can assume a bounded subset of the integers at each time step. In a manner of speaking, this work occupies the ground between continuously varying input shaping and ON–OFF input shaping. This brings about some new features and opportunities particular to this discrete range of inputs that must be considered and are described more fully in the next section.

3.1.2 Open-loop switching control methods

The following method is adapted from the ON–OFF input shaping methods described in section "Cellular actuator concept". A prerequisite for this method to work is that the frequency of the dominant modes of vibration must be known, at least approximately. These can usually be found experimentally without too much difficulty. It

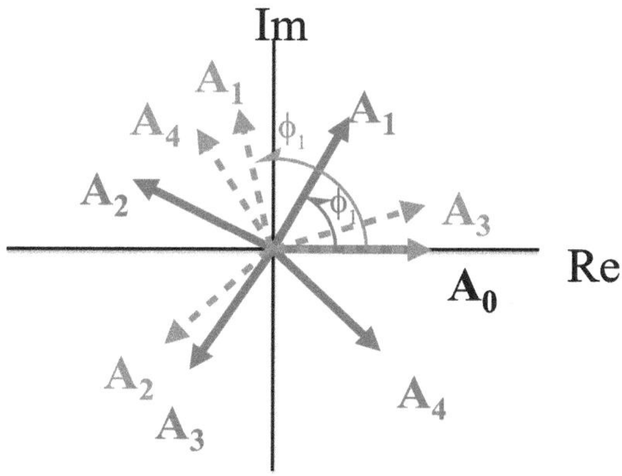

Figure 3.2 Vector diagram showing impulses represented as phasors in the complex plane. The vector diagrams for two frequencies are superimposed on top of one another. The lower frequency is shown with the solid arrows, the higher of the two as the dashed arrows. The phase of the impulse A_1, ϕ_1, is shown explicitly. To be a vibration suppressing command, both sets of impulses must sum to zero in the complex plane.

does not require knowledge of the dynamic system's residues, which is more difficult to determine.

When one of the PZT stacks transitions between states (forward, OFF, reverse), it generates a force, which is developed quickly with respect to the actuator's system dynamics. It can therefore be represented mathematically as an impulse. Each impulse, in turn, is represented by a phasor in the complex plane. In ON–OFF input shaping, this phasor must have magnitude 1 (or 0); in this context it can take on a wider range of values. The direction of the phasor, ϕ_j, is the angle corresponding to the time delay from the initial impulse, t_j, with respect to the natural period of the mode of vibration, T_n, (angular frequency ω_n) being considered. The effects of the impulses on the lightly damped system can be illustrated graphically on a vector diagram, as shown in Fig. 3.2. If all the impulses sum to zero, this is equivalent to saying that the resonant poles are canceled by the zeros arising from the time delay inherent in the command and the residues for these poles will be close to zero. This means that the oscillation in the output, at least according to the model, will be imperceptible.

$$\phi_j = 2\pi \frac{t_j}{T_n} = \omega_n t_j. \tag{3.2}$$

This method assumes that the system is undergoing point-to-point motion, and the switching command is held at a steady state value once completed, and there is no residual oscillation at the time a move occurs. Suppressing oscillation arising from disturbances or nonzero initial conditions would require a feedback method. The response of the system (neglecting damping) once the switching command is completed at the resonant frequency represented on the phasor diagram is proportional to the vector sum of the phasors. If the vector sum of these impulses is zero, there will be no residual oscillation at frequency ω_n after the command is completed.

3.1.2.1 All ON/All OFF control

The switching method of Singhose, Singer, and Seering [238] can be easily applied to a system like the cellular actuator that accepts only ON–OFF commands to each of its individual inputs. Discretizing the command according to the number of inputs to the hardware means that actuator effort will correspond to some number of inputs ON. For point-to-point motions, this means that any steady state or destination value that can be represented by how many of the available inputs are in the ON state at any given instant. If an impulsive switching command is developed for coarse actuation applications (such as spacecraft thrusters [296,148] where the control command consists of burn times and free-floating segments), this same command will suppress oscillation at the frequencies designed for multiple actuators acting identically in concert. To reach the desired steady state value, one merely applies the vibration suppressing input identically to each input involved in the command. For instance, if the actuator is to move to the steady state position reached by turning 3 more inputs on, 3 channels would be selected from those currently inactive, and these 3 units would transition between ON and OFF multiple times according to the command developed, finishing with those 3 ON. This will be referred to as *All ON/All OFF control*, and it is effective at suppressing oscillation induced by the command itself. To move the other direction, units that are active will be selected to deactivate. This is illustrated in Fig. 3.3. Which individual PZT stack is activated at each switching step can be the same or different, as shown in the figure. An advantage of All ON/All OFF Control is that once a vibration suppressing command is calculated, it does not need to be calculated again for different move distances. The same vibration suppression command can be used for a move of any size simply by changing the number of PZT stacks to activate.

3.1.2.2 Discrete switching vibration suppression

In the following section, we define a Discrete Switching Vibration Suppression command, a wider class of commands that suppress vibration than the All ON/All OFF control commands. This takes advantage of the redundancy present in the cellular actuator structure. The wider range of commands result in more flexibility. Since any of these commands can be used to suppress vibration equally well, this can be used to achieve other, secondary goals, by choosing the vibration suppression command that performs better with regard to another goal.

Most prior examples of switching vibration command have a single actuator: a thrust from a rocket, current to a single motor, and so on, so all amplitudes are normalized to one. Traditional input shaping (which convolves the impulse with a continuously variable command) is usually implemented using a continuously variable actuator such as an electric motor, and therefore the size of the impulse amplitudes on the phasor diagram can be non-unity. Since input shaping was originally developed

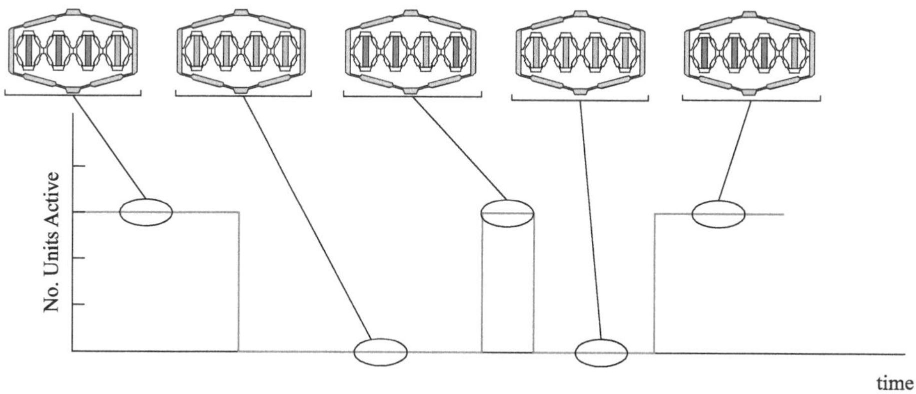

Figure 3.3 Illustration of all-ON all-OFF control for a move from 0 to 4 units ON (green denotes the ON state in both the plot and the illustration). (For interpretation of the references to color in this figure legend, the reader is referred to the web version of this section.)

for cranes, which have low fundamental frequencies [238], damping effects can reduce the amplitude within a single period of vibration. Therefore, they compensate for damping effects by reducing the amplitude of the impulses on the phasor diagram in a corresponding fashion compared to the initial impulse. Other potential uses of non-unity impulses are not explored.

Variable amplitude impulses have a very different meaning in the context of a highly redundant discretized actuator. Rather than "dialing down" a unity amplitude to compensate for damping to whatever value is needed, a variable amplitude impulse can be selected from a finite set whose elements correspond to any number of inputs changing from the OFF state to the ON state. One concludes that there are a wide variety of switching commands that can be sent to the actuator, and many of these may suppress vibration just as well as All ON/All OFF control while arriving at the desired goal.

The simplest modification one might think of can be illustrated as follows. Imagine a cellular actuator that has to increase its effort by a total of 5 inputs active. If there are two frequencies to suppress, this will take a total of 5 impulses. Imagine that these 5 impulses would be selected from a pool of 11 currently inactive inputs. All ON/All OFF control would select 5 units to activate, leaving the other 6 inactive. The 5 units selected would be activated, deactivated, activated again, deactivated, and then activated a final time, at timings specified by the phasor diagram calculation. The remaining 6 would be untouched. However, a small variation, selecting a set of 5 channels to activate or deactivate *for each impulse* would have the same effect, cycling other PZT stacks in and out of the active state in the process. This is a byproduct of the redundant cellular architecture and as we will see, there are a whole host of other commands that will also suppress oscillation in the same way. The following development will generalize the notion of switching commands and provide a mathematical foundation to describe this wider set of commands.

If the PZT stacks are uniform, activation of any of the PZT stacks (channels) within the actuator will produce a uniform change in force, no matter which channel is selected. This was found to hold in practice for the isometric case [219]. Even if the actuator is changing length, based on the relative size of the constituent parts of the actuator and the low stiffness of the outermost layer the dynamics of the actuator can be assumed to be fast compared to the load it is driving, hence the isometric assumption holds with regard to the time it takes to produce the force. We can then normalize the total force produced by the actuator by the force generated by activating a single unit. Normalizing the command by the effect of activating a single channel leads to interpreting the command as a sequence of a subset of integers. Interpreting the control input $u(t)$ as the number of units chosen to be activated that changes the number active m times can be defined to be

$$u(t) = \sum_{j=0}^{m-1} A_j \delta(t_j), \; A_j \in \mathbf{D} \subset \mathbb{Z}, 0 \notin \mathbf{D} \tag{3.3}$$

where A_j is some integer impulse amplitude for impulse j, and $\delta(t_j)$ is the Dirac delta function. Each A_j is normalized with respect to the force provided by a single actuator, which is assumed to have the same magnitude in either direction. Since applying an impulse j at some time corresponds to turning a certain number of actuators on or off, it is limited by practical considerations, namely, how many actuators are, at a given time, in the ON state and therefore available to be turned off, and how many actuators are in the OFF state and are available to be turned on. Thus, the set of possible input amplitudes, **D**, will change every time the activation state changes. Physically speaking, A_j represents how many additional stacks are turned on at instant t_j. A negative value of A_j indicates that stacks are to be turned off.

Just as in All On/All Off control, each A_j can be represented as a phasor in the complex plane, each rotating at a natural frequency of vibration, but this time their amplitude is one of the integers in **D** rather than ± 1. It is assumed that any attenuation due to damping is negligible in the time it takes to complete the command. If damping is significant and the value of the damping coefficient ζ is known, a weighted sum with time delay can be iteratively solved to produce a more accurate vibration suppression command; the same general principle applies.

What happens if there are multiple resonances to be suppressed? The same method works, but a set of impulses must be found so that there is no residual oscillation at *any* of the resonant frequencies. In general, this will require more impulses in the command (i.e., a larger value of m). When there are multiple resonant frequencies to be suppressed, this is equivalent to having a whole set of vector diagrams, one for each natural frequency. Each vector diagram represents the same impulses and they will have the same amplitudes, but they will appear on the diagram to have different phase angles because, although they are applied at the same instant in time, each vector diagram measures phase delay relative to *its own* natural frequency. For a command to suppress oscillation, *each and every* vector diagram must sum to zero for the *same* command. Since each vector diagram has a real and imaginary component, each vector diagram contributes two equations that must be satisfied. There is one final thing that needs to be done in order to solve these equations. The phase variable in each equation refers to a different natural frequency. To produce a unified set of variables across the system of equations, we use the fact that the individual frequencies of each mode of vibration are known, and can write the entire system of equations in terms of the lowest natural frequency of the system, denoted ϕ_j. This gives us the following set of equations:

$$\begin{aligned}
A_0 + A_1 \cos\phi_1 + \cdots + A_{m-1}\cos\phi_{m-1} &= 0,\\
A_1 \sin\phi_1 + \ldots + A_{m-1}\sin\phi_{m-1} &= 0,\\
A_0 + A_1 \cos\frac{f_2}{f_1}\phi_1 + \cdots + A_{m-1}\cos\frac{f_2}{f_1}\phi_{m-1} &= 0,\\
A_1 \sin\frac{f_2}{f_1}\phi_1 + \cdots + A_{m-1}\sin\frac{f_2}{f_1}\phi_{m-1} &= 0,\\
&\vdots\\
A_0 + A_1 \cos\frac{f_n}{f_1}\phi_1 + \cdots + A_{m-1}\cos\frac{f_n}{f_1}\phi_{m-1} &= 0,\\
A_1 \sin\frac{f_n}{f_1}\phi_1 + \cdots + A_{m-1}\sin\frac{f_n}{f_1}\phi_{m-1} &= 0,
\end{aligned} \tag{3.4}$$

where f_i is the natural frequency of the ith mode and there are N_m significant modes. Any solution to this system of equations will result in zero oscillation at each frequency f_i. Since (3.4) contains two equations for every frequency to be suppressed, due to the real and imaginary parts, in general the command must contain 2 impulses occurring at some phase ϕ per frequency to solve the system. When there is just a single mode of vibration to be suppressed, the solution can be found analytically, but with 2 or more, it can only be solved numerically, in general. Specific cases will admit solutions with fewer impulses, and this will be discussed later on.

Given the discretized nature of the actuation, operation for point-to-point motion will be from one number of active PZT stacks, to some other desired value. The change in activation from one steady state value to the next can then be expressed as an integer-valued goal y_g, which corresponds to the total number of inputs turned on for that particular point-to-point command. A negative value of y_g corresponds to turning PZT stacks off. So in addition to solving Eq. (3.4), the command must change the total number of units active by the amount y_g by the time it is finished. This adds an additional constraint that must be satisfied,

$$\sum_{j=0}^{m-1} A_j = y_g. \tag{3.5}$$

Any set of allowable integer amplitudes (within **D**) and corresponding phase values that solves Eqs. (3.4) and (3.5) constitutes a DSVS command. The wide range of discrete switching vibration suppression commands means that different alternatives can be evaluated to select the one that finishes in the fastest time, uses the least amount of energy, has the most gentle transient, and so on.

3.1.2.3 Minimum switching discrete switching vibration suppression

As one might imagine, Eqs. (3.4) and (3.5) will not have a unique solution due to the transcendental functions involved. Examining different combinations of A_j, one finds that many can achieve the goal value, and there may be multiple ϕ_j that suppress vibration at the frequencies considered.[1] With such a large set of possible commands, which one should be chosen? As will be seen later, there are a number of advantages of choosing a command that satisfies the *minimum switching condition*, or

$$\{A_j \in \mathbf{D} \subset \mathbb{Z}\} \text{ such that } \sum_{j}^{m-1} |A_j| \text{ is minimized.} \tag{3.6}$$

3.1.3 Redundantly actuated two-layer flexible cellular actuator

The 6-input hinged rhomboid cellular actuator shown in earlier sections will be used as a case study in the remainder of this section to illustrate the process of designing a discrete switching vibration suppression command for a particular actuator. The command for other cellular actuator modules and combinations thereof can be generated using the same process, and how to do so will follow naturally from the description given here. The key quantities that must be known are the natural frequencies of the device, and the number of inputs (individual PZT stacks, if addressable) it has.

3.1.3.1 Actuator components

The term "redundant actuation" will be used to describe a situation in which there are many actuators that drive the same degree of freedom. This should not be confused with redundant degrees of freedom, such as robot with 7 or more joints. Discrete switching vibration suppression commands were evaluated on the cellular actuator shown in Fig. 1.31 is a redundantly actuated device, because there are 6 PZT stacks, and only one degree of freedom. Energizing one or more PZT stacks causes the actuator to elongate, and discharging them causes it to contract. It is constructed from the same strain-amplified [182,265,267] lead zirconate titanate (PZT) piezoelectric stack actuators (Cédrat corporation APA50XS) as in the devices describe in the previous sections. These have a maximum no load displacement of 80 μm [2]. As in prior sections, the Cédrat units connected in series constitute the first layer of the assembly. There are 6 in the first layer. As this has a rather small free displacement, its strain is amplified by the second layer a rhomboidal strain amplifier with flexure hinge that surrounds the first layers. The design of this second layer rhombus was described in

[1] Delaying one of the impulses corresponding to the least common multiple of all natural frequencies from its original location could produce an equivalent command, but this will likely be so long of a delay as to be only of theoretical interest.

Figure 3.4 Close-up view of the two-layer cantilevered actuator uses to validate the DSVS method.

[267]. It was designed so that the final actuator would have 20% strain. The original version was made from phosphor bronze, but the version used for testing in this section was made from high silicon bronze, as it is easier to procure in sheets. The elastic modulus of the two materials is nearly identical.

Although the damage tolerance properties of the cellular actuator make it particularly appealing, we remind the reader that they are not there merely for redundancy's sake. Multiple PZT stacks are also necessary to produce the amount of displacement and force required, just as many muscle cells are needed in human muscle. So from one perspective, with the cellular actuator technology, we get the damage tolerance and redundancy for free. But the other advantage that comes with this technology architecture is that the redundancy can be used to prevent oscillation at the resonant frequencies of the device using DSVS commands. This redundancy of actuation can be exploited to impose additional constraints on the system while achieving the desired control objective, e.g., position, force.

3.1.3.2 Determining the actuator's dynamic response

Section 3.1.2 made the point that in order to design a DSVS command, the resonant frequencies of all significant vibrational modes must be known. This is a function of not only the actuator's geometry and material properties, but also its end conditions. We will present the actuator in a cantilever configuration with one end fixed to a rigid support and the other end free to move, as shown in Fig. 3.4. The reason the cantilever configuration works well is that it results in an intuitive set of modes with natural frequencies that are distinct enough whose numerical values are in a range that clearly identify the concept. The vibrational modes have a physically intuitive manifestation described below. The frequency response of the actuator was evaluated using finite

Figure 3.5 Significant vibration modes of the cellular actuator. © 2012 IEEE, reprinted with permission

element methods and confirmed by experiment. The description of the experiment and the analysis of the dynamic model accuracy is deferred to Section 3.1.7 so the apparatus can be described first.

The cellular actuator with the flexure hinged rhomboid has two significant modes ($N_m = 2$) that need to be suppressed, hence it is simple, but represents the general case as there is more than just a single mode of vibration. A finite vibration study was conducted in the Dassault Systèmes' COSMOSWORKS (now known as Solidworks Simulation) environment with a fixed end condition at one of the mounting flanges and the other flange free to move. The two vibrational modes determined by the study are illustrated in Fig. 3.5. The first mode, shown on the left, involves the first layer moving back and forth in the axial direction of the actuator, with the portion of the second layer strain amplifier closest to the fixed mounting flange behaving like a leaf spring. In the second mode, shown on the right, the first layer is largely stationary, but small motions of the first layer cause the free mounting flange to oscillate in the axial direction, with the portion of the second layer strain amplifier closest to the free side acting like a leaf spring.

As one might expect, COSMOSWORKS identified other significant modes that were observed in the finite element study, such as a rocking mode about the fixed flange, and a torsional mode, but these would not be excited by activation of PZT stacks in the first layer. These may be of concern if the effect of external disturbances in the direction transverse to the actuation direction are expected, but just as they are not excited by energizing the PZT stacks, the PZT stacks do not have the ability to suppress them. Therefore, these modes are ignored in the development.

3.1.3.3 Discrete positions for the actuator

Because one end of the cellular actuator is fixed to a rigid support, the free end will have a finite relative displacement as the actuator extends. The goal displacement y_g is normalized and can take any natural number from 0 to 6. In order for (3.4) and (3.5) to be satisfied, there must be a total of $m = 2N_m + 1 = 5$ impulses. $m > 5$ can be chosen if some strategy is chosen for determining a solution for an underconstrained system of equations; however, if more impulses are added than are strictly necessary, it is unlikely that this will produce a minimum switching solution. The total number (6) of units present have an impact on the set **D** of impulses allowed. Since the first impulse occurs at $t = 0$, the phases ϕ_j of the remaining four impulses must be found. Because the number of first layer units on cannot exceed 6, the amplitudes A_j must be chosen such that

$$\{\forall j \in [0, 5] \subset \mathbb{Z}, A_j \mid 0 \leq \sum_{k=0}^{j} A_k \leq 6\}. \tag{3.7}$$

This ensures that no commands will be calculated which the actuator is not physically capable of implementing.

3.1.4 Determination of switching pattern

To illustrate how the method described in the previous section can be used in practice, we will apply the method to a positioning system that consists of a cellular actuator with the dynamic properties described in Section 3.1.3. This gives 7 steady state goal positions, corresponding to $0, 1, 2, 3, 4, 5$, and 6 PZT stacks on. Activation of any stack is considered to have the same effect; for positions 1–5, the activation pattern is non-unique, but they are considered to be equivalent.

As there are a whole litany of DSVS commands, this section will compute only the Minimum Switching DSVS (MSDSVS) commands. Furthermore, this section does not try to solve equations (3.4) and (3.5) directly because solving for a combined set of integer-valued (A_j) and real-valued (ϕ_j) solutions does not use well-known methods. Instead, a set of A_j that is minimal is assumed *a priori*, and then a solution is sought for the ϕ_j from all possible sets of A_j that are both minimal and meet the goal displacement at the conclusion of the command. If no solution for ϕ_j that suppresses vibration is found, slightly more switching is allowed and the process continues. Once a solution is found, it is still said to be minimal, at least for practical purposes, because no solution with a smaller number of transitions is available.

3.1.4.1 *Finding the minimum switching solution*

The following describes the algorithm used to compute the MSDSVS move. It is provided with the resonant frequencies that need to be suppressed and computes the timings and integer amplitude values (number of units to switch) at each time. In some cases, the actuator will have to move past the desired goal position and come back in order to suppress vibration. Fig. 3.6 summarizes the process in a flowchart for the case where $m = 2N_m + 1$. Commands with larger values of m (more impulses) in general will not be minimal and would require a modification of the algorithm not presented here. The algorithm is implemented in MATLAB. It does not specifically designate precisely which unit should be activated or deactivated at each instant in time, only how many. This needs to be determined by a separate process and one method for doing this will be presented in a subsequent section.

The algorithm breaks neatly into two parts: The first part determines a set of impulse amplitudes that will reach the goal displacement with the minimum number of changes in state to the inputs. The second part determines the phases that result in a vector sum of zero at each of the resonant frequencies on the vector diagram, or

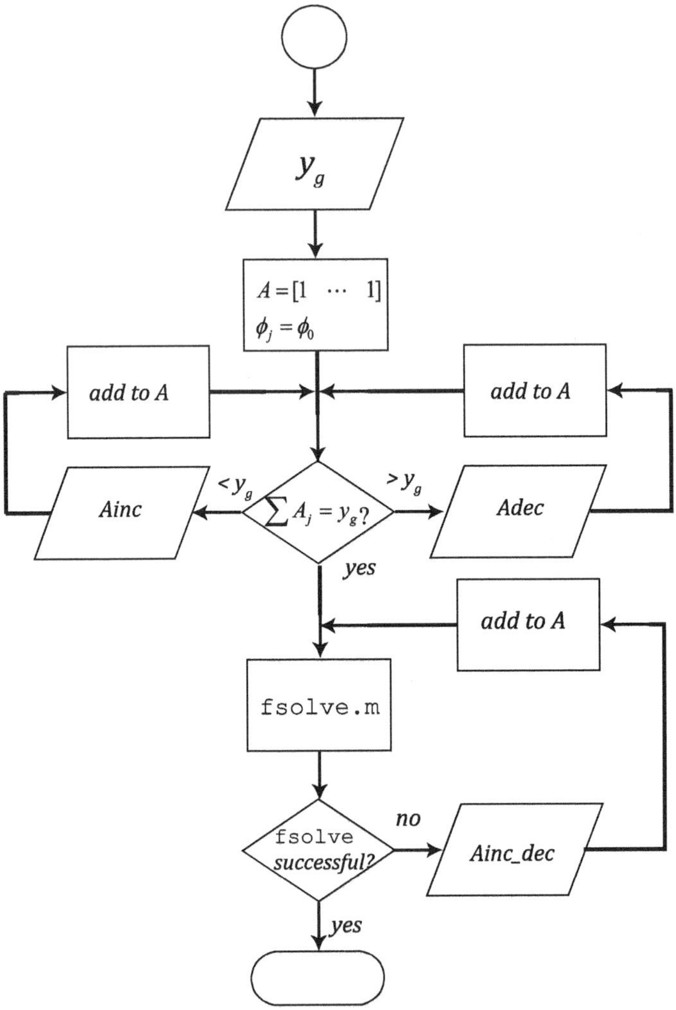

Figure 3.6 Flowchart depicting algorithm to determine switching pattern. © 2012 IEEE, reprinted with permission

zero residual oscillation. **A** is initially set to a vector of length m, where all elements are 1. The vector of phases, ϕ, is set to some convenient initial condition, such as evenly spaced angles between 0 and π. Most often, the sum of all the elements of this initial **A** will not equal the goal displacement, y_g; this is just an initial condition. To correct this, the elements of **A** are either incremented or decremented using the rules shown in Table 3.1. These rules ensure that no element of **A** will ever be zero, which would correspond to deleting an impulse. This would mean there are too few impulses to guarantee that (3.4) and (3.5) are satisfied. This process produces a

Table 3.1 Command selection algorithm rules.

Name	Form	Note
inc	$[0 \ldots 010 \ldots 0]^T$	
Adec	$[0 \ldots 0 -1 0 \ldots 0]^T$	−1 has the same index as some $A_j \neq 1$
Adec	$[0 \ldots 0 -2 0 \ldots 020 \ldots 0]^T$	−2 has the same index as some $A_j = 1$
Adec	$[0 \ldots 0 -2 0 \ldots 010 \ldots 010 \ldots 0]^T$	−2 has the same index as some $A_j = 1$
Ainc_dec	some Ainc + Adec	nonzero elements of Ainc and Adec may not have the same index

monotonic "stairstep" pattern (which contains, by inspection, the minimum number of switches) if $y_g \geq m$. The "stairsteps" may not be same size, however.

This opens up a curious point: there is an intrinsic difference between shorter moves ($y_g < m$) and longer moves ($y_g \geq m$). If there are many vibrational modes to suppress, the sheer number of impulses required for vibration suppression will exceed the goal for shorter moves, and the actuator will have to move back and forth several times to suppress oscillation. For longer moves, from an energetic perspective, vibration suppression is free; the "stairstep" solution requires no additional impulses. One could get vibration suppression by "climbing the stairs" and "moving up a step" at key times rather than moving to the goal in one fell swoop. If $y_g < m$, this method produces one of several possible patterns with as few negative valued input transitions as possible. Negative amplitudes correspond to inputs in the opposite direction of the desired displacement and represent wasted energy in one sense, however, they may be necessary to suppress all vibrational modes, especially for short moves. Once a "minimum switching" set of amplitudes is determined, the vector **A** is passed to a numerical solver which attempts to solve (3.4) by changing the phases ϕ_j. If the solver is successful, the set of amplitudes determined by the first part of the algorithm and the phases returned by the solver constitute the MDSDSVS command.

It may be that even for moves where $y_g \geq m$, for that specific combination of frequencies and amplitude candidates a set of phases cannot be found that results in a zero sum on the vector diagram for each and every frequency. The choices are then to allow some amount of residual oscillation and move to an optimization routine rather than a numerical solver, or to move to another set of impulses that will have more transitions. For our purposes we chose the latter; if the solver is unable to determine a solution, **A** is modified using the final rule shown in Table 3.1. This modifies the individual impulse amplitudes while leaving their sum unchanged, moving to the configuration with the next fewest number of individual input transitions.

It is important to note that the order in which the A_j are placed in the first step is not important, because the solver allows the impulses to "pass by" one another when looking for the solution. For instance, a "stairstep" command for $y_g = 6$ will have one "stair" that is larger than the others. The algorithm does not have to test the case

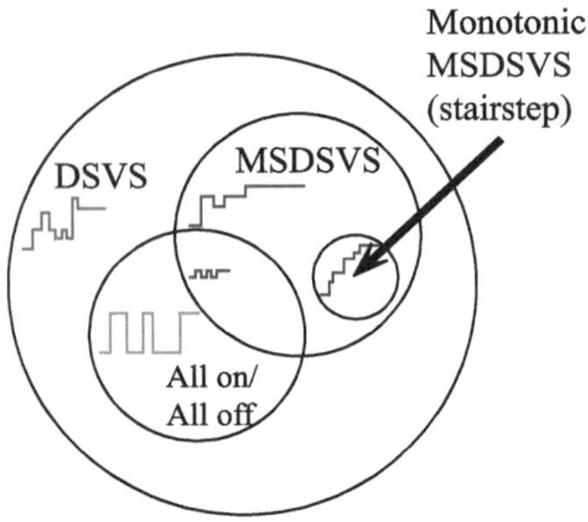

Figure 3.7 Venn diagram illustrating the various Discrete Switching Vibration Suppression Commands.

with the large stair first and the large stair last separately; the solver places the large stair in the sequence in which it needs to be. However, before giving up on a minimum switching candidate, and increasing the number of transitions, it is important to verify that all the combinations that reach the goal have been tried. For $m = 6$ this is easily managed, but for $m > 6$ this can require considerable bookkeeping.

To sum up, the various types of switching commands that can suppress vibration in a piezoelectric cellular actuator are illustrated in the Venn Diagram in Fig. 3.7. Note that DSVS is the most general, encompassing the other categories. Note also that an MSDSVS command is not guaranteed to be monotonic in nature. The move from 0 to 1 unit on, for instance, lies in the intersection between All ON/All OFF control and MSDSVS – it is minimal, but it is not possible for it to be monotonic due to the short length of the move.

3.1.4.2 Numerical solver and implementation details

To generate commands for the cellular actuator, the amplitudes A_j and phases ϕ_j are found by a custom MATLAB function, `arb_novib2f.m`, which uses MATLAB's (R2009b) native `fsolve.m` function. `fsolve.m` employs an algorithm based on that proposed by Powell [199]. With two significant resonant modes ($N_m = 2$), and 6 inputs, 5 impulses are required for a minimum switching solution that achieves zero residual vibration. The algorithm begins with

$$\mathbf{A} = [A_0\ A_1\ \ldots\ A_5] = [1\ 1\ 1\ 1\ 1]. \tag{3.8}$$

`arb_novib2f.m` then follows the procedure in the previous section to generate a sequence of impulses that results in a final position of y_g. $Ainc$ is some permutation of the elements of $[\ 0\ 0\ 0\ 1\ 0\ ]^T$, $Adec$ is some permutation of the elements of $[\ 0\ -2\ 0\ 1\ 0\ ]^T$. Once y_g is achieved, `fsolve.m` attempts to find ϕ_j that result in zero residual vibration. If it is unsuccessful, $Ainc\_dec$ of the form $[\ 0\ -2\ 1\ 1\ 0\ ]^T$ is added and `fsolve.m` attempts to find a solution. If it is possible to achieve a vibration-free move with fewer than 5 impulses, `fsolve.m` may calculate two identical ϕ_j. In this case, the two impulses are removed and replaced by a single impulse whose amplitude is the sum of the two amplitudes.

3.1.5 Illustrative example of switching algorithm

To aid in understanding how the algorithm functions a situation described below is presented that exhibits the various situations that occur during execution of the algorithm. It is purely hypothetical and should not be thought of to be normative or to correspond to a particular set of frequencies. Not every set of frequencies and goal displacements will require going through each and every step described here. Fig. 3.8 shows an example of the process conducted by `arb_novib_2f.m`. In this case, let's say $y_g = 6$. The default amplitude vector $\mathbf{A} = [\ 1\ 1\ 1\ 1\ 1\ ]$ results in a final position $\sum_{j=0}^{4} A_j$ of 5, not 6, so the pattern must be modified. Firstly, to satisfy (3.5), the vector $[\ 0\ 0\ 0\ 1\ 0\ ]$ is added to $\mathbf{A}$, causing the second to last impulse to have an amplitude of 2, which satisfies the desired y_g with a monotonically increasing command, so it is guaranteed to have the minimum number of switches. Next, `fsolve.m` attempts to find ϕ_j that satisfy (3.4). For purposes of illustration, let us say that in this case, `fsolve.m` does not return a valid solution. `fsolve.m` does not enforce $\phi_j < \phi_{j+1}$, so it is not strictly mathematically necessary to try different permutations of $\mathbf{A}$. However, from a numerical perspective, the presence of local minima and singularities may affect the reachability of a solution from a given set of initial conditions. Unfortunately, for systems of transcendental equations a definitive statement about non-existence is not always available. So for practical purposes we permit ourselves to stop looking. Since the monotonic solution was not successful, we must introduce an amplitude of -1. In this example, we add $[\ 1\ 0\ -2\ 0\ 1\ ]$ to $\mathbf{A}$. This introduces two more switches (the smallest amount we can add) while leaving the goal displacement unchanged. `fsolve.m` then attempts to find a solution. To show some other situations the algorithm must account for, let us say that it is once again unsuccessful. We than have the choice of introducing an additional negative impulse by repeating the previous step, or we can increase the absolute value of an existing negative impulse. In this example, we choose the latter, and add $[\ 0\ 1\ -1\ 0\ 0\ ]$. Let us say that this time `fsolve.m` returns a solution. This must be checked to make sure that it does not violate (3.7). When the A_j are arranged in order of ascending ϕ_j, one sees that this is so. This is then designated as the minimum switching command.

The implementation of the command determined in this example proceeds as follows: at $t = 0$ two PZTs would be turned on, then two of those remaining off are turned on at $t = \frac{\phi_1}{\omega_1}$. At $t = \frac{\phi_2}{\omega_1}$ two of the four turned on at previous timings would be turned off. At this point there are two PZTs on and four off. Two of those off are turned on, and the remaining two turned on at the timings corresponding to ϕ_3 and ϕ_4 to complete the command. This results in a total of 10 "switches." It should be noted that as an implementation of this command is constructed, one needs to keep track of which inputs are on and which are off at each stage of the command, so that an input that is already on is not commanded to activate, and vice versa!

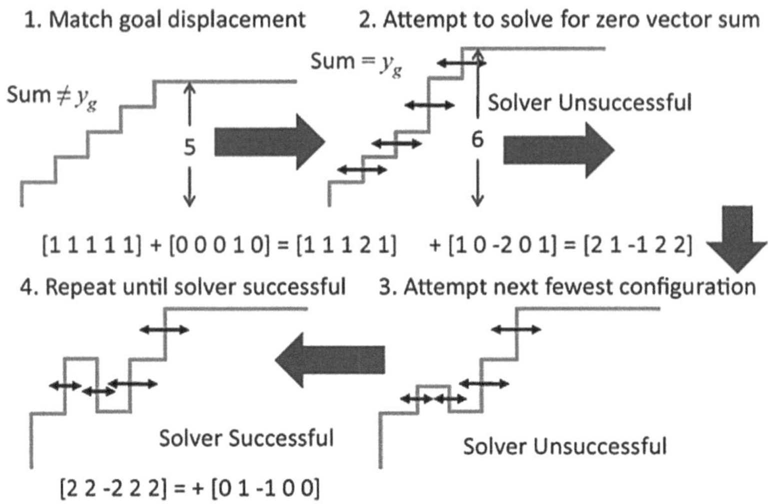

Figure 3.8 Diagram of how `arbnovib2f.m` finds a command that meets the goal y_g and suppresses the two modes of vibration. © 2012 IEEE, reprinted with permission

3.1.5.1 Existence and uniqueness considerations

The algorithm here looks for a solution until one is found. If the algorithm fails to come up with a solution, it "cuts bait" and changes the amplitudes to something just a little less desirable (introducing more PZT stack transitions to reach the goal) and tries again. But how are we to know that there wasn't some solution using the set of impulse amplitudes we tried and it's just in some sort of narrow region and we just missed that narrow range of initial conditions that would allow Powell's (or other Newton-based) method to get to it? The answer is – we don't. Since (3.4) is a system of nonlinear equations, no statements of existence, non-existence, or uniqueness can be made using classical mathematical methods. Solutions to systems of nonlinear equations are still an area of active research. Like many other real-world search and root-finding problems with strongly nonlinear systems of equations, we can perform a certain number of restarts but at some point we consider the search space "covered." The good news is that as systems like these grow larger and larger, we get more and more "stairstep" candidates to try and the likelihood of finding a minimum solution increased.

Existence questions such as these come hand-in-hand with uniqueness questions. Say that a minimum switching command was found. Might there be more? If one managed to find all the minimum switching commands one could sort through them to find the one that performs the best on some other criterion: minimizing time to the goal, minimizing the oscillation during the transient, trying to even out the time between switches as much as possible, for example. While this is an interesting area of study, we content ourselves with having found a minimum switching command for the applications immediately envisioned for cellular actuators.

Under certain conditions, the Interval Newton Method [8] may be used to make an analytical statement about the existence or uniqueness of a solution for a given set of amplitudes. This depends on the ability to invert the Jacobian of (3.4), which is singular at every multiple of $\frac{\pi}{4}\frac{f_k}{f_1}$ radians, where f_k is the highest natural frequency, so its application will apply to a small local region rather than globally. Because a solution may be near a singularity, small changes in the natural frequencies of the system may mean that the new solution found by the algorithm is not in the neighborhood of the old solution. In fact, a candidate set of amplitudes may produce a solution for a given set of natural frequencies, and fail to produce a solution for neighboring frequencies. For this reason, it is important to verify all MSDSVS patterns experimentally.

If one is unwilling to accept even a few more switches than absolute minimum and wants to determine definitively that all possibilities of phase angles have been exhausted before moving on to another set of candidate amplitudes the Interval Newton Method may be of some use. The problem is that with the sine and cosine function being equal to one another at each odd multiple of $\pi/4$, the Jacobian loses rank at this point. This means that existence and uniqueness of a solution can be checked using the Interval Newton Method for at best a range of one quarter period of the highest natural frequency. The Jacobian can also lose rank from one frequency to another for specific values of ϕ, limiting the ranges which can be evaluated still further. The user is then left to stitch together all these different regions, to say nothing of the phase values where the Jacobian is singular. Given the range of frequencies that could conceivably occur with all types of cellular actuators, this did not appear to have large benefits in practice and no implementation of the Interval Newton Algorithm was undertaken in the code. Nevertheless, it opens up an interesting research direction that could improve the process if an algorithm that would check existence could be developed. Some adaptation of the Interval Newton Method may succeed in future work.

3.1.6 Experimental setup

In this section, MSDSVS commands are calculated for the 6-input flexure hinged cantilevered actuator system for each of the 7 discrete positions (including 0 on) it can reach at steady state. The motion of the output flange is measured and the oscillation quantified. This system is representative of any servo positioning application using discrete cellular actuators in which large amounts of oscillation after a certain time would be unacceptable. Although the model would predict that no oscillation would occur after the command is completed, in practice, some of the non-ideal effects not captured by the model mean that there is still some oscillation present. However, it is substantially smaller than that produced by a step transition.

Once the relevant resonant frequencies to be suppressed and the desired final position (0–6 first units actuators on) have been found, they are provided to `arb_novib2f.m`, which returns a table with the number of units on after each switch in the sequence and the time that each switch occurs. Defining some terminology may be helpful here: a "command" will be used to refer to sequence of values of the total number of units active at any given time, whereas a "pattern" refers to *exactly which* units are active at any given time. This needs to be converted into a binary command to each channel – a pattern must be found that matches the command. In order to do this, the current state (active or inactive) of each channel must be known. A PZT stack that has already been energized cannot be energized a second time to move the actuator farther. So in implementation a list must be maintained of each unit's state. For an actuator with only 6 channels, this is not difficult to do heuristically, but for larger systems this may open up some research questions. In this example, the binary inputs were determined manually, being careful that the aggregate of all first layer units at any given time match the command returned by `arb_novib2f.m`.

There is a practical matter regarding the choice of how to distribute the MSDSVS command to binary signals to each of the units. If the choice of binary sequence stipulates that each input must change state an odd number of times, the pattern for the forward direction can be used to command a motion of the length in the opposite direction, merely by inverting all input lines in the sequence. This is called a "reversible" pattern. This has numerous practical benefits, in particular that a single pattern can be stored for both directions. This was immensely beneficial in the experiment conducted, because it made it easy to return to position 0 (all units inactive) following a move.

The DSVS command is implemented in ANSI C running on a Silicon Laboratories C8051F120DK microcontroller development kit, running at 25.4 MHz. Patterns for each of the 6 moves are stored in header files and can be modified by changing the header file. This makes it easy to use the same code for cellular actuators with varying properties. The command for each first layer unit is generated by each of the 6 Programmable Counter Array (PCA) modules, which toggles an output pin (CEX) state when a match occurs between a given module's internal register and the PCA clock, set to be the system clock/12. This allows implementation of a command which suppresses resonant frequencies as low as 32.3 Hz without rollover. The microcontroller is set to interrupt each time a PCA match occurs on any module, and the subsequent match times are loaded into the various PCA registers. If an output line is supposed to toggle at a given instant, the PCA compare register is loaded with the next transition time at each match interrupt. If it is supposed to maintain its value at the next switching time, the PCA compare register is loaded with the maximum time value, preventing a match from occurring on that module when the next switching time is reached. When the final impulse occurs, the PCA clock is stopped, and all output lines are held at their current values until the next move command occurs. Move commands are invoked by a serial command from a PC over RS-232 using the microcontroller's UART and PuTTY open-source telnet software.

Each microcontroller output is wired to a custom drive circuit introduced in Appendix A.3 based around an Avago ASSR302-C solid state switching IC, which has two bridges per chip. This is a commercially available, easily procured, low cost component, easily scalable to devices using a large number of cellular actuators. The "low" side of the PZT stack actuator in each first layer unit is always connected to ground. When the upper bridge is activated, the "high" side of the PZT is connected to ground. When the lower bridge is activated, the "high" side of the PZT stack is connected

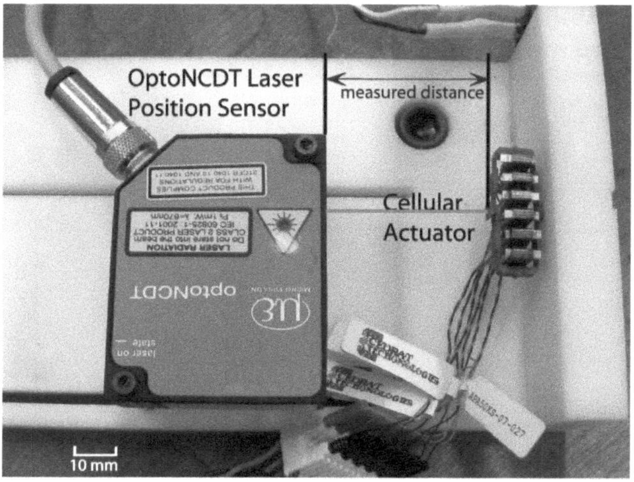

Figure 3.9 MSDSVS experimental setup. © 2012 IEEE, reprinted with permission

to the source voltage, imposing a voltage across the PZT stack and inducing a displacement. A pair of bridges per stack is necessary because the PZT stack actuators are capacitive; simply disconnecting a stack actuator from the source voltage will not cause it to return to its unforced length. It must be grounded so that it can discharge. The ASSR302-C has a delay of 70 μs and an R_{dson} of 30 Ω (typical). The drive voltage is provided by an American Power Design A5 series 5 W high voltage DC/DC converter. It supplies 33.3 mA at 150 V. A delay circuit with a time delay of 40 μs prevents cross-conduction.

One end of the cellular actuator is fixed to a rigid support. The position of the other end is measured by a Micro-Epsilon optoNCDT ILD 2200-20 laser position sensor with a sampling frequency of 10 kHz. For the verification of the MSDSVS algorithm on cellular actuator hardware, the signal from the laser position sensor was recorded with a LeCroy Waverunner 44MXi mixed signal digital oscilloscope. A photo of the experimental setup is shown in Fig. 3.9.

3.1.7 Experimental results

The frequency response of the cellular actuator was obtained using an HP 3562a spectrum analyzer to generate a sinusoidal input and compare it with the output of the laser position sensor. For the purpose of obtaining the frequency response, the cellular actuator was driven with an analog signal through a Cédrat CA45 amplifier, which has an unloaded bandwidth of 10 kHz. A transfer function for a fourth order lightly damped system with two significant modes of vibration with time delay was fit to the data. The frequency response is shown in Fig. 3.10.

Since the displacement generated by the actuator results from deformation only, and not from rigid body displacement, it can be represented as the sum of N_m lightly

Table 3.2 Parameters for experimental frequency response fit.

Mode	R	ω [rad/s]	ζ
1	1.46E–3	478.61	0.0400
2	1.22E–3	2180.3	0.0175

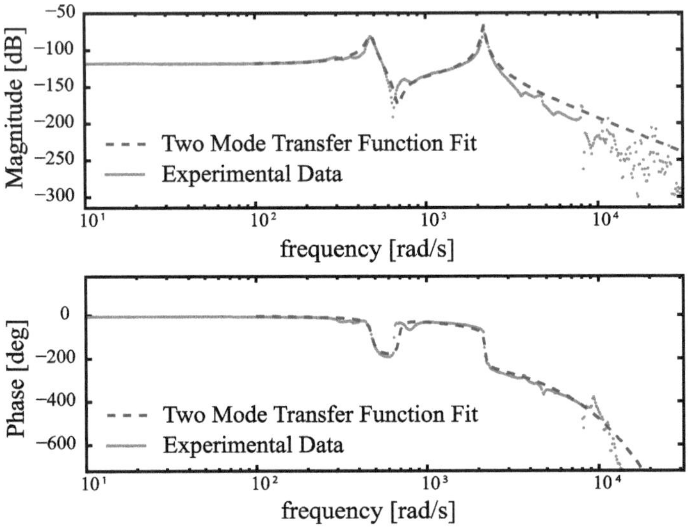

Figure 3.10 Frequency response of cellular actuator. © 2012 IEEE, reprinted with permission

damped second order systems:

$$G(s) = \sum_{i=1}^{N_m} \frac{R_i}{s^2 + 2\zeta_i \omega_i s + \omega_i^2} \tag{3.9}$$

where for mode i, R_i is the residue, ω_i is the natural angular frequency, ζ is the damping ratio, and s is the Laplace variable. The results from the experimental transfer function fit for the 2 most significant modes are in Table 3.2. The time delay in the system was determined from the fit to be 3.9 ms. The higher resonances beyond the second peak have lower magnitudes due to system roll-off and justifiably neglected in the model. Although they are not used in the computation or implementation of the MSDSVS control command, the residues are listed as well.

While any DSVS command should provide a point-to-point move with no residual vibration after the move is completed, different DSVS commands have different properties with regard to the transient motion, robustness to modeling uncertainty and energy expended. It turns out that MSDSVS tends to have favorable properties with regard to these, as exhibited by the results below. In the following, MSDSVS is contrasted with the All ON/All OFF control method. The MSDSVS and All ON/All OFF

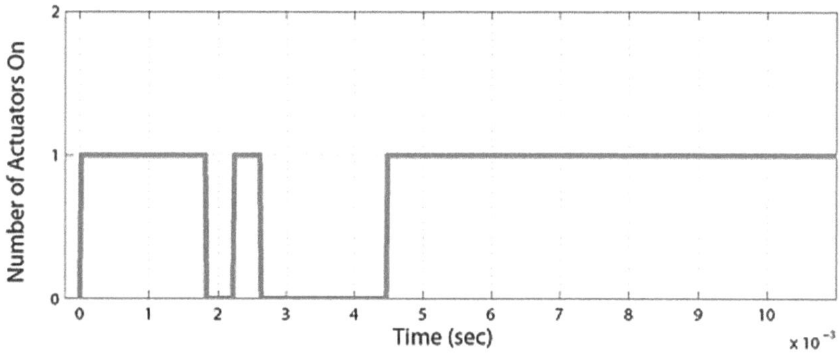

Figure 3.11 Step, All ON/All OFF, and MSDS Commands to Cellular Actuator, $y_g = 1$. Note that for motion to this position, the All ON/All OFF and MSDS Commands are the same. © 2012 IEEE, reprinted with permission

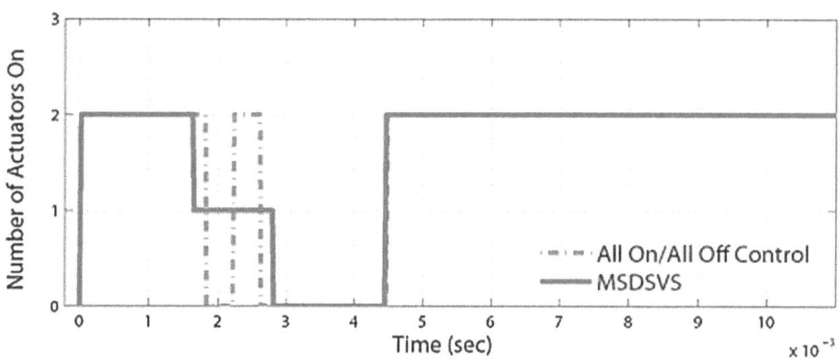

Figure 3.12 Step, All ON/All OFF, and MSDS Commands to Cellular Actuator, $y_g = 2$. © 2012 IEEE, reprinted with permission

commands computed for the 6-input two-layer actuator are shown in Figs. 3.11–3.16. Each was applied to the cantilevered 6 input cellular actuator using the microcontroller and drive circuit. Since 5 impulses are needed to suppress the two frequencies, following the discussion in Section 3.1.4 only $y_g = 5$ and $y_g = 6$ can have monotonic commands. A monotonic pattern was found for $y_g = 5$, which had $A_j = 1 \, \forall j$, but for $y_g = 6$, the solver did not converge and two additional unit transitions had to be introduced. The commands found for each of the 6 move distances are shown in Figs. 3.11–3.16.

Fig. 3.17 shows the response to the Step, All ON/All OFF, and MSDSVS commands for the largest move (from 0 to all 6 units ON). The step response shows considerable oscillation (>50% overshoot), due to the lightly damped nature of the system. All commands are completed by 10 ms, so the remaining period of time shows the decay of any residual oscillation once the command is completed. Of the three responses, the response to the All ON/All OFF command has the largest transient. A similar trend was observed for the other move distances.

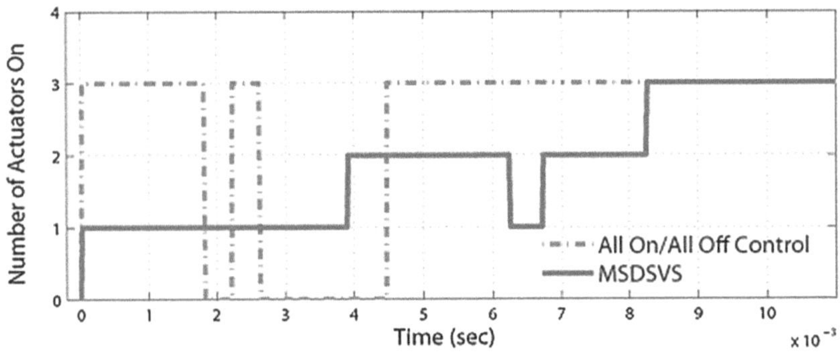

Figure 3.13 Step, All ON/All OFF and MSDS Commands to Cellular Actuator, $y_g = 3$. © 2012 IEEE, reprinted with permission

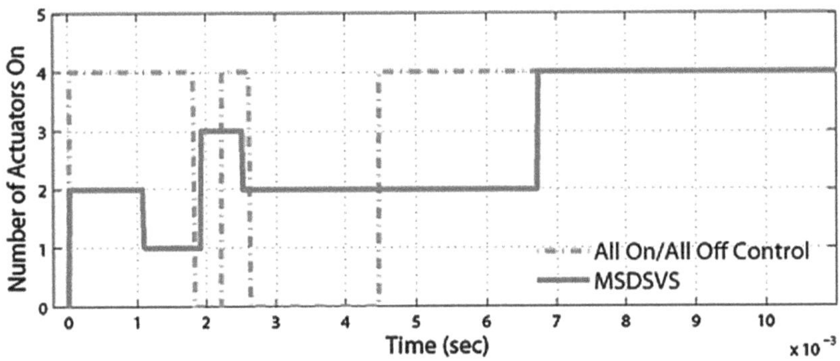

Figure 3.14 Step, All ON/All OFF, and MSDS Commands to Cellular Actuator, $y_g = 4$. © 2012 IEEE, reprinted with permission

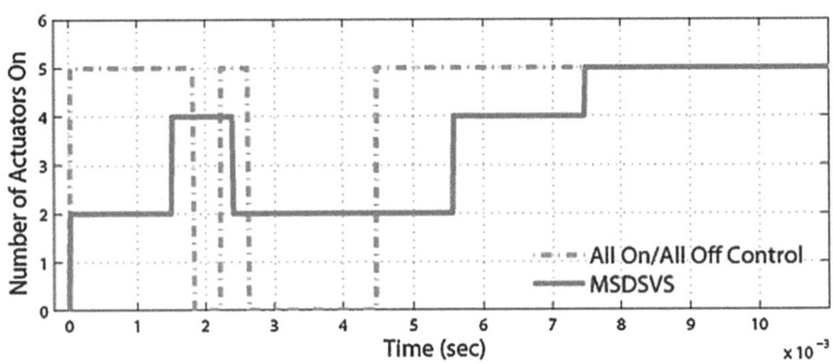

Figure 3.15 Step, All ON/All OFF, and MSDS Commands to Cellular Actuator, $y_g = 5$. © 2012 IEEE, reprinted with permission

When either MSDSVS or All ON/All OFF control is applied to the cellular actuator, the largest component of the FFT (Fig. 3.18) and the RMS oscillation (Fig. 3.19)

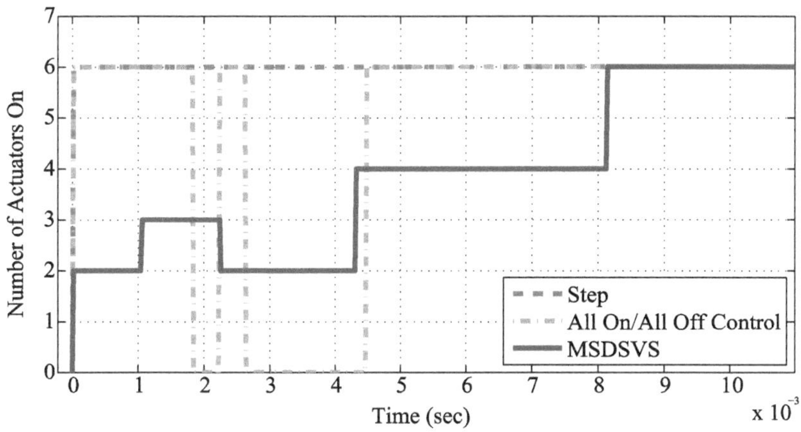

Figure 3.16 Step, All ON/All OFF, and MSDS Commands to Cellular Actuator, $y_g = 6$. © 2012 IEEE, reprinted with permission

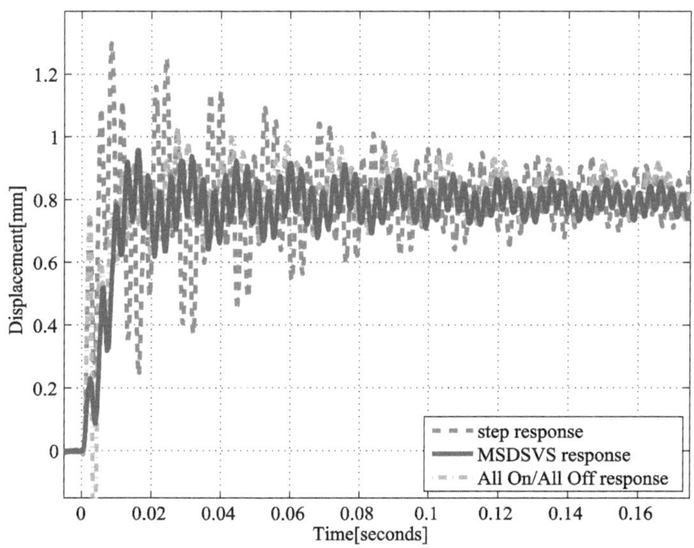

Figure 3.17 Dynamic response of Step, All ON/All OFF, and MSDSVS Commands, $y_g = 6$. © 2012 IEEE, reprinted with permission

are reduced by 50% or more in nearly all cases. In most cases, MSDSVS performed slightly better. For $y_g = 6$, the 2% settling time for a step command was measured to be 699 ms. For the all on all off command, the settling time was 477 ms. For MSDSVS, the settling time was 529 ms. With fine-tuning to compensate for all the non-ideal behaviors, it is expected that the settling time for the vibration suppressed commands can be reduced even further.

Looking at this figure, we see that although the vibration suppression commands are effective at greatly reducing oscillation for the point-to-point move, it does not

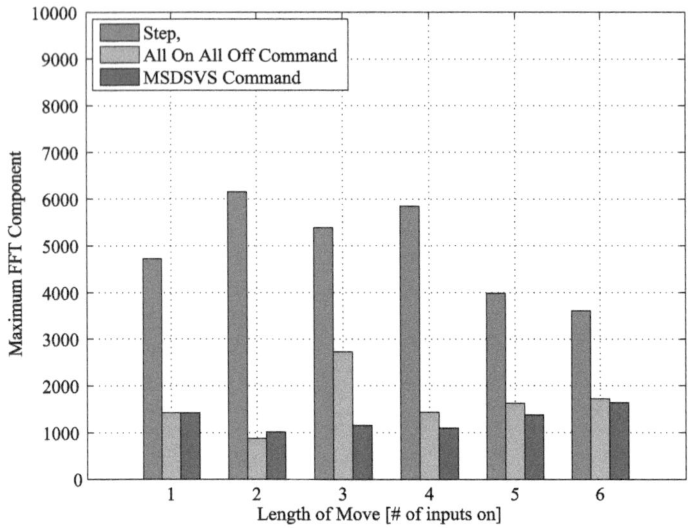

Figure 3.18 Magnitude of largest FFT component, normalized by move distance. © 2012 IEEE, reprinted with permission

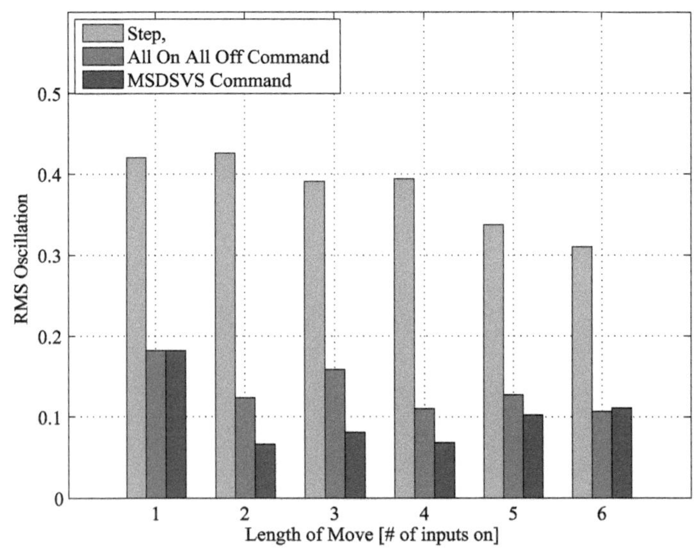

Figure 3.19 RMS oscillation, normalized by move distance. © 2012 IEEE, reprinted with permission

match the simulation. The simulation predicted that there would be *no* oscillation remaining. Why does the method not behave as well as predicted? As with many other situations in soft robotics, the limitations of the model must be kept in mind, with a healthy presumption of uncertainty in the model. In the case of the cellular actuator, these limitations and sources of uncertainty are as follows. Both All On/All Off and

(A) (B) (C) (D)

Figure 3.20 Frames from high-speed video showing the motion of the actuator: (A) actuator before the command is applied, (B) maximum excursion when all 6 inputs are activated at once, (C) actuator at maximum excursion under an MSDSVS command from position 0 to 6, and (D) actuator at maximum excursion under All ON/All OFF control from position 0 to 6.

MSDSVS methods are based on linear analysis, and they suppress oscillation in an actuator that is described accurately by a linear model. In actuality, some non-negligible nonlinear effects are present. The natural frequency of the cellular actuator changes slightly as it extends. This can be seen in Fig. 3.20. Comparing the shape of the actuator with zero units active in Fig. 3.20A with the shape of the actuator when all 6 units are active Fig. 3.20B–C, there is a discernible difference in angle of the rhombus and the length. This changes the natural frequency by a perceptible amount (a nonlinear effect). Although it is more evident from Fig. 3.17, it is possible to see the increased overshoot of the step command compared to the vibration suppressing commands in Fig. 3.20 if you look closely. This phenomenon will be more pronounced when a longer chain of actuators is used, rather than a single one.

The values in Table 3.2 are only exact for the intermediate position corresponding to the DC offset of the spectrum analyzer, +2 V, during the test because the frequency sweep with the spectrum analyzer caused oscillation about this intermediate position. The command was designed based on the frequency for this intermediate length, not the natural frequency of the actuator when it was extended. This provides a plausible explanation for why the move from 0 to 3 units on has the least residual oscillation: The resonant frequency measured is for the average excursion, which occurs close to the position corresponding to three units on. Fine-tuning of the command based on the final position reached to better match the frequency at that position may improve the performance. When the command is developed, it is assumed that the circuit responds so quickly compared to the cellular actuator that any delays in the circuit are negligible. This is not completely true. The frequency of the second mode is high enough to where the dead time to prevent cross-conduction and time delay in the system begins to become significant. What's more, the time constant of the charging of the PZT and switching circuit is beginning to become significant at the frequency of the second mode as well. In addition, the assumption that an ON command has the same force magnitude as an OFF command, may not be entirely accurate for the cellular actuator. If these properties are well known, adjustments can be made to compensate for them when generating the command. Finally, during the frequency response testing, it was observed that when the actuator was excited with a pure sinusoid at the frequency of the first resonant mode, the output was not a pure sinusoid. Instead, persistent oscillation at the frequency of the second mode was superimposed over the response at the excitation frequency, indicating a nonlinear dynamic effect which is not captured by the linear analysis. This is illustrated in Fig. 3.21.

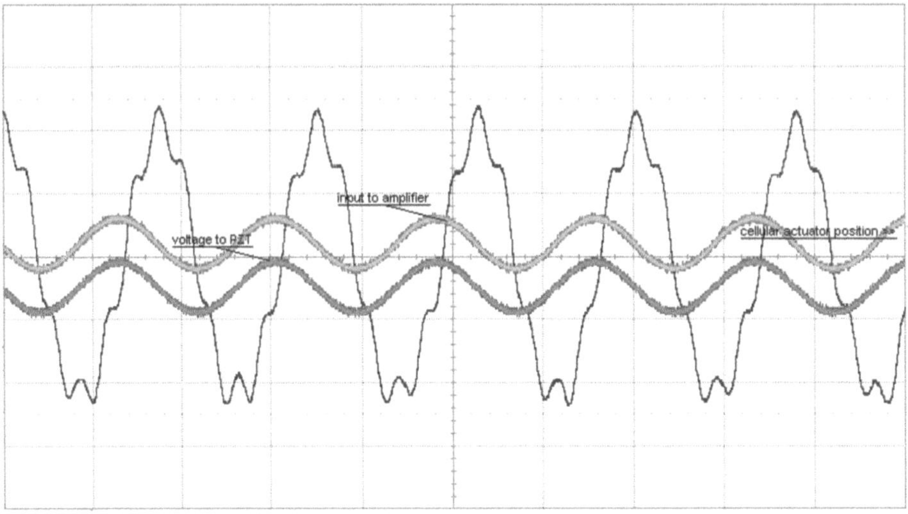

Figure 3.21 Oscilloscope capture of the cellular actuator receiving a sinusoidal input. The green signal is from a function generator, the yellow is the high voltage signal across the PZT stack from the amplifier. The blue curve is the displacement of the cellular actuator, measured by the OptoNCDT laser position sensor. Notice that although the input voltage is sinusoidal, the displacement is not. (For interpretation of the references to color in this figure legend, the reader is referred to the web version of this section.)

Those interested in using cellular actuator technology can take heart, however, for the single 6-input cellular actuator represents something of an extreme case. Because of the high loads involved in positioning or manipulating objects, most robotic devices in practice will use large numbers of cellular actuators. The camera positioner introduced in Section 4.4, for instance, has a lower dominant natural frequency.

3.1.8 Non-ideal effects and command robustness

If both MSDSVS and All ON/All OFF control both reduce oscillation, one may wonder whether there are any advantages in using one over the other (or any DSVS command, for that matter). The advantage of All ON/All OFF control is that the command is nearly always shorter in duration, and thus, in theory, it can reach a vibration-free response sooner than MSDSVS. However, there are a number of advantages to MSDSVS. MSDSVS, since it has fewer transitions, uses less energy than the All ON/All OFF solution. This corresponds to a smaller control effort being used. In this particular application, there is a certain amount of Joule heating that occurs when the PZT is energized and discharged. This can be expressed as

$$J_E = \int_0^{t_{charge}} {i_C}^2 R_{dson}\, dt \tag{3.10}$$

where J_E is the energy dissipated, and i_C is the current, and t_{charge} is the charging/discharging time. In order to determine the amount of energy used for a move

Table 3.3 Number of switches required to reach y_g for each command.

y_g	Step	All ON/All OFF	MSDSVS
1	1	5	5
2	2	10	6
3	3	15	5
4	4	20	8
5	5	25	7
6	6	30	8

Table 3.4 Overall performance comparison between commands (RMS and FFT are normalized).

Command	Avg RMS	Avg FFT	Avg Energy/Move Dist [mJ]
Step	0.380	4950	11.9
All ON/All OFF	0.135	1630	30.0
MSDSVS	0.102	1280	18.6

using a given command, the voltage supplied to the input of the DC/DC converter was measured using a LeCroy ADP305 differential probe. The current to the DC/DC converter was measured with a LeCroy AP015 current probe. After subtracting steady state leakage currents, the product of the two signals was integrated numerically over the time duration of the move. Fig. 3.22 shows how the difference in energy dissipated grows quickly with an increase in the number of inputs. Table 3.3 shows the number of switches required to reach a given position, which directly correlates to the results shown in Fig. 3.22. The term "switch" in this context means a transition of a single PZT stack, so, for instance, implementing an impulse of $A_j = 2$ would count as two "switches," since 2 stacks must be energized. Table 3.4 summarizes the information in Figs. 3.18–3.22. The MSDSVS command reduces the RMS oscillation by 73% and the amplitude of the largest frequency component by 74%. All ON/All OFF control reduces oscillation by nearly the same amount, but the MSDSVS command uses 37% less energy than the All ON/All OFF command and shows better robustness properties. Minimizing the number of cycles also has benefits for the life of the actuator. Piezoelectric bimorphs have been known to delaminate at 10^8 cycles [262].

Due to variations in manufacturing, material properties, and the aforementioned change in natural frequency with length, one may wonder how well the command developed will do across a batch of cellular actuators or across different operating conditions. MSDSVS seems to be more robust to changes in frequency than other DSVS commands. Robustness is often characterized in terms of the residual amplitude of oscillation, A_r, produced by a given All ON/All OFF or MSDSVS command when one assumes that some unmodeled resonance with frequency ω is present. If the amount of oscillation at this frequency (it is not a bad idea to pick the worst case one that could reasonably occur physically) is considered "not too bad", then the command is

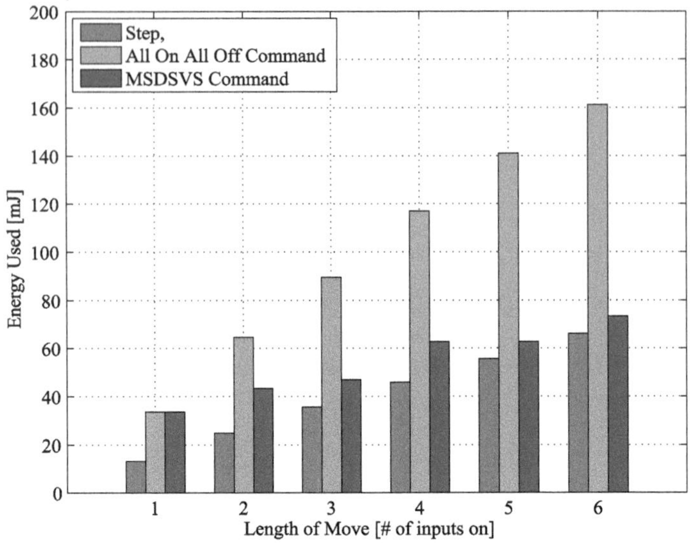

Figure 3.22 Energy consumption per move. © 2012 IEEE, reprinted with permission

said to be "robust". In the command shaping literature, this is found by looking at the maxima of the sensitivity plot, a plot of the amplitude of oscillation created at steady state by the command assuming lightly damped poles at all frequencies, normalized by the oscillation produced by a step input to the plant used to design the command. Moreover, command shaping is a double-edged sword; the command attenuates energy in the command near the frequencies that are being suppressed, but may *amplify* perturbations at other frequencies. Most often these are far from the resonances of the plant, but it pays to understand how the command behaves to avoid nasty surprises. The sensitivity plot helps in this regard.

$$A_r = \frac{|\sum_{k=0}^{m-1} A_k e^{\frac{j\omega\phi_k}{\omega_1}}|}{\sum_{k=0}^{m-1} A_k} \times 100\% \tag{3.11}$$

The sensitivity plot (A_r vs ω) of the MSDSVS command and the All ON/All OFF command is shown in Fig. 3.23. MSDSVS shows much less amplification, in general, if there is an unmodeled lightly damped mode. Because of the periodic nature of the vector diagram summation, both types of commands will produce "wells" at higher frequencies than the ones they are specifically designed to suppress as a byproduct of the process. After all, $1 + e^{st}$ has an infinite number of roots along the imaginary axis. These extra "wells" on the sensitivity plot may be a boon in the event of an unmodeled resonance anywhere near them. The All ON/All OFF command has two such frequencies near 590 and 720 Hz, but looking across the entire frequency range shown, MSDSVS performs better.

To better understand this, let us simulate a situation where the frequency used to design the command is inaccurate, say due to temperature variation, manufacturing

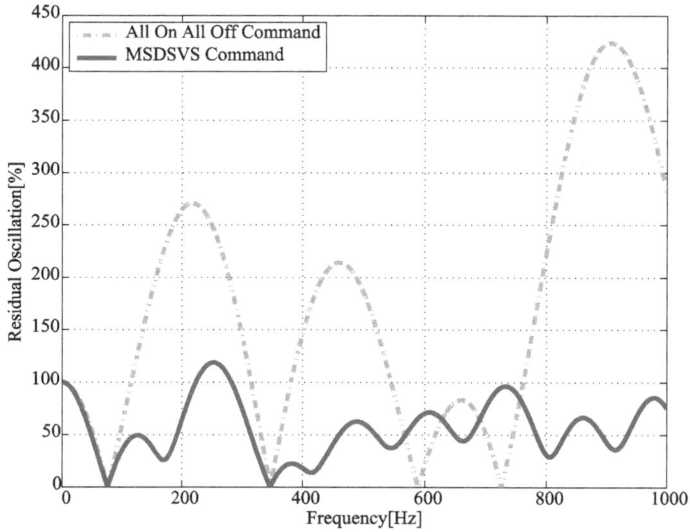

Figure 3.23 Sensitivity plot for All ON/All OFF control and MSDSVS. © 2012 IEEE, reprinted with permission

variations, or simply bad record keeping between resonant frequency evaluation and command design. Fig. 3.24 shows the time response of the cellular actuator to an All ON/All OFF command and an MSDSVS command when both commands were designed for an actuator with a first mode natural frequency of 45.7 Hz, and the cellular actuator's true first natural frequency is 76.1 Hz. The move distance shown is $y_g = 6$. Despite the fact that the wrong command is used, MSDSVS still exhibits fair vibration suppression properties, even with a frequency that is off by 40%, whereas All ON/All OFF control exhibits almost no vibration suppression.

To see how this performs in a real physical situation, two masses were added to the cellular actuator by clipping lead split shot sinkers (normally used for fishing) to the actuator as shown in Fig. 3.25. This would change the resonant frequency considerably. The command was not modified to account for this added mass, i.e., they are the same as shown in Figs. 3.11–3.16. Once again, this means that the frequency attenuated by the command does not match the resonant frequency of the plant. Fig. 3.26 shows the response for $y_g = 6$. The natural frequencies of the modified actuator are $f_1 = 56.8$ Hz and $f_2 = 341$ Hz. In this case, both All ON/All OFF control and MSDSVS still provide some benefit. The amount of residual oscillation for the All ON/All OFF and MSDSVS methods are comparable, as predicted by Fig. 3.23 when the first mode is lowered.

3.1.9 Discussion

MSDSVS is effective at suppressing the oscillation engendered by the input command to the actuator. However, the transfer zeros created by the switching command are not

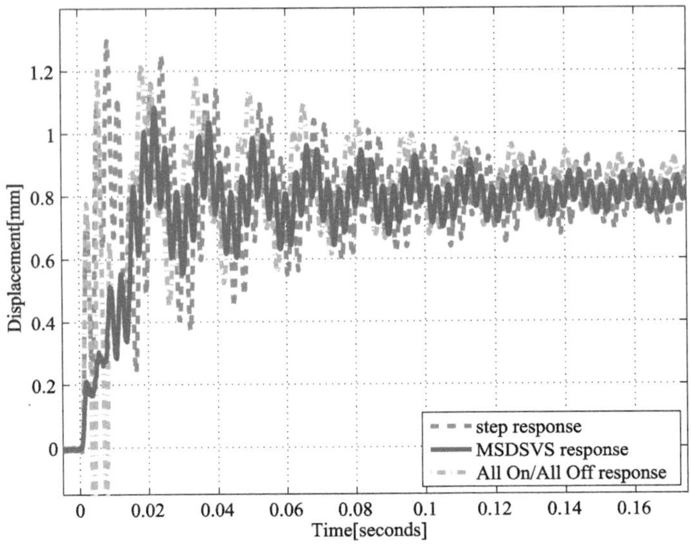

Figure 3.24 Cellular actuator response when the command is designed based on an incorrect natural frequency. © 2012 IEEE, reprinted with permission

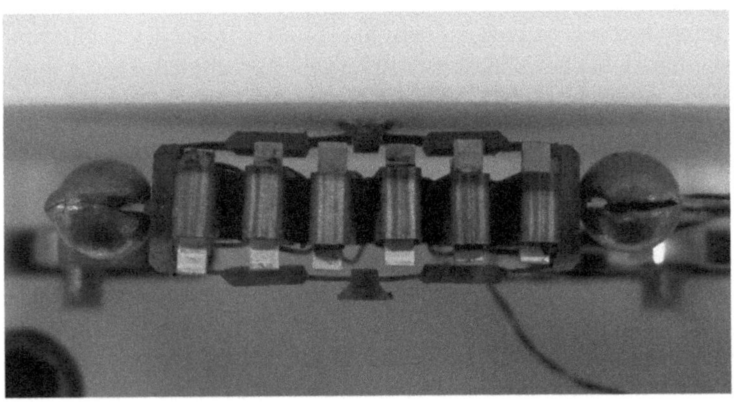

Figure 3.25 Split shot sinkers attached to the cellular actuator to shift the resonant frequency of the first mode.

in the path of the disturbance inputs and do nothing to prevent oscillation arising from other excitation in the environment [64]. This is because it is fundamentally an open loop method. To reject the effects of unknown disturbances, some measurements and estimate of the system state and appropriate feedback control law will be required.

Intersample discretization, proposed by Schultz and Ueda [216], is a method that uses switching inputs between near neighbors in place of zeroth order hold during the sample. This shows some promise for using switching inputs to gradually drive oscillation out of the system. Generating a feedback command that uses switching to drive out oscillation will likely involve some sort of predictive element over intervals

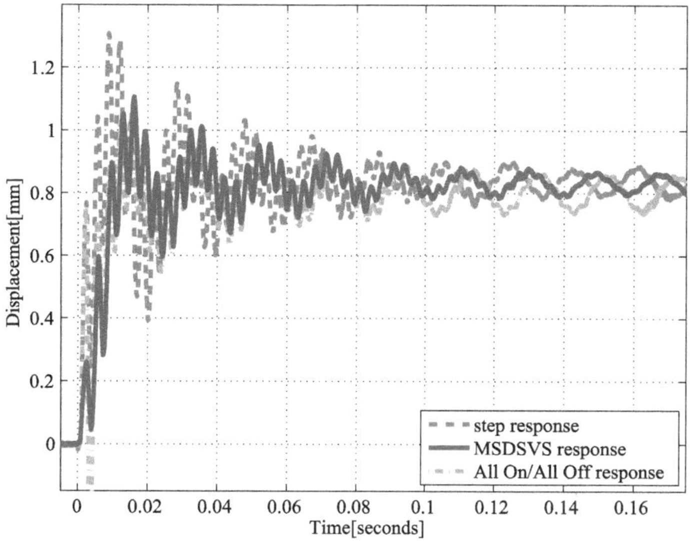

Figure 3.26 Cellular actuator response when the natural frequency is changed by adding mass to the actuator. © 2012 IEEE, reprinted with permission

of time combined with state estimation. This is largely an open problem. Some clues lie in the literature on quantized control, most of which trace its lineage to the work of Delchamps [52]. Most of these are focused on analyzing reachability of various points in the state space and proofs of stability of various types of quantizers [30,26, 195,194]. Abdelwahed et al. are more directed toward applications [5], but real-time implementation of a feedback control system along the lines of the cellular actuator is yet to be demonstrated.

3.2 Broadcast control for cellular actuator arrays

3.2.1 Cellular control system

This section focuses on position control of a cellular actuator array. We assume that each individual cell takes bistable ON–OFF states and provides an uniform displacement η regardless of the stress applied to the cell:

$$y_t^i = \begin{cases} \eta, & \text{if the } i\text{th cell is ON,} \\ 0, & \text{if the } i\text{th cell is OFF,} \end{cases} \tag{3.12}$$

where y_t^i is the displacement of the ith cell at discrete time t. Consider an array structure where N cells are connected in series. Suppose that N_t^{ON} cells are ON and N_t^{OFF}

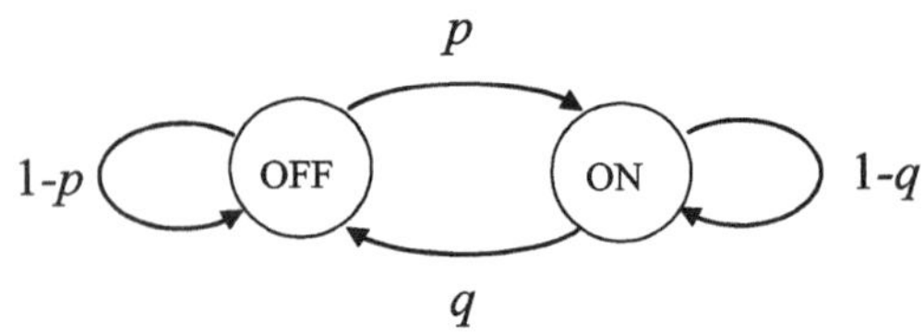

Figure 3.27 Single cell.

cells are OFF at time t. The aggregate output y_t is given by

$$y_t = \sum_{i=1}^{N} y_t^i = \eta \cdot N_t^{ON}. \tag{3.13}$$

The gross stroke (output range of the system) is then given by $L = \eta \cdot N$. The number of OFF cells, $N_t^{OFF} (= N - N_t^{ON})$, does not contribute to the aggregate output.

Each cell has a decision-making unit that changes the transition probability from one state to the other by receiving a broadcast command. Let p $(0 \leq p \leq 1)$ be the transition probability from OFF to ON, and q $(0 \leq q \leq 1)$ be the transition probability from ON to OFF as shown in Fig. 3.27. Assume that the transition is performed in discrete time step, hence the behavior of the cell is modeled as a discrete-time, non-homogeneous Markov process.

The ensemble behavior of N cells can be represented as a Markov process [72] shown in Fig. 3.28. The conditional mean of the number of ON cells and that of OFF cells are given by the state transition equation:

$$\begin{bmatrix} E[N_{t+1}^{ON} | N_t^{ON}, N_t^{OFF}] \\ E[N_{t+1}^{OFF} | N_t^{ON}, N_t^{OFF}] \end{bmatrix} = \begin{bmatrix} 1-q & p \\ q & 1-p \end{bmatrix} \begin{bmatrix} N_t^{ON} \\ N_t^{OFF} \end{bmatrix}. \tag{3.14}$$

The broadcast control is to modulate the transition probabilities p and q as a function of the broadcast command. Hence this Markov process is not homogeneous.

For example, the probability density function for the number of activated cells at time $t+1$ when all the N cells are OFF at time t is given by

$$\Pr\{N_{t+1}^{ON} = x | N_t^{OFF} = N\} = \binom{N}{x} p^x (1-p)^{N-x}, \tag{3.15}$$

where $\binom{N}{x} = \frac{N!}{(N-x)!x!}$.

In practice, however, some fractions of the cells are non-functional, i.e., dead cells. Suppose that N_{dead}^{ON} cells are dead and stay in the ON state for the next transition and that N_{dead}^{OFF} are dead, taking the same OFF state. For simplicity, we consider the case where dead cells stay either in the ON or OFF state but in the intermediate states. The

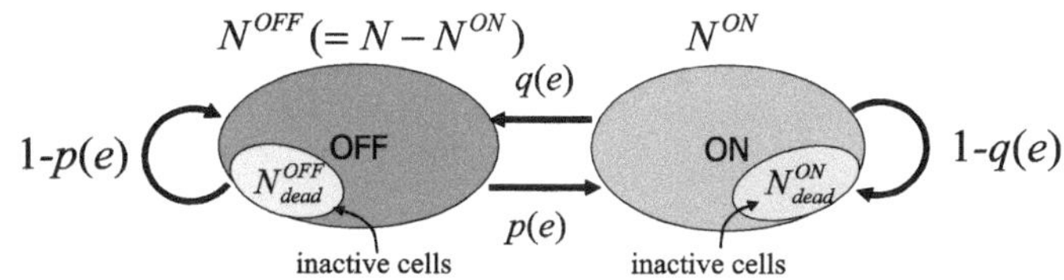

Figure 3.28 Aggregate Markov model.

above state transition equation is then modified to

$$\begin{bmatrix} E[N_{t+1}^{ON}|N_t^{ON}, N_t^{OFF}] \\ E[N_{t+1}^{OFF}|N_t^{ON}, N_t^{OFF}] \end{bmatrix} = \begin{bmatrix} 1-q & p \\ q & 1-p \end{bmatrix} \begin{bmatrix} N_t^{ON} - N_{dead}^{ON} \\ N_t^{OFF} - N_{dead}^{OFF} \end{bmatrix} + \begin{bmatrix} N_{dead}^{ON} \\ N_{dead}^{OFF} \end{bmatrix}. \tag{3.16}$$

Note that $0 \le N_{dead}^{ON}, N_{dead}^{OFF}$ and $0 \le N_{dead}^{ON} + N_{dead}^{OFF} \le N$ hold obviously. N_{dead}^{ON} and N_{dead}^{OFF} may vary, but we treat them as constant values assuming that the variation is slow.

In the following, we modulate the transition probabilities p and q as a function of the broadcast error. Hence this Markov process is not homogeneous. Without the loss of generality, the concept can be used for force control, where each cell produces a discrete force against a compliant element, which will be discussed in a later section. Also, the robustness, or non-uniformity, of the cells will be discussed in a later section.

3.2.2 Broadcast feedback for cellular control system

3.2.2.1 Open-loop characteristics

Consider a feedforward (open-loop) control problem at first. Without loss of generality, we can assume that all the N cells are OFF at time t in the following analysis. Let $r = \eta N_d$ $(0 \le r \le L)$ be a reference input, where N_d is the number of ON cells for r. By setting the transition probability p of the individual cells as

$$p = \frac{r}{L} \quad (0 \le p \le 1), \tag{3.17}$$

the expected aggregate output is given by

$$E[y_{t+1}|y_t = 0] = \eta \cdot E[N_{t+1}^{ON}|N_t^{ON} = 0] = \eta \cdot p \cdot N = r, \tag{3.18}$$

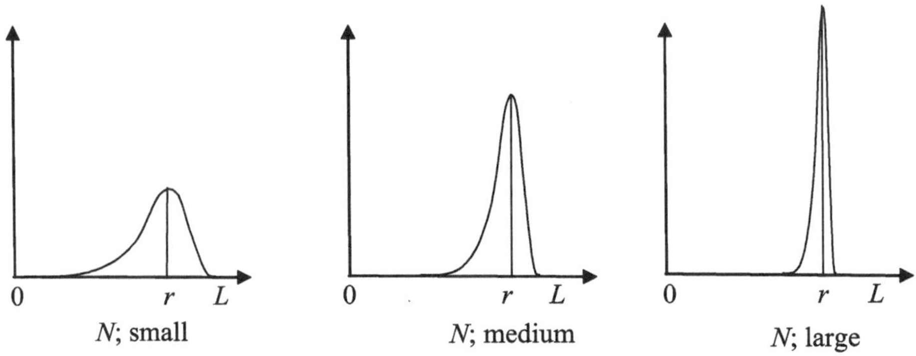

Figure 3.29 Probability distribution for different N.

which agrees with the reference output. More importantly, the variance of the output is given by

$$\begin{aligned} Var[y_{t+1}|y_t = 0] &= E[(y_{t+1} - E[y_{t+1}|y_t = 0])^2] \\ &= \eta^2 \cdot E[(N_{t+1}^{ON} - E[N_{t+1}^{ON}|N_t^{ON}])^2] \\ &= \eta^2 \cdot N \cdot p(1-p) = \frac{L^2}{N} p(1-p). \end{aligned} \tag{3.19}$$

Note that the variance reduces as the number of cells increases. In other words, the variance reduces as the given gross stroke length L is divided into more cellular units, each producing a finer output η. Fig. 3.29 shows plots of the output probability distribution for different N. The standard deviation $\sigma_{t+1} = \sqrt{Var[y_{t+1}|y_t = 0]}$ reduces in proportion to $1/\sqrt{N}$. This property implies that the cellular control system having more cells turns out to be more predictable. As N tends to infinity, it can be driven to produce a desired output with an arbitrary accuracy in the mean square sense by broadcasting the error or state transition probability.

3.2.2.2 Closed-loop control by broadcast feedback

The drawbacks of the above broadcast open-loop control are:

- For a finite number of cells, N, the output inevitably deviates from the reference having a probability distribution with a finite variance.
- The exact number of the usable cells must be known.

The latter drawback will be an important issue particularly for a large scale cellular system, where it is difficult to maintain all the cells functional. Some of the cells may die or do not respond to the inputs. It is desired if the control system works without knowing the exact number of functional (healthy) cells. In an attempt to overcome these difficulties, the "Broadcast Feedback" shown in Fig. 3.30 is considered.

The output is measured at discrete time $t = 0, 1, 2, \ldots$. The central controller generates a *universal* control command based on the aggregate output error, i.e., the

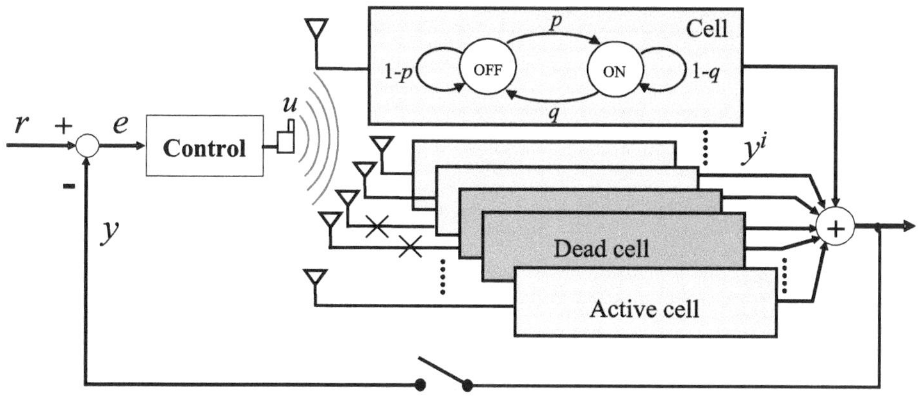

Figure 3.30 Broadcast feedback for a cellular control system with distributed decision-making units.

difference between the reference input and the current aggregate output, $e_t = r - y_t$, and broadcasts it to all the cells. The universal control command u is in terms of this global error; it maps the error to the probabilities in each cell, $u(e_t) : e_t \to p, q$. The output-feedback problem for a stochastic nonlinear system has been studied [53]. In the broadcast feedback approach, the output of the system is the ensemble of each individual cell that takes bistable ON–OFF states.

Figs. 3.31A–B show the change of ON cells and error e_t when $e_t > 0$. If $e_t > 0$, more cells must be turned ON. The updated probability p_{t+1} is calculated in all the cells, which independently make stochastic decisions. As a result, ΔN_{t+1}^{ON} cells are turned on among N_t^{OFF} cells that were OFF at time t. Let us assume that the transition from ON to OFF is prohibited when $e_t > 0$. This prohibition simplifies the problem and avoids *zero dynamics* as will be described in Section 3.2.3.2. This changes the number of ON cells to

$$N_{t+1}^{ON} = N_t^{ON} + \Delta N_{t+1}^{ON}, \tag{3.20}$$

and the output and the error to

$$y_{t+1} = \eta \cdot N_{t+1}^{ON}, \tag{3.21}$$

$$e_{t+1} = r - y_{t+1}, \tag{3.22}$$

respectively.

The expected value, i.e., conditional mean, of ΔN_{t+1}^{ON} is obtained from the binomial probability distribution as

$$\Delta \bar{N}_{t+1}^{ON} = E[\Delta N_{t+1}^{ON} | N_t^{OFF}] = p_{t+1} \cdot N_t^{OFF}, \tag{3.23}$$

which brings the error to

$$E[e_{t+1}|e_t, r] = p_{t+1}(r - L) + (1 - p_{t+1})e_t. \tag{3.24}$$

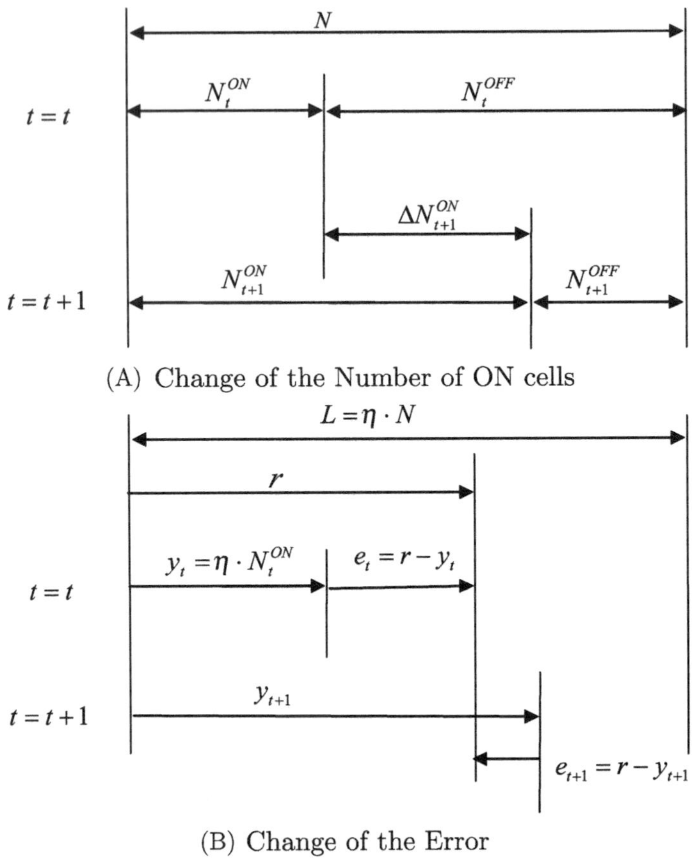

Figure 3.31 Compensation for the error by closed-loop control.

From (3.20) to (3.23), the variance of ΔN_{t+1}^{ON} is also obtained from the binomial probability distribution as

$$Var[\Delta N_{t+1}^{ON} | N_t^{OFF}] = p_{t+1}(1 - p_{t+1}) \cdot N_t^{OFF}. \tag{3.25}$$

The variance of error at $t + 1$ is then given by

$$\begin{aligned} Var[e_{t+1} | e_t, r] &= \eta^2 N_t^{OFF} p_{t+1}(1 - p_{t+1}) \\ &= \frac{1}{N} L(L - r + e_t) p_{t+1}(1 - p_{t+1}). \end{aligned} \tag{3.26}$$

Note that the variance gets smaller as the number of cells increases. Similar results can be obtained for $e_t < 0$.

3.2.3 Stability analysis of broadcast feedback

3.2.3.1 Asymptotic stochastic stability

The stability of the broadcast feedback control system consisting of N cells with an aggregate output feedback is analyzed. The control system is distributed with no intercellular communications, i.e.,

- The central controller does not know the number of active or dead cells such as N^{ON}, N^{OFF}, N_{dead}^{ON}, and N_{dead}^{OFF}. It broadcasts only the error between the aggregate output and reference.
- The local control at each cell uses its own internal state and broadcast information.

Therefore, the system is described by the inhomogeneous Markov process with the state transition equation (3.16) where state transition probabilities are functions of the broadcast output error e. In general, stability and performance can be improved by observing internal states of the system. For example, if the number of non-functional cells is observed and effectively incorporated in the transition probabilities, the expectation of the error of the next time step may be set to zero, resulting in the optimal control. However, the complexity of the system increases, which is not acceptable for a large-scale cellular control system.

We assume that the sampling rate of the broadcast feedback is sufficiently slow compared to the cell dynamics, so that each cell completes transition within the sampling period. The basic motivation of cellular architecture, where the actuator material is divided into many small units, is to increase speed of response as well as to overcome the complex hysteresis and nonlinear properties of the material. The above assumption is based on that the cell dynamics is much faster than the broadcast feedback using the aggregate output, therefore hysteresis and other nonlinear properties are negligible. The assumption is reasonable when PZT-based cellular actuators as shown in Section 1.31 are used. Note that Section 3.4 discusses how hysteresis can be compensated when relatively slow actuator materials, such as SMA, are used.

The broadcast feedback control system is stochastic. The state transition is known only in the stochastic sense. Therefore, stability theory for deterministic systems is not applicable. The following stability theory using a stochastic Lyapunov function by Kushner [143,144] is applicable to this problem.

Theorem 1 (Asymptotic Stochastic Stability of Discrete Parameter System [143, 144]). *Let $V^S(x)$ be a scalar-valued, non-negative, continuous function, satisfying $V^S(0) = 0$, $V^S(x) > 0$, $x \neq 0$. $V^S(x)$ has continuous first derivatives in the bounded set $Q_m = \{x : V^S(x) < m\}$, $m < \infty$. Let $x_0, x_1, \ldots$ be a scalar-valued discrete-parameter Markov process, where x_0 is the initial condition in Q_m. If a non-negative, real, scalar function $k(x_t)$ exists, such that the difference between $V^S(x)$ at time t and the conditional mean $E[V^S(x_{t+1})|x_t]$ at time $t+1$ is bounded as*

$$E[V^S(x_{t+1})|x_t] - V^S(x_t) \triangleq -k(x_t) \leq 0 \tag{3.27}$$

in Q_m, then x_t converges to

$$x_t \to P_m = Q_m \cap \{x : k(x) = 0\} \tag{3.28}$$

with a probability no less than $1 - V^S(x_0)/m$.

$V^S(x)$ is called a stochastic Lyapunov function. Assume that reference r is constant. Change of r merely means the change of the coordinate origin, i.e., $x_t = e_t = r - y_t$. Let us consider $V^S = e_t^2$ for a candidate of the stochastic Lyapunov function. The change to the Lyapunov function candidate, ΔV_t^S, is calculated as follows:

$$\begin{aligned}
\Delta V_t^S &= E[V_{t+1}^S|e_t] - V_t^S \\
&= Var[e_{t+1}|e_t] + E[e_{t+1}|e_t]^2 - e_t^2 \\
&= -k(e_t) \le 0,
\end{aligned} \tag{3.29}$$

where $E[e_{t+1}^2|e_t] = Var[e_{t+1}|e_t] + E[e_{t+1}|e_t]^2$ has been applied. Note that the variance appears in (3.29), indicating the effect of the variance on the stability condition. If the process is deterministic and, thereby, the variance is zero, the stability condition has no difference from that of a deterministic Lyapunov function. Due to the stochastic nature of the process, the left-hand side of the above inequality condition is larger with the added variance term. Therefore, more strict (conservative) stability condition must be met for the stochastic process. It is obvious that $Var[e_{t+1}] \to 0$ (see (3.26)) and $e_{t+1} \to E[e_{t+1}]$ if $N \to \infty$, resulting in deterministic analysis shown in Appendix A.6.

When the inequality condition, (3.29), is satisfied, the process is called a nonnegative supermartingale, for which the Lyapunov function is guaranteed to converge to a nonnegative limit with probability 1. See [143,144,62] for more details and proof.

3.2.3.2 Unilateral transition control

In order to simplify the problem, the transition from ON to OFF is prohibited when $e_t > 0$, and the transition from OFF to ON is prohibited when $e_t < 0$ as follows:

$$p(e) = \begin{cases} 0 & (e \le 0), \\ p(e) & (e > 0), \end{cases} \tag{3.30}$$

$$q(e) = \begin{cases} q(e) & (e < 0), \\ 0 & (e \ge 0). \end{cases} \tag{3.31}$$

We call this control law a unilateral transition control, which modifies the single cell dynamics as shown in Fig. 3.32. The above transition probabilities split (3.16) into two 1-dimensional random walk models. By substituting (3.30) and (3.31) for (3.16), the following discrete systems are obtained:

$$E[e_{t+1}|e_t] = e_t - \eta p(e_t)(N_t^{OFF} - N_{dead}^{OFF}), \quad e_t > 0, \tag{3.32}$$

$$E[e_{t+1}|e_t] = e_t + \eta q(e_t)(N_t^{ON} - N_{dead}^{ON}), \quad e_t < 0. \tag{3.33}$$

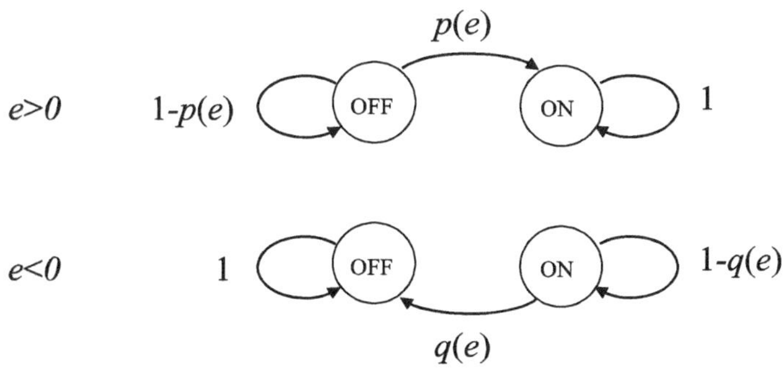

Figure 3.32 Single cell with unilateral transition control.

As a result, the problem to be solved is how to design $p(e)$ and $q(e)$ that guarantee the convergence of the error in terms of (3.32) and (3.33). Note that the stationary condition will be preserved once $e_t = 0$ holds since $p(0) = q(0) = 0$ provides

$$E[e_{t+1}|e_t = 0] = 0. \tag{3.34}$$

The cellular control system with the unilateral transition does not have *zero dynamics* that is unobservable from the aggregate output. This legitimates the concept of the broadcast feedback that merely broadcasts the aggregate output error. Consider the following Lemma 1.

Lemma 1. *Let* $\Delta e_t = e_t - e_{t-1}$ *and* $\Delta y_t^i = y_t^i - y_{t-1}^i$ $(i = 1, \ldots, N)$. *Observation of* $\Delta e_t = 0$ *implies that no cell changed its internal state from time* $t - 1$ *to* t, *i.e.,*

$$|\Delta e_t| = 0 \Rightarrow \Delta y_t^i = 0 \; \forall i. \tag{3.35}$$

Proof. Suppose $\exists i$ such that $y_{t-1}^i = 0$ and $y_t^i = \eta$, i.e., cell i made a transition from OFF to ON. This cannot happen unless $p > 0$. This implies $e_{t-1} > 0$ and $\Delta e_t < 0$ since $\Delta y_t^i \geq 0$ for all i only when $p > 0$ and $q = 0$. Similarly, $\Delta e_t > 0$ if $\exists i$ such that $y_{t-1}^i = \eta$ and $y_t^i = 0$. Therefore, the following proposition holds:

$$\exists i \text{ s.t. } \Delta y_t^i \neq 0 \Rightarrow |\Delta e_t| \neq 0. \tag{3.36}$$

The contraposition of this proposition gives Lemma 1. □

Since there is no zero dynamics, a Lyapunov function based on only the aggregate output e is enough to analyze the stability of the entire cellular system. Unlike the above state transition law, if bilateral state transitions are allowed, there is a chance that prolonged oscillations may occur. Furthermore, proof of stability including internal stability becomes more complicated.

3.2.3.3 Stability conditions

By substituting (3.26), (3.32), and (3.33) for (3.29), we obtain

$$\Delta V_t^S = \begin{cases} \{\eta(1-p) - 2e_t + p(\eta N - \eta N_d - \eta N_{dead}^{OFF} + e_t)\} \\ \times p(\eta N - \eta N_d - \eta N_{dead}^{OFF} + e_t), \quad e_t > 0, \\ \{\eta(1-q) + 2e_t + q(\eta N_d - \eta N_{dead}^{ON} - e_t)\} \\ \times q(\eta N_d - \eta N_{dead}^{ON} - e_t), \quad e_t < 0. \end{cases} \tag{3.37}$$

The following transition probabilities hold $\Delta V_t^S < 0, \forall t$:

$$p(e) = \begin{cases} 0, & e \le \eta/2, \\ 0 < p(e) < \min(\frac{2e-\eta}{\eta(N-N_d-N_{dead}^{OFF}-1)+e}, 1), & e > \eta/2, \end{cases} \tag{3.38}$$

$$q(e) = \begin{cases} 0 < q(e) < \min(\frac{-2e-\eta}{\eta(N_d-N_{dead}^{ON}-1)-e}, 1), & e < -\eta/2, \\ 0, & e \ge -\eta/2, \end{cases} \tag{3.39}$$

where the subscript "t" is omitted. Since m is arbitrary, any trajectory from any error e_0 reaches $P_m = \{e : |e| \le \eta/2\}$ with probability 1 when $t \to \infty$. Note that the radius of P_m is less than the resolution of the output η.

The stability condition given by (3.38) and (3.39) does not meet the conditions for the cellular control system since each cell should know N^{ON} and N_{dead}^{ON} for $e_t < 0$, or N^{OFF} and N_{dead}^{OFF} for $e_t > 0$. The following theorem provides the solution:

Theorem 2 (Error Broadcast for the Cellular Control System). *Suppose that a broadcast feedback controller performs proportional control of the output y, i.e., only the error e_t is broadcast. If each cell has the following set of transition probabilities* as a function of the broadcast error *(see Fig. 3.33), the error asymptotically converges to $P_m = \{e : |e| \le \eta/2\}$ with probability* 1*:*

$$p(e) = \begin{cases} 0, & e \le \eta/2, \\ 0 < p(e) < \min(\frac{2e-\eta}{\eta(N-1)}, 1), & e > \eta/2, \end{cases} \tag{3.40}$$

$$q(e) = \begin{cases} 0 < q(e) < \min(\frac{-2e-\eta}{\eta(N-1)}, 1), & e < -\eta/2, \\ 0, & e \ge -\eta/2, \end{cases} \tag{3.41}$$

where only the nominal number of the cell N in the system and the displacement of a single cell η $(= L/N)$ are required.

Proof. Recall that the actual number of ON cells is always less than the number of cells included in the system: $N \ge N^{ON} - N_{dead}^{ON} = N_d - N_{dead}^{ON} - e/\eta$ for $e < 0$, then (3.40) is obvious. Similarly, (3.41) is obvious since $N \ge N^{OFF} - N_{dead}^{OFF} = N - N_d - N_{dead}^{OFF} + e/\eta$ for $e > 0$. □

Corollary 1 (Convergence of internal states). *Consider a cellular control system with a unilateral transition law by error broadcast, e.g., (3.40) and (3.41). Suppose a*

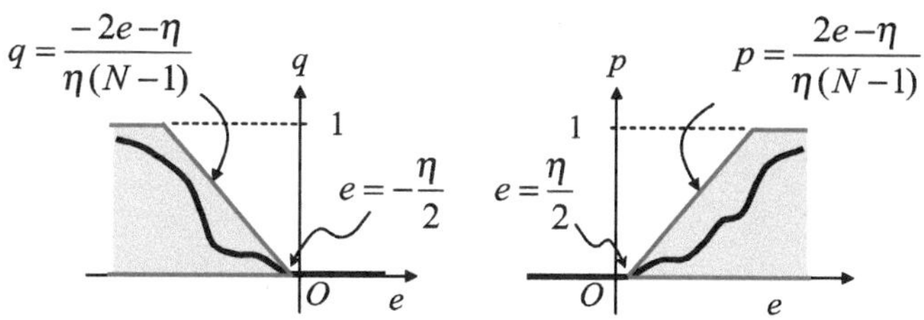

Figure 3.33 Stable transition probabilities.

stochastic Lyapunov function L^S of the system is nonnegative supermartingale. This implies not only $e_t \to P_m$ with probability one, but also implies that the internal state of every cell converges with probability one, i.e., $\Delta y_t^i \to 0\ \forall i$.

The proof is a direct consequence of Theorem 2 and Lemma 1.

Remark 1 (Robustness against cell failure). *Recall (3.40) and (3.41) hold for any N_{dead}^{ON} and N_{dead}^{OFF}. Then, we can say that* the system is stable for any percentage of dead cells *for any feasible r if the system is stable for N (100% cells are active).*

Remark 2 (Convergence when N is large enough). *By letting $N \to \infty$ ($\eta \to 0$), (3.40) and (3.41) can be approximated to the transition probabilities from the deterministic analysis of the expectation.*

The convergence of the expectation can be analyzed by assuming $e_t \to E[e_t]$ when $N \to \infty$. Let $V^D = e_t^2$ be a Lyapunov function, where the superscript "D" denotes the deterministic analysis. $\Delta V_t^D = V_{t+1}^D - V_t^D$ is given as follows:

$$\begin{aligned}\Delta V_t^D &= e_{t+1}^2 - e_t^2 \\ &= \begin{cases} p(\eta N - \eta N_d - \eta N_{dead}^{OFF} + e_t) & \\ \times\{-2e_t + p(\eta N - \eta N_d - \eta N_{dead}^{OFF} + e_t)\}, & e_t > 0, \\ q(\eta N_d - \eta N_{dead}^{ON} - e_t) & \\ \times\{2e_t + q(\eta N_d - \eta N_{dead}^{ON} - e_t)\}, & e_t < 0. \end{cases}\end{aligned} \tag{3.42}$$

$\Delta V_t^D < 0$ holds if

$$\begin{aligned} p(e) &= \begin{cases} 0, & e \le 0, \\ 0 < p(e) < \min(\frac{2e}{\eta(N-N_d-N_{dead}^{OFF})+e}, 1), & e > 0, \end{cases} \end{aligned} \tag{3.43}$$

$$\begin{aligned} q(e) &= \begin{cases} 0 < q(e) < \min(\frac{-2e}{\eta(N_d-N_{dead}^{ON})-e}, 1), & e < 0, \\ 0, & e \ge 0. \end{cases} \end{aligned} \tag{3.44}$$

Recall that $L \ge \eta(N_d - N_{dead}^{ON}) - e$ and $L \ge \eta(N - N_d - N_{dead}^{OFF}) + e$ always hold. Then, we have the following transition probabilities for the cellular control system without using N^{ON}, N^{ON}_{dead}, N^{OFF}, or N^{OFF}_{dead} explicitly:

$$p(e) = \begin{cases} 0, & e \leq 0, \\ 0 < p(e) < \min(2e/L, 1), & e > 0, \end{cases} \tag{3.45}$$

$$q(e) = \begin{cases} 0 < q(e) < \min(-2e/L, 1), & e < 0, \\ 0, & e \geq 0. \end{cases} \tag{3.46}$$

For large N, $e_t \to 0$ with probability 1 since $P_m = \{0\}$. In addition, only the nominal gross stroke L is required for the design of the transition probabilities.

Remark 3 (No overshoot response). *No overshoot response of $E[e_t]$ for any negative initial error can be obtained if $0 < q(e) < -e/L$ $(e < 0)$ for (3.45), i.e., $E[e_t] \to -0$, $t \to \infty$, since the equilibrium $e = 0$ is asymptotically stable and we have $e_0 < E[e_1] < \cdots < E[e_t] < E[e_{t+1}] < \cdots < 0$. Note that this convergence is irrelevant to the design of $p(e)$. Similarly, $E[e_t] \to +0$, $t \to \infty$ can be obtained without overshoot for any positive initial error by $0 < p(e) < e/L$ $(e > 0)$ for (3.46). This convergence is irrelevant to the design of $q(e)$.*

3.2.4 Simulation: uniform cellular array

3.2.4.1 Simulation model

A positioning control is applied to an actuator in which N (= 1000) of the MEMS–PZT cellular actuators introduced in Section 1.1.3 are connected in series. The gross stroke of the actuator is $L = 19.6$ [mm] since the length of each cell is 280 [μm] and the displacement is $\eta = 19.6$ [μm] (7% of 280 μm). The total length of the actuator when fully extended is 0.28 [m], and the length when fully contracted is 0.2604 [m]. The feedback controller broadcasts the error between the reference and the current displacement, $e_t = r - y_t$.

The broadcast signal e_t is updated in every 0.01 [s] so that decision-making in each individual cell is performed in sync with this update. Since the response of the PZT actuator in excess of 5 kHz is fast enough, the dynamics of PZT actuator is negligible compared to the dynamics of the decision-making. Sampling delay $T = 0.01$ [s] and white noise are added to the observation of e.

3.2.4.2 Transition probability design

Assume that the number of the cells ($N = 1000$) is large enough. From (3.45) and (3.46), a practical design for $p(e)$ and $q(e)$ that satisfies the condition of stability is given as follows:

$$p(e) = \begin{cases} 0, & e \leq 0, \\ \min(g_p e/L, 1), & e > 0, \end{cases} \tag{3.47}$$

$$q(e) = \begin{cases} \min(-g_q e/L, 1), & e < 0, \\ 0, & e \geq 0, \end{cases} \tag{3.48}$$

where g_p and g_q are control gains that satisfy $0 < g_p, g_q < 2$.

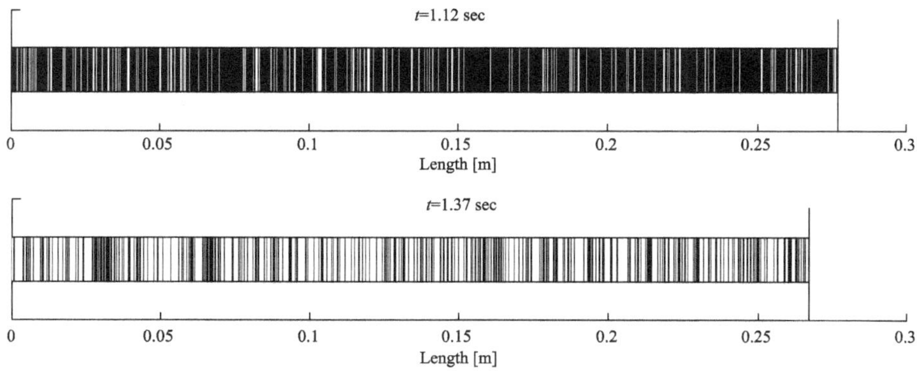

Figure 3.34 Simulation snapshots: (white) ON cell, (black) OFF cell.

3.2.4.3 Position control by broadcast feedback

The probabilities given by (3.47) and (3.48) are applied to each cell. Step response and trajectory tracking for a 2 Hz sinusoidal function are performed for $g_p = g_q = 0.8$, $g_p = g_q = 1.5$, and $g_p = g_q = 2.2$. Recall that the convergence is guaranteed by $0 < g_p, g_q < 2$, and no-overshoot response is achieved by $0 < g_p, g_q < 1$. In addition, the performance is examined when 30% of cells become inactive to confirm the robustness against the failure. We assume inactive cells stay in the OFF state and never change to ON; $N_{dead}^{ON} = 0$, $N_{dead}^{OFF} = 300$. Note that both trajectories are designed to stay within the working range obtained by 70% of (700) active cells.

Fig. 3.34 shows snapshots of the simulation. Internal state of a cell is represented by black when the cell is OFF, and by white when the cell is ON.

3.2.4.4 Simulation results

Fig. 3.35 shows the results of step response and trajectory tracking for $g_p = g_q = 0.8$, $g_p = g_q = 1.5$, and $g_p = g_q = 2.2$ when 100% of cells are active (healthy) and 30% of cells are dead. The solid line shows the current displacement and the dashed line shows the reference. The displacements for full contraction and full extension are indicated by two solid lines. When 30% of cells becomes inactive, the displacement for full contraction changes ($y = 0.266$ [m]). As observed in Fig. 3.35A, stable no-overshoot responses are achieved for $g_p = g_q = 0.8$. Responses become oscillatory by setting $g_p = g_q = 1.5$, but still stably converge to the references as shown in Fig. 3.35C. These damped oscillations resemble a standard second-order system but are actually of overcompensation and switching between the dynamics given by (3.32) and (3.33). The responses become more oscillatory and do not converge to the references by setting $g_p, g_q \geq 2.0$ as shown in Fig. 3.35E. Note that $0 < g_p, g_q < 2.0$ is merely the sufficient condition, and the violation of this condition does not always lead to instability.

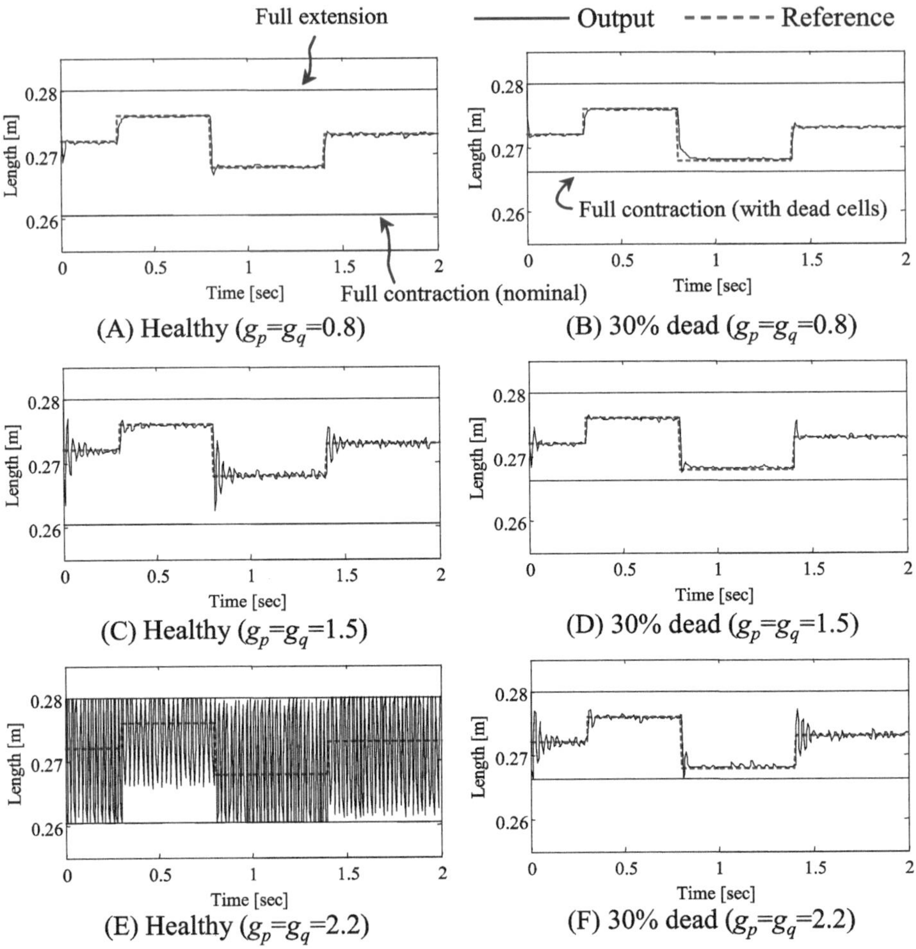

Figure 3.35 Simulation results: step response.

As described in Section 3.2.3.3, increasing the number of inactive cells does not affect the stability of the system as shown in Fig. 3.35B, D, and F; however, this makes the system more conservative, resulting in sluggish response. The oscillatory responses in Fig. 3.35C and E are suppressed as shown in D and F. Especially, the instable responses in Fig. 3.35E are improved to be stable. It is easy to show that the condition of stability is satisfied for these cases by calculating the equivalent gains, i.e., $\tilde{g}_p = \tilde{g}_q = 2.2 \times 700/1000 = 1.54 < 2.0$.

Fig. 3.36 shows the results of sinusoidal trajectory tracking. Strictly speaking, since the stability shown in Section 3.2.3.3 is guaranteed only when the reference r is constant, convergence to a continuously changing reference is not obvious even though the frequency of reference is much slower than the response of actuator. Therefore, in Fig. 3.36, we merely confirm the applicability of the proposed control scheme for

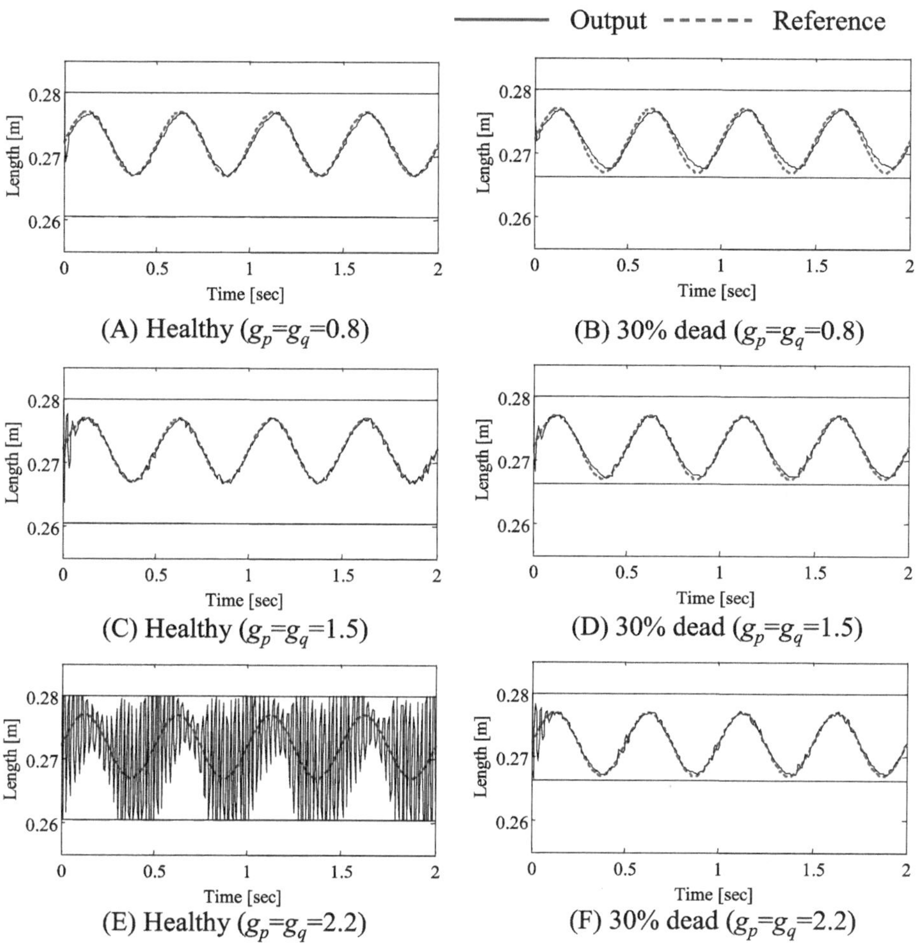

Figure 3.36 Simulation results: sinusoidal trajectory tracking.

continuous tracking. As can be observed in the figure, the responses for the sinusoidal trajectory show the same tendency as shown in Fig. 3.35. The output tracks to the reference stably except when all the cells are healthy and $g_p = g_q = 2.2$ as shown in Fig. 3.35E. Even when 30% of the cells are dead, the output keeps tracking the trajectory stably although the response becomes sluggish.

3.2.5 Simulation: non-uniform cellular array

In the previous section, the displacement of each actuator was assumed to be uniform. The transition probability at each cell was also assumed to be the same. The broadcast control method can tolerate a certain level of non-uniformities, and applicable to controlling an actuator array with non-uniform cellular length and non-uniform transition probability. This section demonstrate that even in the presence of these non-uniformities, the aggregate output of the cellular units follows a reference robustly.

3.2.5.1 Non-uniformity of cells

Since cellular actuator systems consist of a vast number of small cellular units, several problems due to the limitation of manufacturing tolerances are considered unavoidable. One of these problems is the difficulty in maintaining the uniformity of the cells, such as response time to the signal, displacement, force, and life cycle. The cellular actuator system is expected to sustain a sufficient response capability even in the presence of these non-uniformities. Several major problems can be addressed as follows.

Failure of cells. As discussed in the previous section, it is difficult to maintain all the cells functional; some of the cells may die or not respond to the inputs. This problem may be due to the creep of the material, disconnection of the power lines, or break in the receiver circuit. In addition, the number of the *dead cells* may vary during the operation. It is notable that a muscle can function robustly and stably although a significant fraction of the cellular units are not functional.

Non-uniformity of cell performance. Even with the recent rapid progress in micro manufacturing technologies, it is difficult to produce many micro actuator units that have uniform displacement or force. A certain degree of variations is unavoidable. In contrast, the length of sarcomere in biological muscle system is not strictly uniform, varying from 2 to 3 μm, and it is considered difficult for the central nervous system to know the whole distribution of the length. However, the muscle control system seems to be working without major problems regarding this point.

Non-uniformity of embedded transition probabilities. Similarly, it is difficult to let all the cells have uniform local controllers that generate uniform transition probability from the broadcast signal; the generated probabilities at local controllers may have some fluctuations due to noise, offset, or signal attenuation. Needless to say, these problems can be observed in biological systems, which are affected by thermal noise.

3.2.5.2 Broadcast control of non-uniform actuator arrays

A position control of a series of the MEMS–PZT cellular actuators with 7% strain [271] is examined for $N = 25$ and $N = 1000$. The length of a single cell is 280 μm and the displacement is $\eta = 19.6$ μm (7% of 280 μm). For example, for $N = 1000$, the gross stroke of the actuator is $L = 19.6$ mm. The total length of the actuator array when fully extended is 0.28 m, and the length when fully contracted is 0.2604 m. The broadcast signal e_t is updated in every 0.01 s so that decision-making in each individual cell is performed in sync with this update. Since the response of the PZT actuator in excess of 5 kHz is fast enough, the dynamics of PZT actuator can be neglected compared to the dynamics of the decision-making. A step-type reference displacement is given and the response is examined.

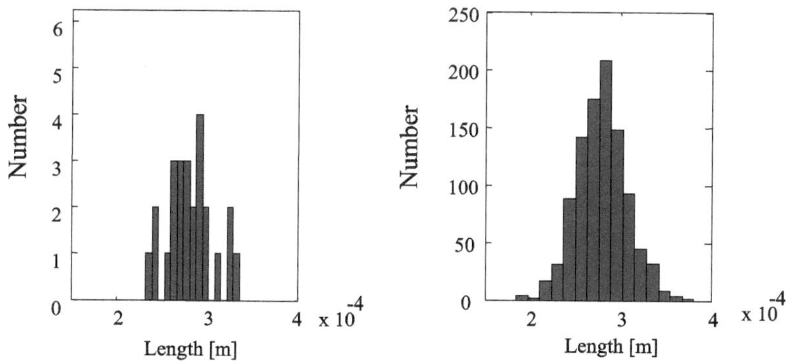

(A) N=25 (B) N=1000

Figure 3.37 Distribution of the cell length. © 2007 IEEE, reprinted with permission

The following three cases are examined:

1. $N = 25$. Non-uniform transition probability. Uniform cell length.
2. $N = 25$. Uniform transition probability. Non-uniform cell length.
3. $N = 1000$. Non-uniform transition probability. Non-uniform cell length. 200 units (20%) are dysfunctional, staying in OFF state.

Fig. 3.37 shows the distribution of cell length where the mean of the length is 280 μm. The non-uniformity of the transition probability due to noise is represented by

$$p^i(e) = \tilde{g}_p(e) + v_p^i, \tag{3.49}$$

$$q^i(e) = \tilde{g}_q(e) + v_q^i, \tag{3.50}$$

where $\tilde{g}_p(e)$ and $\tilde{g}_q(e)$ are deterministic functions of e uniformly given to all the cells. v_p^i and v_q^i represent white noise in the ith cell. Noise with mean 0 and variance 4.0×10^{-4} is applied.

By considering the uniformity of cell length, (3.40) and (3.41) are modified as follows:

$$\bar{p}(e) = \begin{cases} 0, & e \leq \eta/2, \\ \bar{p}(e) = \min(\frac{g_p e - \bar{\eta}}{L - \bar{\eta}}, 1), & e > \bar{\eta}/2, \end{cases} \tag{3.51}$$

$$\bar{q}(e) = \begin{cases} \bar{q}(e) = \min(\frac{-g_q e - \bar{\eta}}{L - \bar{\eta}}, 1), & e < -\bar{\eta}/2, \\ 0, & e \geq -\bar{\eta}/2, \end{cases} \tag{3.52}$$

where $\bar{\eta}$ is the mean of η $(i = 1, \ldots, N)$. $g_p = g_q = 1.5 < 2$ were chosen as to satisfy the condition of stochastic stability. Fig. 3.38 shows the regions of $p(e)$ and $q(e)$ suggested by (3.40) and (3.41).

The following simulation results demonstrate that the transition control given by (3.51) and (3.52) has robustness against the cell non-uniformities. The results are com-

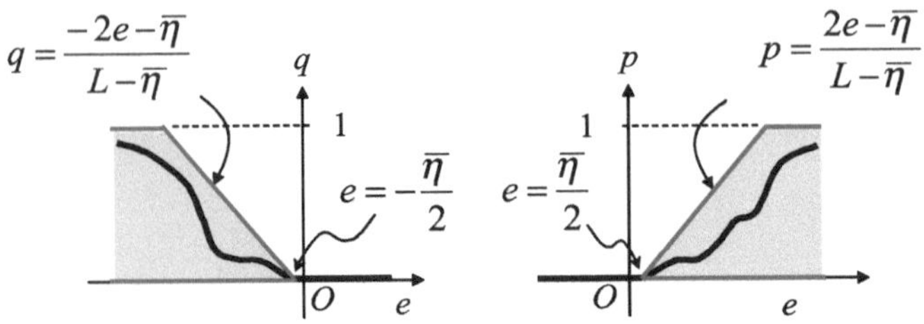

Figure 3.38 Stable transition probabilities. © 2007 IEEE, reprinted with permission

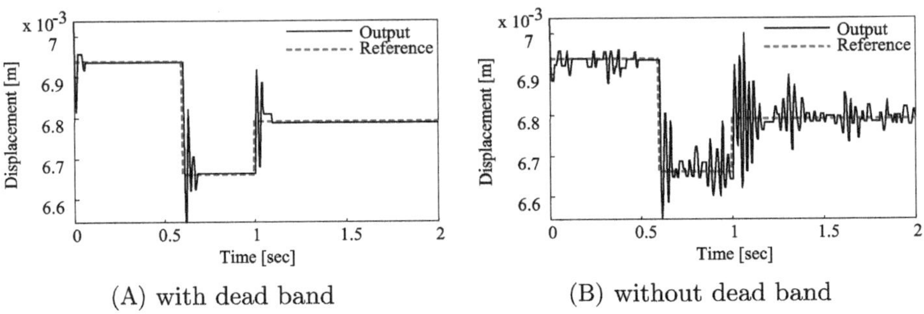

Figure 3.39 Simulation results for $N = 25$ with non-uniform transition probability. © 2007 IEEE, reprinted with permission

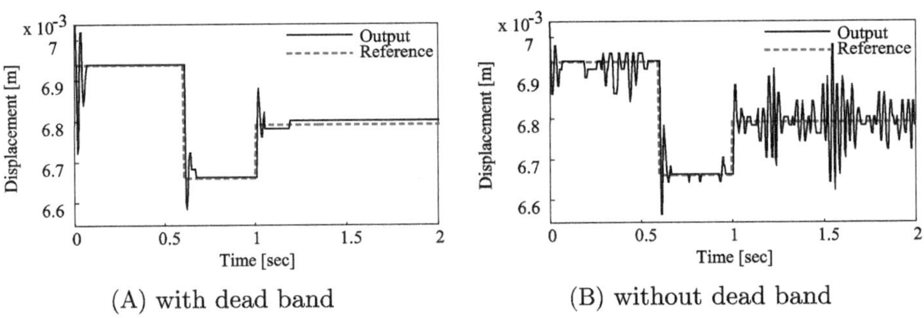

Figure 3.40 Simulation results for $N = 25$ with non-uniform cell length. © 2007 IEEE, reprinted with permission

pared with the ones generated by using (3.47) and (3.48) without a dead band. Note that the control law given by (3.47) and (3.48) demonstrated its efficacy for a uniform actuator array with $N = 1000$ [263]. Although the presented transition control is slightly more conservative than (3.47) and (3.48), since a small dead band is required, it takes the stochastic variance and quantization effect into account and guarantees the convergence for any N.

Figs. 3.39 to 3.41 show the simulation results that correspond to Cases 1–3. As shown in Fig. 3.39A, the output stably tracks the given trajectory even the transition probability is affected by noise. The transition control shows a potential to cope with the non-uniformity of cell length as shown in Fig. 3.40A. In contrast, the response

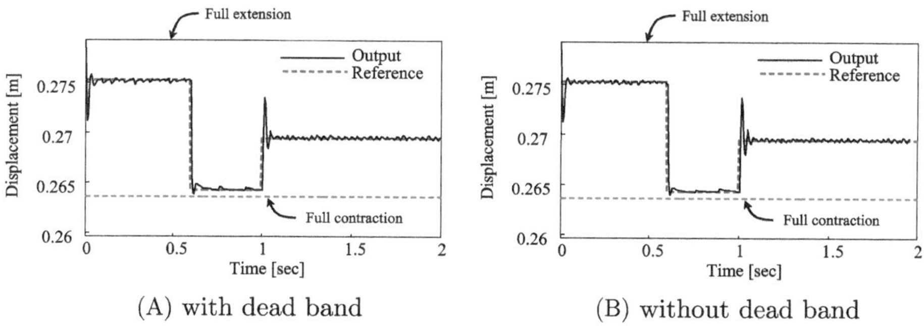

Figure 3.41 Simulation results for $N = 1000$ with non-uniform cell length, non-uniform transition probability, and 20% of dead cells. © 2007 IEEE, reprinted with permission

becomes oscillatory without the dead band in the transition control. Fig. 3.41 shows another result of the cellular control system. That is, if the number of cells N is large enough, the effect of the variance by non-uniformities as well as by stochastic transition control becomes negligible, resulting in high robustness. See [272] for more analysis.

3.2.6 Discussion

A broadcast feedback approach has been proposed for a large-scale stochastic cellular control system. The broadcast feedback controller uniformly broadcasts the aggregate error signal to all the cellular units that modulates the transition probabilities in individual cells. Stochastic properties of the cellular control system have been presented for both open-loop and closed-loop control. Based on stochastic Lyapunov analysis, the stability conditions have been derived in terms of the modulation of transition probabilities. The unilateral transition control has been shown to be sufficient to eliminate zero dynamics. In addition, the robustness to cell failure has also been proved. The validity of the proposed control structure has been confirmed by simulation with an artificial cellular actuator consisting of many segments of smart actuator material.

The framework of broadcast feedback may be applicable to other distributed stochastic control systems. Azuma et al. address the controllability of stochastic multi-agent systems [14]. Wood et al. discuss stochastic modeling of emergent behaviors of a cell population in angiogenesis [288]. Julius et al. apply a hybrid control approach to stochastic control of lactose regulation of *Escherichia coli* [122].

Investigation of the effects of actuator hysteresis is presented in Sections 3.3 and 3.4. Consideration of a time-varying reference as well as time-varying number of non-functional cells is necessary for generalization. The extension to continuous systems without the synchronization clock for the cells is important for the design of real cellular units. Furthermore, observation of internal states without increasing the complexity of the system is worth a try for improving the performance. For this observer-based control, two problems have to be solved: (i) how to design the observer for stochastic systems and (ii) how to guarantee the stability of the overall system including the observer and broadcast controller.

3.3 Hysteresis loop control of hysteretic actuator arrays

3.3.1 Segmented binary control for hysteretic cellular actuator units

Consider a material having a pronounced hysteresis loop, as illustrated in Fig. 3.42. The process is bistable, taking a binary output value, ON or OFF, when the input is higher or lower than a threshold value: u_N for the ascending branch and u_F for the descending branch as shown in the figure. PZTs, shape memory alloys (SMAs), and other smart materials exhibit this type of nonlinearity distributed across the material body.

Feedback control of this type of nonlinear distributed system is a challenging problem. In the past, many advanced control methods have been studied. Song et al. in [239], for example, has an extensive literature survey of SMA and hysteresis modeling and control during the last three decades. Recently, Segmented Binary Control (SBC) has been found to be an effective approach to overcoming bistable, hysteric, distributed nonlinearities [225]. SBC is a technique wherein a single actuator material is segmented into many discrete cellular units, each of which can be fully contracted or fully relaxed. Thus, the overall system output is formed from the collection of individual unit outputs; the aggregate output is proportional to the number of the cellular units that are turned on, if each unit produces the same magnitude of displacement; see Fig. 3.43. In SBC, a local controller responsible for each cellular unit regulates the binary state of its own segment. It merely drives the state of the unit towards ON or OFF, rather than stopping it at a specific point between the two. Therefore, it is simple and straightforward.

This segmented binary control architecture provides a means of overcoming not only distributed nonlinearity, but also time delay in the system response, when combined with another technique described next. Suppose that a unit is being turned off and moving from S_3 to S_0 in Fig. 3.43. In order to move in the opposite direction, the unit must move horizontally across the hysteresis band. No output change is observed until it arrives at the ascending branch between S_0 and S_1. This latency time can be eliminated by using a technique named Hysteresis Loop Control (HLC).

In HLC, each unit, after making an ON–OFF transition, is pushed toward a hold position, S_1 or S_3. u_{FR} is the input for S_1 and u_{NR} for S_3, respectively. Note that a unit at these states is ready for making an immediate transition, having no latency time for moving across the hysteresis band. As illustrated in Fig. 3.43, prohibiting the units that are still in a transition phase (unit A) and calling upon only the ready-state units (unit B), the latency time can be eliminated. An output change can be observed immediately after the ready-state units have reacted to the central controller's command. This kind of coordination is impossible for a single bulk process, but is feasible

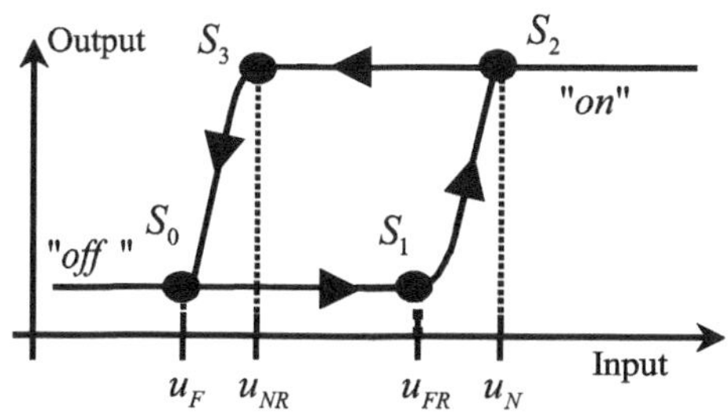

Figure 3.42 Input–output characteristics of hysteretic materials. © 2011 ASME, Wood, L., reprinted with permission

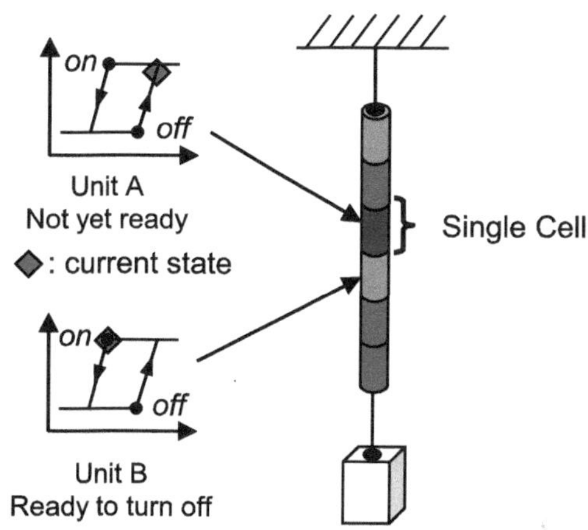

Figure 3.43 Coordinated control of multitude of cellular units. © 2011 ASME, Wood, L., reprinted with permission

for segmented cellular systems. Furthermore, this coordination can be implemented easily and effectively, as addressed in the following section. This requires the local controller of each cellular unit to simply prohibit switching of the ON–OFF command unless it is in a ready state. This resembles "refraction" in neuroscience; a neural cell does not fire for some period after it has fired last time. This prohibitory rule is called a refraction rule.

Note that for relatively slowly responding materials, such as SMAs, a local controller monitors and maintains the temperature of each SMA segment according to four user-specific temperatures: hot temperature (S_2), hot-to-cold transition temperature (S_3), cold temperature (S_0), cold-to-hot transition temperature (S_1), which has been reported effective to improve the speed of response of the entire system [225]. On the other hand, for fast responding materials, such as PZTs, local controllers can be designed even more simpler; bistable ON–OFF operation of the charge amplifier may be sufficient.

The displacement control of a single SMA wire actuator that is uniformly segmented into N units as shown in Fig. 3.43 is considered. The total displacement of the

SMA wire is given as the summation of the individual displacements:

$$y = \sum_{i=1}^{N} \epsilon_i l_i \tag{3.53}$$

where $\epsilon_1 = \epsilon_2 = \cdots = \epsilon_N = \epsilon$ is the strain and $l_1 = l_2 = \cdots = l_N = l$ is the length of each of the uniform segments. The initial strain is given by a preloaded tension of the wire as shown in Fig. 3.43.

3.3.2 Implementation of hysteresis loop control of an SMA unit

The hysteresis loop control (HLC) is implemented to improve the speed of response of a segmented SMA actuator [270]. For SMA wire actuators, Fig. 3.43 can be represented as Fig. 3.44 that illustrates the concept of strain–temperature control for hysteresis compensation. Let T be the temperature of a segment of SMA. In Fig. 3.44, T_F is cold temperature, T_{FR} is cold-to-hot transition temperature, T_N is hot temperature, and T_{NR} is hot-to-cold transition temperature, respectively. As also shown in Fig. 3.43, the four states, S_0, S_1, S_2, and S_3, are defined along the strain–temperature hysteresis loop.

Let i_C be the current charged to a target segment, and i_{Cd} be a constant current for heating the segment. The lowest-level temperature control of SMA is performed by ON–OFF switching, i.e., $i_C = i_{Cd}$ for heating up, and simply $i_C = 0$ for cooling down. Fig. 3.45 illustrates this control. States S_0 and S_1 are classified as ON states, and S_2 and S_3 are OFF states. The HLC consists of relatively simple four temperature control modes as shown below:

Mode 1 (Heating up from S_1 to S_2) $i_C = i_{Cd}$ when $T < T_N$. Move to Mode 2 if $T \geq T_N$.

Mode 2 (Keeping ready state S_3) $i_C = \begin{cases} i_C = i_{Cd} & T < T_{NR} \\ i_C = 0 & T \geq T_{NR} \end{cases}$. Hold T_{NR} until an OFF command is given.

Mode 3 (cooling down from S_3 to S_0) $i_C = 0$ when $T > T_F$. Move to Mode 4 if $T < T_F$.

Mode 4 (Keeping ready state S_1) $i_C = \begin{cases} i_C = i_{Cd} & T < T_{FR} \\ i_C = 0 & T \geq T_{FR} \end{cases}$. Hold T_{FR} until an ON command is given.

Note that the transition from ON to OFF can be executed during Mode 1. Similarly, the transition from OFF to ON can be executed during Mode 3. However, these transitions would create minor hysteresis loops with potentially longer transition times. Therefore, the "refraction" rule should be implemented.

3.3.3 Transition probability distribution and hysteresis loop

The transition between ON and OFF in each cell is calculated based on probability density functions that take the current control mode and temperature as inputs [270].

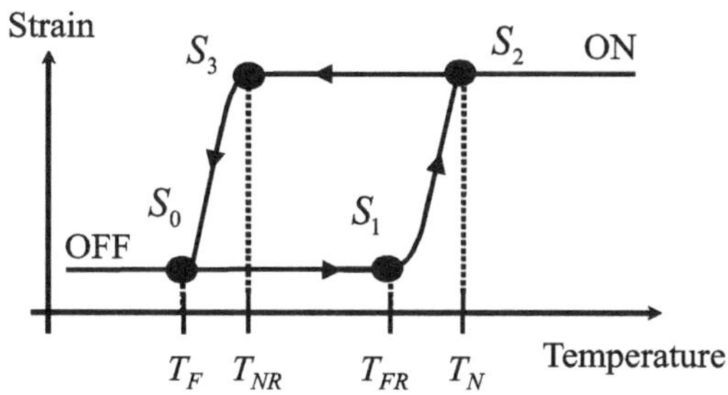

Figure 3.44 Hysteresis loop control of SMA actuator unit.

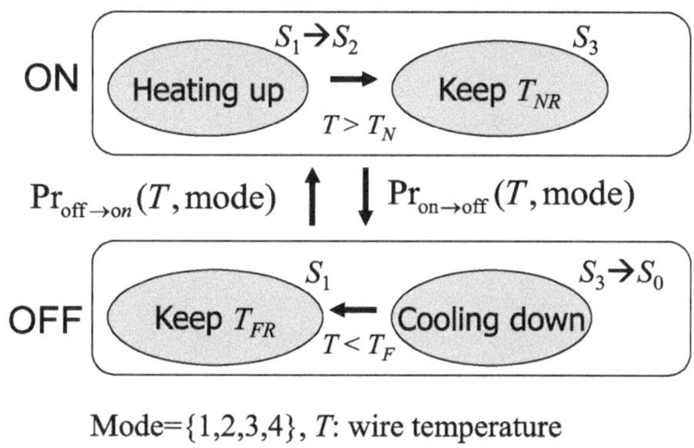

Figure 3.45 State transition of hysteresis loop control.

Let $p = \mathrm{Pr}_{\text{off}\rightarrow\text{on}}$ be the transition probability from OFF to ON. Similarly, let $q = \mathrm{Pr}_{\text{on}\rightarrow\text{off}}$ be the transition probability from ON to OFF. For example, p and q may be given by using a sigmoid function as follows:

$$p(T, \text{Mode}) = \begin{cases} \frac{1}{1-e^{|T-T_F|/g_p}}, & \text{Mode} = 2 \text{ or } 3, \\ \frac{1}{1-e^{-|T-T_F|/g_p}}, & \text{Mode} = 1 \text{ or } 4, \end{cases} \tag{3.54}$$

$$q(T, \text{Mode}) = \begin{cases} \frac{1}{1-e^{-|-T-T_N|/g_q}}, & \text{Mode} = 2 \text{ or } 3, \\ \frac{1}{1-e^{|T-T_N|/g_q}}, & \text{Mode} = 1 \text{ or } 4, \end{cases} \tag{3.55}$$

where g_p and g_q are gains.

The probability distribution functions are mapped along with the hysteresis loop controlled by HLC (see Fig. 3.46). For example, the probability of ON (or keeping ON state) becomes greater proceeding along the hysteresis loop, and takes the largest value at T_N, which enables the recruitment of "ready" cells at S_1. Note that the cells in transition from S_1 to S_2 are already recruited and ON. Therefore these cells do not change their behavior. Similarly, the probability of OFF (or keeping OFF state) becomes larger toward T_F enabling the recruitment of "ready" cells at S_3.

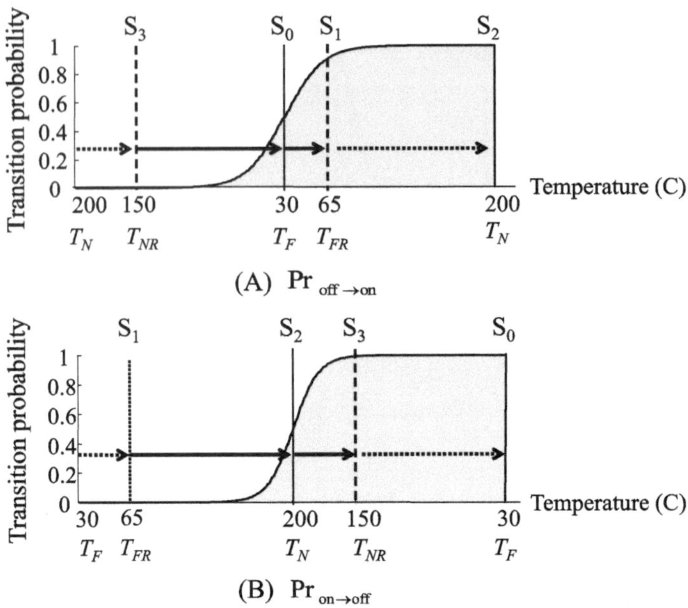

Figure 3.46 Mapping of transition probability profiles on hysteresis loop.

3.3.4 Localized stochastic transition

As described in Section 3.2, each cellular actuator has a decision-making unit which decides whether to accept or ignore the broadcast control command u in a random manner. Since the HLC takes only ON or OFF state, the input to a local HLC is sufficient to take two values. Here $u_C = 0$ is used to command the OFF state and $u_C = 1$ to command the ON state. For example, when the broadcast command is $u > 0$ and the current HLC's state is OFF (therefore the current command is $u_C = 0$), the u_C changes to $u_C = 1$ according to the transition probability $p = \mathrm{Pr}_{\mathrm{off}\to\mathrm{on}}$ when

$$f_{CDF}(g_d|\mathrm{rand}\ u|) \geq 1 - p(T, \mathrm{Mode}). \tag{3.56}$$

rand is a normally distributed random number with mean 0 and variance 1, $g_d > 0$ is a gain and $f_{CDF}(x)$ is the is the cumulative distribution function (CDF) of the normal distribution with mean 0 and variance 1,

$$f_{CDF}(x) = \frac{1}{\sqrt{2\pi}} \int_{-\infty}^{x} e^{-t^2} dt. \tag{3.57}$$

Similarly, the command $u_C = 1$ changes to 0 when

$$f_{CDF}(g_d|\mathrm{rand}\ u|) \geq 1 - q(T, \mathrm{Mode}). \tag{3.58}$$

Each local HLC performs the temperature control described in Section 3.3.2 based on the command $u_C = \{0\ (\text{OFF state}), 1\ (\text{ON state})\}$. The left-hand side of (3.56) gen-

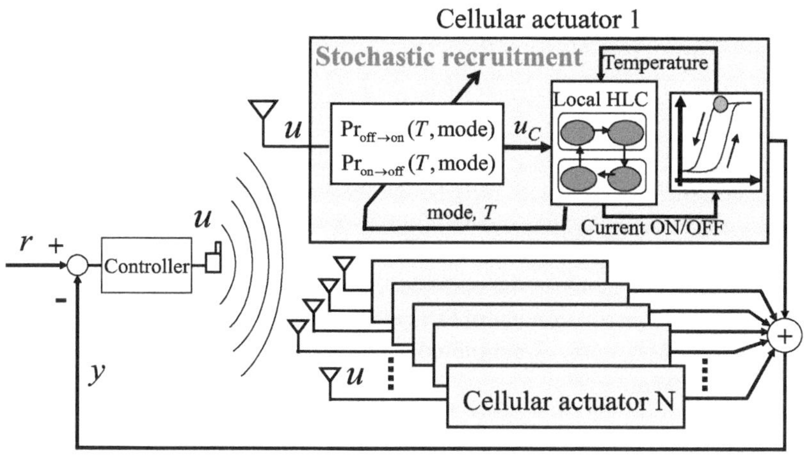

Figure 3.47 Broadcast feedback with localized stochastic recruitment (BFSR) for SMA cellular actuator array. © 2006 IEEE, reprinted with permission

erates a cumulative probability with which the transition probability from OFF state to ON monotonously increases as the magnitude of the u increases. The right-hand side of (3.56) represents the threshold determined based on the current internal state. The threshold decreases as the current state is near the "ready" one. The transition from ON of OFF is described by (3.56) in a similar fashion. In the following simulation, (3.56) and (3.56) are evaluated in every sampling period.

3.3.5 Broadcast control approach to the coordination of hysteric cellular actuator array

A simple P-type controller is applied that broadcast input based on the error between the desired and current displacement:

$$u = k_p(r - y) \tag{3.59}$$

where k_p is a gain. This general control signal includes the magnitude of the error and the direction for error compensation. The control signal is broadcasted to all the cellular units uniformly. Fig. 3.47 shows a diagram of the broadcast feedback control combined with hysteresis loop control named Broadcast Feedback with localized Stochastic Recruitment (BFSR).

3.3.6 Centralized cell coordination

For the sake of performance comparison, three control structures that employ centralized coordination of cellular actuator units are considered. Individual SMA cells are controlled by HLC. Ideally, as implemented in [225], a centralized controller should gather all the information of individual cells, including the temperature and mode of

HLC in operation as well as whether each cell is ready or not. This fully-centralized coordination can recruit appropriate cells to activate or deactivate, and would maximize the performance. However, as a drawback, this architecture requires significant resources, and therefore does not seem practical to apply to a cellular control system with a vast number of actuator units. The centralized cellular coordination considered in this section is assumed to only know which cells are currently ON or OFF, and does not gather information on each individual HLC such as temperature or modes. In other words, the centralized controller does not know which cells are ready or not.

The central controller has an internal variable, N_a, that represents the total number of cells to be activated. The total number of ON cells, or N_a, represents the steady-state length of the current SMA wire. A simple controller updates N_a at every sampling time based on the positioning error

$$\Delta N_a = k_c(r - y) \tag{3.60}$$

where k_c is a feedback gain. When the dynamics of the cellular control system is negligible, the proof of the exponential convergence of N_a that realizes r is straightforward similar to the deterministic analysis in Appendix A.6. The next Section 3.4 provides the proof. One might think that $k_c = 1/l$ is an intuitive choice to realize r in next sampling period. While this gain guarantees convergence, a little more aggressive choice of k_c usually works better due to slow response and hysteresis of SMA actuator units.

The following three structures are considered based on how N_a cells are selected, and ΔN_a cells are newly turned on when $\Delta N_a > 0$ in next sampling period or turned off when $\Delta N_a < 0$.

Centralized binary-scheme recruitment (CBR). Fig. 3.48 shows a diagram of the control structure. In this structure, the N segments are regrouped into $\log_2 N$ (base 2 logarithm of N) groups as shown in Fig. 3.49 by following the binary scheme [225]. The centralized controller orderly recruits N_a cells through a $\log 2\, N$-bit address line. The binary-scheme grouping has eliminated the redundancy in actuation and enables fully deterministic recruitment. N discrete displacements can be specified by minimum $\log 2\, N$-bit addressing.

Centralized sequential recruitment of uniform segments (CSR). By allowing for a full N-bit address line equivalent to the number of cells, centralized coordination can be composed without loosing the redundancy in actuation. Fig. 3.50 shows a diagram of the control structure. In this structure, the recruitment is orderly and sequential as shown in Fig. 3.51 by directly sending a total of N binary commands, $u_{C1}, u_{C2}, \ldots, u_{CN}$, to the cellular system. The same update law in (3.60) is used to generate N_a. The central controller sequentially turns on ΔN_a cells out of inactive cells when $\Delta N_a > 0$, or turns off then $\Delta N_a < 0$.

Centralized random recruitment of uniform segments (CRR). The block diagram of the third structure is shown in Fig. 3.52. The central controller randomly select ΔN_a cells out of inactive cells and turns them for $\Delta N_a > 0$, or turns off for $\Delta N_a < 0$ as shown in Fig. 3.53.

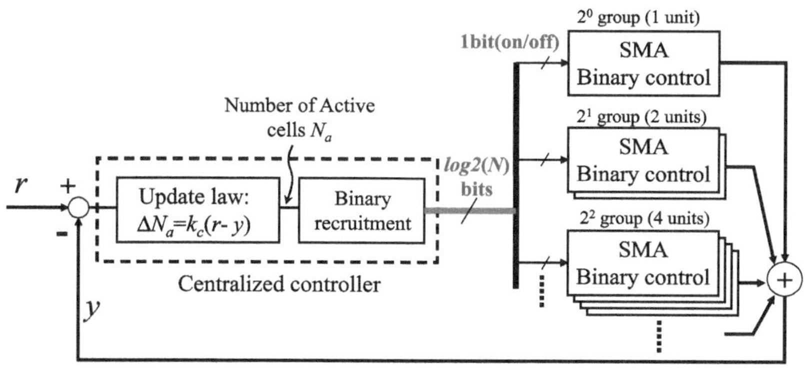

Figure 3.48 Centralized binary-scheme recruitment (CBR). © 2006 IEEE, reprinted with permission

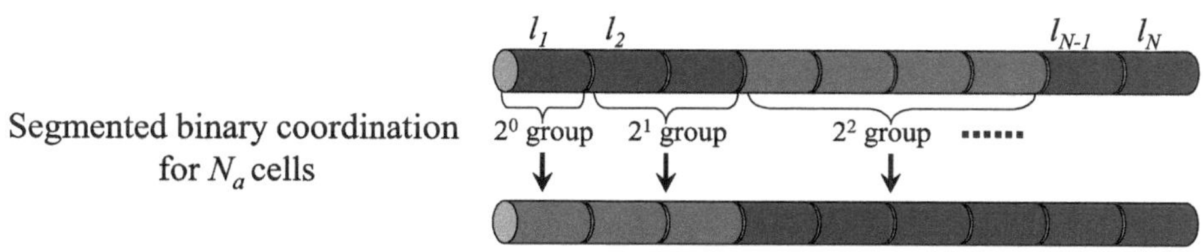

Figure 3.49 Binary recruitment. © 2006 IEEE, reprinted with permission

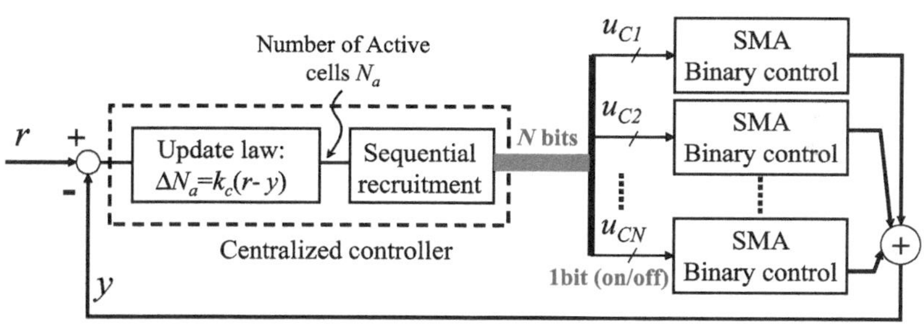

Figure 3.50 Centralized sequential recruitment of uniform segments (CSR).

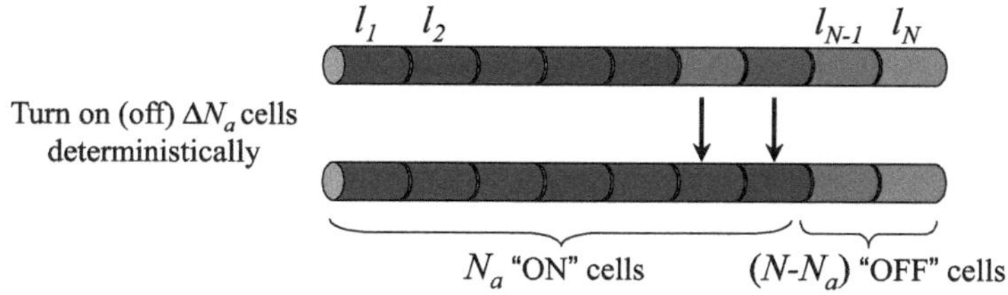

Figure 3.51 Sequential recruitment.

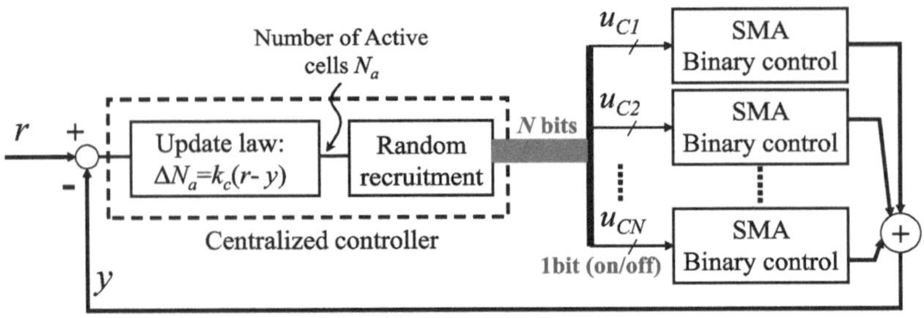

Figure 3.52 Centralized random recruitment of uniform segments (CRR). © 2006 IEEE, reprinted with permission

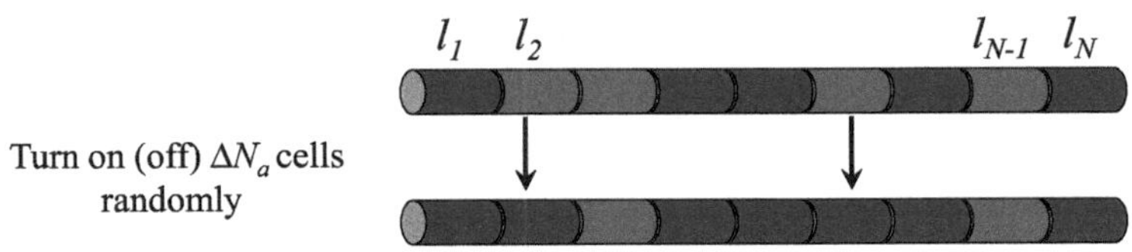

Figure 3.53 Random recruitment © 2006 IEEE, reprinted with permission

3.3.7 Simulation environment

A displacement control of an SMA wire segmented into 100 cells; $N = 100$ is considered. Step response and sinusoidal trajectory tracking are examined. A performance comparison is made between the proposed broadcast control architecture and three other methods with centralized coordination to demonstrate that (i) redundancy and randomness improve performance, and (ii) centralized coordination is unnecessary.

The diameter of the SMA wire is 0.18 [mm] and the length of each segment is 50 [mm]. A mathematical model of SMA proposed by Madill et al. [162] is used. The parameters of SMA regarding mechanical and thermal properties are determined based on [162] accordingly. The preloaded tension of the wire is 250.0 [MPa] and assumed to remain constant during the operation.

The parameters of HLC are given as $T_N = 200$ [°C], $T_{NR} = 150$ [°C], $T_F = 30$ [°C], and $T_{FR} = 60$ [°C]. The current input is set as $i_{Cd} = 1.5$ [A]. In this simulation, no forced-cooling device such as a Peltier device is used. The cooling process takes longer than that of heating. The gain for the P-type broadcast control is $k_p = 1350$. Also for the localized stochastic transition. $g_d = 0.125$, $g_p = 150$, and $g_q = 100$ are used. The simulation model is implemented by using MATLAB/Simulink as shown in Fig. 3.54. The sampling period is set 0.01 [s]. Fig. 3.55 shows a screen shot of the simulator.

The desired displacement r for the step response is given as

$$r(t) = \begin{cases} 0.12, & 0 \le t < 15, \\ 0.065, & 15 \le t < 30, \\ 0.08, & 0 \le t < 60, \end{cases} \tag{3.61}$$

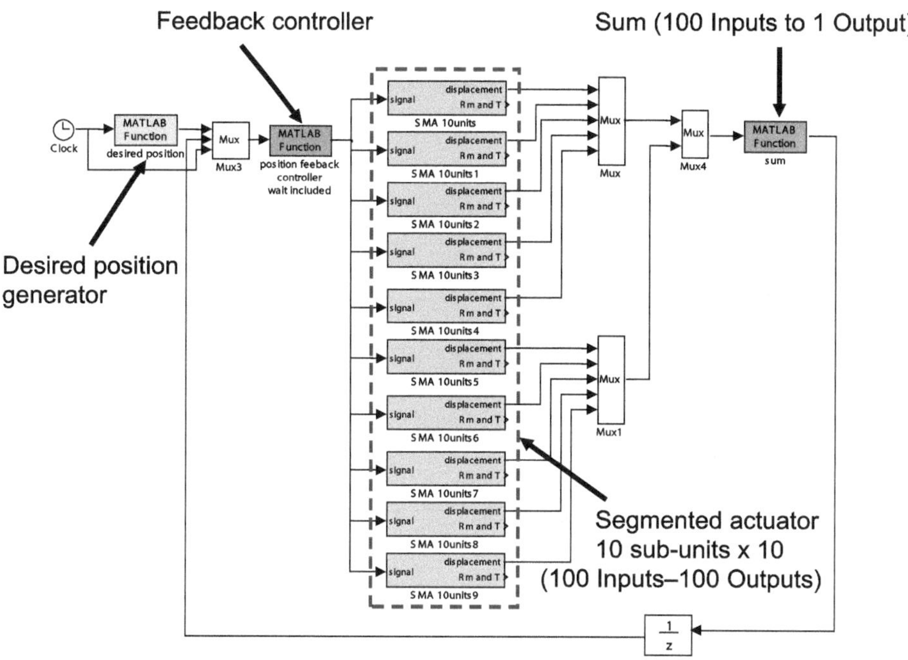

Figure 3.54 Simulink model of SMA array control. © 2006 IEEE, reprinted with permission

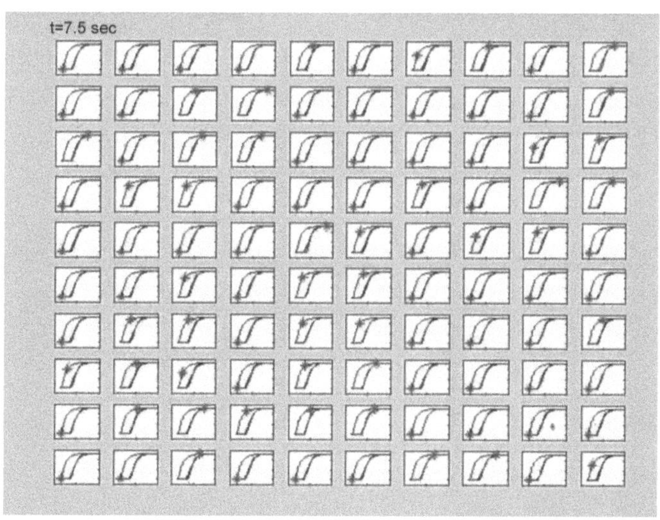

Figure 3.55 Screen shot of simulation.

where t is time. Also, a sinusoidal trajectory of 0.07 [Hz] (0.44 rad/s) for trajectory tracking is given by

$$r(t) = 0.1 - 0.02\sin(0.44t) + 0.06e^{-t/2}. \tag{3.62}$$

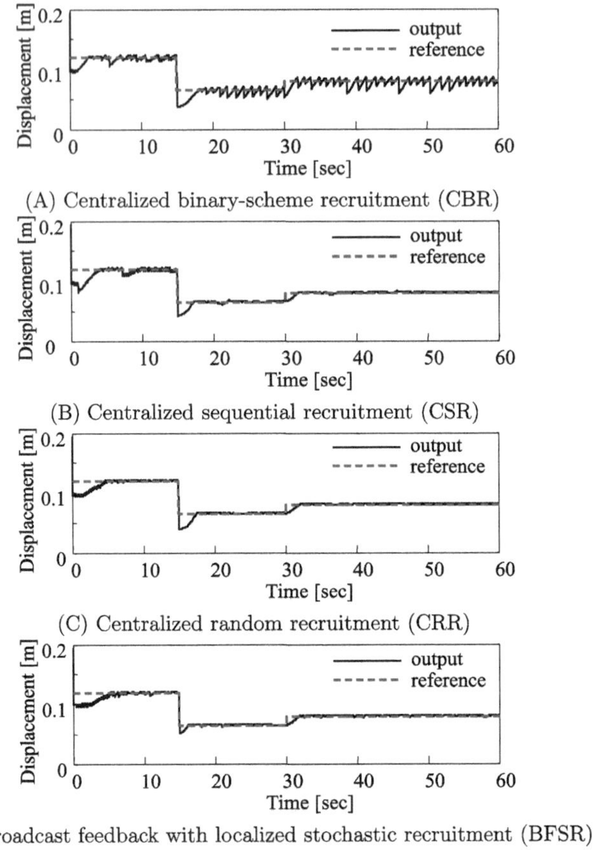

Figure 3.56 Step response. © 2006 IEEE, reprinted with permission

For the centralized control structures, the control gain $k_c = 400$ is used. For the CRBS, a 7-bit address line is used to address 7 groups (1, 2, 4, 8, 16, 32, and 37 cells) since $N = 100$.

3.3.8 Simulation results

Figs. 3.56 and 3.57 show the results of step response and sinusoidal trajectory tracking. Randomly selected $N/2$ cells are set to ON and the rest of them are set to OFF as the initial condition. In each plot, the solid line shows the current displacement and the dashed line shows the reference.

As shown in Fig. 3.56, the settling time of the initial step response error is 2–3 [s], and then the output continues to track the desired position. The responses of the CRR and BFSR are smooth and satisfactory. On the other hand, the CBR exhibits the least smooth profile, followed by the CSR. The same tendency can be observed in Fig. 3.57.

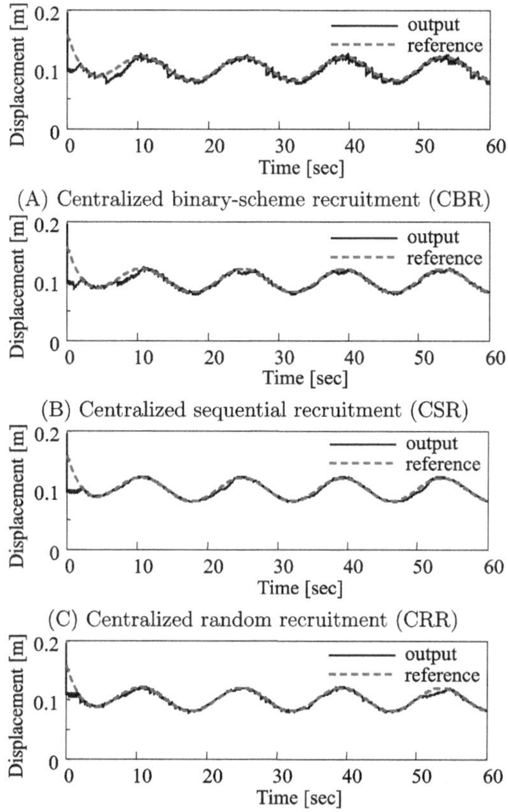

(A) Centralized binary-scheme recruitment (CBR)

(B) Centralized sequential recruitment (CSR)

(C) Centralized random recruitment (CRR)

(D) Broadcast feedback with localized stochastic recruitment (BFSR)

Figure 3.57 Sinusoidal trajectory tracking.

Recall the same update law (3.60) with the same gain k_c are applied to CBR, CSR, and CRR. Therefore, the differences in the resultant trajectories are due to the differences in recruitment.

A comparison between CBR and CSR indicates that the redundancy in actuation can compensate for the slow response of individual cells and improve the performance. In CBR, the performance of precise tracking essentially relies on the speed of response of the specific 2^0 group. On the other hand, in CSR, however, other cells can alternate the role since all cells have the same stroke. A comparison between CSR and CRR also indicates that random recruitment is more effective in activating "ready" cells than sequential recruitment where neighboring cells may have been recruited in recent sampling periods therefore are less likely to be "ready" to make a transition.

The main argument of this section can be found in the comparison between the CRR and BFSR. These two methods achieve almost the same performance. That is, as far as a vast number of redundant actuators are involved, random recruitment can be done locally; no centralized controller is necessary that gathers the information of ON and OFF states of individual cells. Individual cells do not need to communicate to each other. Rather, they can make a transition based on their own transition probability and the broadcast command. The observation is consistent with what is discussed in the previous Section 3.2 even though the responses of individual cells are not negligibly fast but slow and hysteretic.

3.3.9 Discussion

The broadcast feedback approach with localized stochastic recruitment is applicable to a cellular control system with pronounced hysteresis. Each of the cellular actuator units is controlled by HLC. The transition probability density functions of each cell are mapped along with the hysteresis loop to improve its response time. It has been shown through simulation that centralized coordination is unnecessary and local stochastic recruitment is sufficient.

This section does not argue that localized stochastic recruitment is superior to centralized recruitment in terms of control performance. Even though states of individual cells are not available for centralized recruitment, a more sophisticated coordination may be implemented, e.g., specifying a cell in which sufficient time has been passed after the last state transition from its time history stored in the memory. The centralized recruitment approach cannot avoid the trade-off between performance and hardware resources. This problem is essential for the control of a vast number of hysteretic cellular actuators. The cellular actuator concept aims at providing a practical solution by utilizing redundancy and randomness, and also obtaining a comparable performance to the centralized recruitment approach.

In Section 3.4, a rigorous mathematical formulation and analytical proof of stability based on supermartingale theory are given to validate the approach presented in this section.

3.4 Supermartingale theory for broadcast control of distributed hysteretic systems

3.4.1 Anonymous control and stochastic recruitment

One drawback of the coordinated Hysteresis Loop Control (HLC) for reducing latency time is that the central controller needs to know the status of every cell in the collective so that it knows which cells are ready to make a transition immediately [225]. To know the status of each cell and make coordinated control decisions, each cell must report to the central controller whether it is ready at each time step. As the number of cellular units grows to yield high resolution in the actuator output, this local-to-central communication may become complex and impractical.

Though the status of each cell is needed for coordination, the output of any particular cell does not need to be regulated specifically, only the collective system output. Any cell contributes to the collective output in the same way and a particular collective output can be caused by any of the cells being ON as long as the correct number is ON, i.e., the cells contribute to the collective equivalently and thus anonymously.

Anonymous contribution means that a central controller does not really need to command each cell specifically, but can instead issue a global command so that a certain fraction of cells may change output. The challenge is to use global commands that are brief, requiring little bandwidth, and that require very little interpretation/computation at the cells. An approach that the authors' group has previously proposed is to design the cells to behave stochastically according to a signal that is broadcasted to all cells [263,273,272]. Each cell makes a stochastic decision to change its output with probability p, from OFF to ON, and probability q, from ON to OFF; see Fig. 3.27. In broadcast control, these state transition probabilities are modulated by the central controller. If it is desired that more cells turn to ON, the central controller broadcasts a higher value of probability p, in order to recruit more cells to turn to ON, and vice versa. This stochastic recruitment control is totally anonymous and broad in nature, thus meeting our requirements. However, previous results were derived only for the case of simple ON–OFF units having no dynamics and hysteresis.

This section presents stochastic broadcast control of hysteretic cellular systems that can effectively eliminate the latency time associated with the hysteresis of the system. Combining the refraction rule and the stochastic broadcast control, the multitude of cellular units can be coordinated to improve speed of response, which is infeasible for a single bulk process. In the following, a mathematical process model and control law will be given, followed by a stability analysis and simulation experiment.

3.4.2 System representation

3.4.2.1 Local behaviors of hysteresis loop controllers

Each cellular unit has a local controller for regulating its state within the hysteresis loop; see Fig. 3.58. Each local controller makes a decision stochastically based on the state transition probabilities, p and q, broadcasted by the central controller. In Hysteresis Loop Control, the state of each unit is moved toward a hold state (S_1 or S_3) after each transition, so that its next transition can be made promptly without traversing the hysteresis band. Suppose that a unit is held at S_1. When the local controller has decided to transition, it first moves to S_2, changing its output from OFF to ON, and then continues to move to the hold state S_3. The former transition is called an output transition and the latter is called a loading transition, since it loads the cell for its next transition.

Let t_{12} be the time for a forward output transition from S_1 to S_2 and t_{23} be the time for a loading transition from S_2 to S_3, while t_{30} and t_{01} are transition times from S_3 to S_0 and S_0 to S_1, respectively. We assume that each local controller does not alter its decision to transition after the decision has been made. That is, once a controller decides to leave S_1, for example, it always reaches S_2 and at that time the local output changes from OFF to ON. The central controller, too, does not change its broadcast signal while the individual units are in transition either from S_1 to S_2 or from S_3 to S_0.

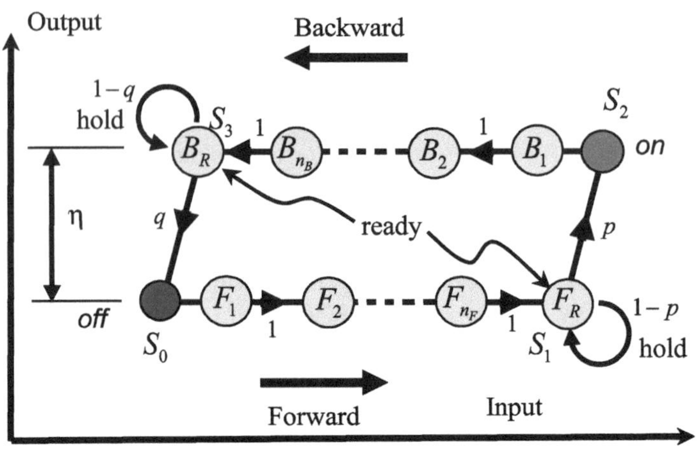

Figure 3.58 Markov chain representation of single cell local control system [289,290].

The central controller monitors the collective outputs in terms of aggregate value y_t, and updates its broadcast signals, p and q, after those units have arrived at a new state, S_2 or S_0. Therefore, the sampling interval of the central controller is set to

$$t_s \geq \max\left[t_{12}, t_{30}\right]. \tag{3.63}$$

It is interesting to note that, although the sampling rate determined by the above condition tends to be low, the overall broadcast feedback control system is proved to be stable, as will be addressed in the later section.

The loading transition times, t_{01} for forward and t_{23} for backward transitions, are discretized as multiples of sampling interval:

$$(n_F - 1)t_s < t_{01} \leq n_F t_s, \quad n_F = 1, 2, \ldots, \tag{3.64}$$

$$(n_B - 1)t_s < t_{23} \leq n_B t_s, \quad n_B = 1, 2, \ldots. \tag{3.65}$$

If the loading times are nonzero but are not longer than the sampling interval, $0 < t_{01} \leq t_s$, $0 < t_{23} \leq t_s$, integers n_F and n_B are set to 1, namely, $n_F = 1$ and $n_B = 1$. If each unit has no hysteresis, loading times become zero, $t_{01} = 0$ and $t_{23} = 0$, and all units move directly to ready states after output transitions. For this special case we set $n_F = 0$ and $n_B = 0$. Otherwise, n_F-steps of intermediate state transitions are involved in forward loading and n_B-steps are involved in backward loading.

Fig. 3.58 summarizes the behavior of individual cellular units with a state transition network consisting of intermediate states $F_1, \ldots, F_{n_F}$ and $B_1, \ldots, B_{n_B}$, and ready states F_R and B_R. Loading transitions are deterministic, as indicated by probability 1 at each arc. Output transitions are allowed only from the ready states (a refractory rule), and the transition probabilities p and q may vary as specified by the central controller. Thus each local control system is a non-homogeneous Markov chain.

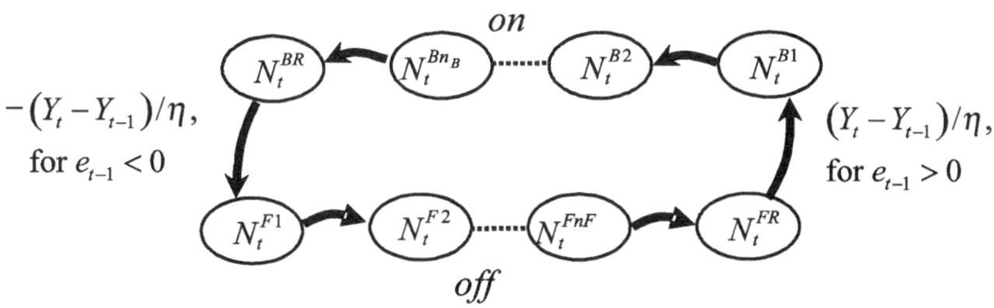

Figure 3.59 Distribution of cells and their transitions [290].

3.4.3 Aggregate state, internal dynamics, and observability

Controlling the collective output of a system of cells is of interest. The states and outputs of the individual cells are inconsequential. Since we can only measure the global output, y_t, it cannot be expected that we are able to infer in which node of the Markov chain in Fig. 3.58 each individual cell exists. Since each cell contributes anonymously and in the same way as any other cell would, the most we can tell is the number of cells that are ON and OFF. Since the number of ON and OFF cells determines the aggregate output, the number is all that we wish to regulate and we have no interest in the outputs of particular cells.

Given a control, p and q, the probability distribution over the aggregate number of ON and OFF cells during the next time step can be completely described in terms of the numbers of forward and backward units ready for changing outputs, N_t^{FR} and N_t^{BR}, respectively. Likewise, the number of cells in the ready node F_R during the next time step, N_{t+1}^{FR}, can be described by the number of cells in the F_{n_F} node, $N_t^{F_{nF}}$, which is then determined by the number of cells in the F_{n_F-1} node, $N_t^{F_{nF}-1}$, and so on. It is clear then that all of the information required to describe the future behavior of the collective output is given by the number of cells in each node. Thus, the aggregate system state can be defined as

$$\boldsymbol{X}_t = \left[N_t^{F_1} \ldots N_t^{F_{nF}}\ N_t^{FR}\ N_t^{B_1} \ldots N_t^{B_{nB}}\ N_t^{BR}\right]. \tag{3.66}$$

Note that the elements in vector $\boldsymbol{X}_t$ sum to N:

$$N = N_t^{FR} + N_t^{BR} + \sum_{i=1}^{n_F} N_t^{F_i} + \sum_{i=1}^{n_B} N_t^{B_i}. \tag{3.67}$$

If the total number of the cellular units, N, is known, one of the elements in vector $\boldsymbol{X}_t$ can be derived from the others, hence it can be eliminated from the state vector. Counting the number of cells in the same node, the distribution of the N cells over the $(n_F + n_B + 2)$ nodes can be obtained as shown in Fig. 3.59.

Defining a state space as a collection of all possible values of the aggregate state vector in (3.66):

$$State = \Big\{ \boldsymbol{X} \,|\, 0 \le N^{FR} \le N, 0 \le N^{BR} \le N, 0 \le N^{F_i} \le N,$$
$$0 \le N^{B_i} \le N, 1 \le i \le n_F, 1 \le j \le n_B \Big\} \quad (3.68)$$

and we can treat the aggregate system as a vector Markov process in S

$$\mathrm{P}_r \left(\boldsymbol{X}_{t+1} | \boldsymbol{X}_t, \boldsymbol{X}_{t-1}, \ldots \right) = \mathrm{P}_r \left(\boldsymbol{X}_{t+1} | \boldsymbol{X}_t \right). \quad (3.69)$$

Note that the N cellular units are independent, identical Markov processes. Therefore, this vector Markov process is a type of Markov population model with time-varying transition probabilities.

While (3.66) contains all of the information required for describing the future behavior of the system, it is not possible to infer the number of cells in each node based on output observation, y_t, alone. In other words, the aggregate system state cannot fully be determined by observations of aggregate output, that is, the system is unobservable. To see this, consider the case where a certain number of cells transition from F_R to B_1 and simultaneously the same number of cells transition back from B_R to F_1. There is no change in the output but the system state clearly changed. This is one type of internal dynamics that is not visible from the aggregate output.

Though the output is not enough to determine aggregate system state if transitions are allowed in both directions at the same time, it is possible to make the system deterministically observable if we limit allowable transitions. If a unilateral control strategy, (3.30) and (3.31), from [263] and also presented in Section 3.2, is used then the system becomes observable. That is, the change in output $y_t - t_{t-1}/\eta$ completely describes the number of cells that transitioned from F_{n_F} to B_1 or from B_{n_B} to F_1. Namely, for e_{t-1} we can write

$$\begin{aligned}
N_t^{F_1} &= 0, \\
N_t^{F_2} &= N_{t-1}^{F_1}, \\
&\vdots \\
N_t^{F_{n_F}} &= N_{t-(n_F-1)}^{F_1}, \\
N_t^{FR} &= N - y_t/\eta - \sum_{i=1}^{n_F} N_t^{F_i},
\end{aligned} \quad (3.70)$$

and

$$\begin{aligned} N_t^{B_1} &= (y_t - y_{t-1})/\eta, \\ N_t^{B_2} &= N_{t-1}^{B_1}, \\ &\vdots \\ N_t^{B_{n_B}} &= N_{t-(n_B-1)}^{B_1}, \\ N_t^{BR} &= y_t/\eta - \sum_{i=1}^{n_B} N_t^{B_i}. \end{aligned} \tag{3.71}$$

A similar set of expressions can be written for $e_{t-1} < 0$. Therefore, all the components of the aggregate state vector $\boldsymbol{X}_t$ can be constructed from the previous $\tau = \max(n_B, n_F)$ time steps of collective output observations

$$\mathbf{Z}_t = \left[y_t \dots y_{t-\tau}\right]^T. \tag{3.72}$$

Note that if initial conditions are unknown, $\boldsymbol{X}_t$ can be fully determined after $t \geq \tau$. This implies that the aggregate system is observable under the unilateral transition constraint. The unilateral control assumption can be well justified for engineering applications like SMA actuators because it is wasteful to command any cells opposite to the direction of the desired output.

3.4.4 Control

3.4.4.1 Stochastic control law

The objective of this section is to modify the stochastic control law described in Section 3.2 in terms of transition probabilities, p and q, to be broadcasted by the central controller, that will bring the error $e_t = r - y_t$ to zero in a stochastic sense, given a constant reference input r. Various stability and convergence criteria can be applied to this problem [144]. For example, if the probability of not reaching the reference by time $t_f > 0$, i.e., $\Pr(e_t \neq 0, \forall t \leq t_f)$, approaches zero as t_f tends to infinity, it represents a type of stability. For non-hysteric, cellular stochastic control systems it has been proven that this condition can be achieved with a broad class of control laws [185]. As long as a control law allows us to hit the target with a probability that is lower bounded by a uniform limit, namely, $\Pr(e_{t+1} = 0 | e_t; p, q) \geq \varepsilon > 0, \forall e_t, \forall t$, then $\Pr(e_t \neq 0, \forall t \leq t_f) \leq (1 - \varepsilon)^{t_f} \to 0$. Many control laws can, in fact, satisfy this stability condition [263,185]. Depending on specific control laws, however, the convergence process will be different.

Fig. 3.60 illustrates the distribution of next-step error $\Pr(e_{t+1} | \boldsymbol{X}_t)$ at each time slice. It is desired that this distribution of error may be reduced quickly, as illustrated by the envelope in the figure. In this section, the mean squared error $E[e_{t+1}^2 | \boldsymbol{X}_t; p, q]$ as a metric of this error distribution is considered, and a feedback control law based on a stability criterion using the second-order moment is derived [144]. In the case

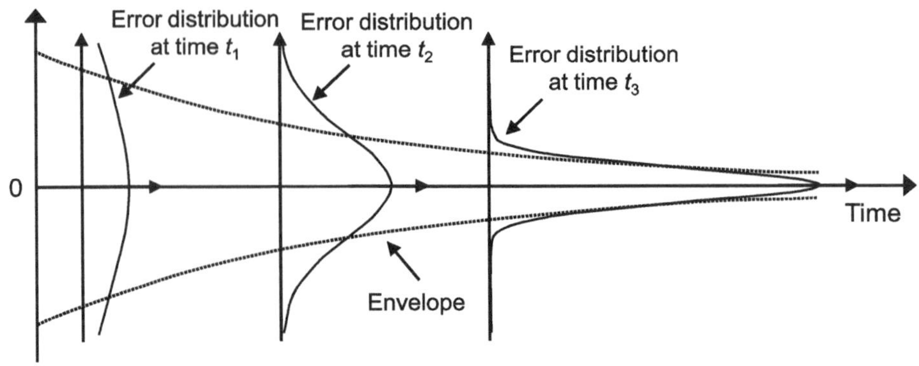

Figure 3.60 Conceptual diagram of converging error distribution.

of general hysteric systems, the behavior is complex, and it is difficult to guarantee that the mean squared error monotonically reduces at every time step. Exploiting the preloading scheme and the refraction rule, however, we can guarantee smooth and effective convergence in a straightforward manner. The specific control law will be derived below.

Given the current state $\boldsymbol{X}_t$ which is derived from the observation $\boldsymbol{Z}_t$, the expected next-step squared error can be written as

$$\begin{aligned} J(p, q; \boldsymbol{X_t}) &\triangleq E[e_{t+1}^2 | \boldsymbol{X}_t] \\ &= \mathrm{Var}[y_{t+1} | \boldsymbol{X}_t] + (E[y_{t+1} | \boldsymbol{X}_t] - r)^2 \end{aligned} \tag{3.73}$$

where broadcast probabilities p and q are used by all of the cellular units at time t. A control law can be obtained by finding p and q that reduce the next-step squared error as much as possible. Such a control law is merely a local, incremental control law. Nonetheless, useful global properties, e.g., supermartingale convergence and decaying rate, can be derived from the incremental expected squared error, as will be discussed later in this section.

The mean squared error is broken down to the output variance term and the output offset term, as shown in the second line of Eq. (3.73). Both terms can be computed explicitly as functions of p and q. Without loss of generality, we can assume $e_t > 0$ and set $q = 0$, which yields

$$\begin{aligned} E[y_{t+1} | \boldsymbol{X}_t] &= E[y_{t+1}^1 + \cdots + y_{t+1}^N | \boldsymbol{X}_t] \\ &= y_t + \eta p N_t^{FR}, \end{aligned} \tag{3.74}$$

where the second line follows from the fact that only cells in F_R can transition. Likewise, the variance becomes

$$\begin{aligned} \mathrm{Var}[y_{t+1} | \boldsymbol{X}_t] &= \mathrm{Var}[y_{t+1}^1 + \cdots + y_{t+1}^N | \boldsymbol{X}_t] \\ &= \eta^2 (p - p^2) N_t^{FR}, \end{aligned} \tag{3.75}$$

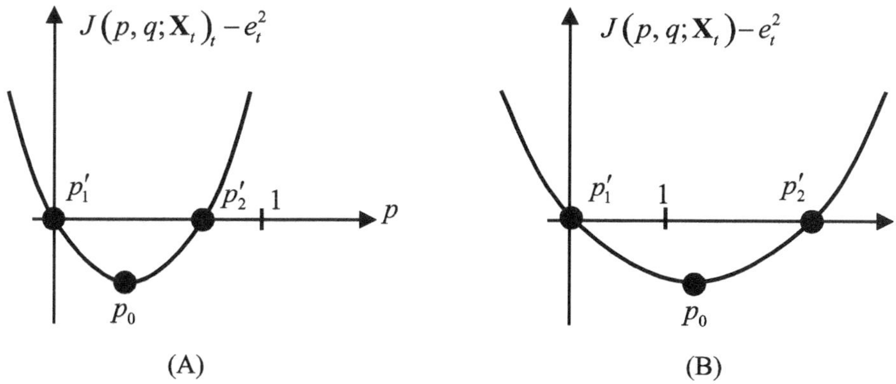

Figure 3.61 Sketch of $J(p, 0; X_t) - e_t^2$ when $N_t^{FR} > 1$ and: (A) $N_t^{FR} \leq e_t/\eta$; (B) $N_t^{FR} > e_t/\eta$.

where the second line follows from the binomial distribution of the cellular units to transition. A similar set of expressions can be written for $e_t < 0$.

Substituting the mean and variance of the next-step output y_t given by (3.74) and (3.75) into (3.73) yields

$$J(p, 0; X_t) = \eta^2 N_t^{FR}(N_t^{FR} - 1)p^2 + \eta N_t^{FR}(\eta - 2e_t)p + e_t^2. \tag{3.76}$$

The main objective here is to minimize J in (3.76) subject to the range of meaningful probabilities $0 \leq p \leq 1$. Eq. (3.76) is a convex parabola if $N_t^{FR} > 1$ and a line for $N_t^{FR} = 0, 1$. The minimum of the parabola can be found by setting $dJ(p, 0; X_t)/dp = 0$, which yields

$$p_0 = \frac{2e_t - \eta}{2\eta(N_t^{FR} - 1)}. \tag{3.77}$$

Fig. 3.61A shows a representative sketch of $J(p, 0; X_t) - e_t^2$ for $N_t^{FR} > 1$ when the minimum of the parabola is at $p_0 < 1$ while Fig. 3.61B shows $J(p, 0; X_t) - e_t^2$ when $p_0 > 1$. The expression $J(p, 0; X_t) - e_t^2$ represents a performance metric. It is the reduction of the expected squared error that can be achieved during the next time step. The parabolic curve intersects with the horizontal axis at p'_1 and p'_2. It is always possible to attain zero expected improvement in the squared error, by choosing a control $p = p'_1 = 0$, or, if $p'_2 \leq 1$, by choosing $p = p'_2$.

Let p_{min} be the value of $0 \leq p \leq 1$ that minimizes $J(p, 0; X_t)$. Given that $N_t^{FR} > 1$, from Fig. 3.61A and $e_t = r - y_t$, when $1 \leq N_t^{FR} \leq e_t/\eta$ and $e_t/\eta > 1/2$, $p_{min} = p_0$. For $1 \leq N_t^{FR} \leq e_t/\eta$, all available units should be turned on so $p_{min} = 1$. If $N_t^{FR} = 0$, then the broadcast does not matter and no units can activate. Thus, the complete control law for $e_t > 0$ is

$$p_{\min} = \begin{cases} \frac{2e_t - \eta}{2\eta(N_t^{FR} - 1)}, & N_t^{FR} > 1 \text{ and } \frac{1}{2} \le e_t/\eta < N_t^{FR}, \\ 1, & 1 \le N_t^{FR} \le e_t/\eta \text{ and } e_t/\eta \le \frac{1}{2}, \\ 0 \le p_{min} \le 1, & N_t^{FR} = 0, \\ 0, & \text{otherwise.} \end{cases} \tag{3.78}$$

A similar law for $e_t < 0$ immediately follows.

A few useful insights can be obtained from the above control law that minimizes the expected squared error (3.73). First, the solution is a trade-off between the output variance $\mathrm{Var}[y_{t+1}|\boldsymbol{X}_t]$ and the squared output offset $(E[y_{t+1}|\boldsymbol{X}_t] - r)^2$ expected for the next step. Choosing p that brings the mean offset to zero does not necessarily reduce the expected squared error most effectively in the current step. It is clear from the variance $\mathrm{Var}[y_{t+1}|\boldsymbol{X}_t] = \eta^2(p - p^2)N_t^{FR}$ that, as p gets closer to either 0 or 1, the next step output becomes more predictable. For example, if the number of ready units, N_t^{FR}, is the same as the error e_t/η, the system can reach the reference point deterministically in one step. The above control law tends to pick a value for p that produces more predictable, low variance output for the next step, which is better in reducing the expected squared error.

The one-step convergence scenario, i.e., the transition with $p = 1$ for $e_t = \eta N_t^{FR}$, also implies that a small error e_t does not necessarily mean that the system is close to convergence. If e_t is approximately one-half the available ready cells, $e_t \approx \eta N_t^{FR}/2$, the variance in the next step output is highest, meaning that the target reference is more likely missed. This intricate behavior requires rigorous analysis of convergence and decaying rate, which are the topics of the following sections.

3.4.4.2 Convergence

The objective of this section is to guarantee that the error e_t converges with probability 1 despite the intricate behavior due to hysteresis. Further, we want to analyze the decay rate of convergence. This can be addressed based on the expected squared error used in the above control law derivation. Consider

$$E[e_{t+1}^2|\boldsymbol{X}_t] \le \bar{\alpha} e_t^2. \tag{3.79}$$

If there exists a parameter $\bar{\alpha}$ in $0 \le \bar{\alpha} < 1$ that satisfies (3.79) for all $t \ge 0$, the process is a non-negative supermartingale, and provides an exponential expected decay rate [185]. Unfortunately, in the case of hysteretic cells, there are periods when no cells are ready to transition, $N_t^{FR} = 0$ or $N_t^{BR} = 0$, during which no decay is expected, $E[e_{t+1}^2|\boldsymbol{X}_t] = e_t^2$, or $\bar{\alpha}$. This invariant situation does not continue forever, since preloading is completed within $\tau = \max(n_F, n_B)$ time steps.

Lemma 2. *If $e_t > 0$, then expected squared error decreases, $E[e_{t+1}^2|\boldsymbol{X}_t] < e_t^2$, at least once in every $\tau = \max(n_F, n_B)$ time steps and, at that time, there is a uniform bound $\tilde{\alpha}$ such that, if $e_t \neq 0$,*

$$\tilde{\alpha} \leq \frac{E[e_{t+1}^2|\boldsymbol{X}_t]}{e_t^2} < 1, \tag{3.80}$$

where $\tilde{\alpha}$ does not depend on t.

A detailed proof is shown in Appendix A.7.

Instead of evaluating the decay rate in every time step, using the decay rate of τ time steps yields the following convergence theorem.

Theorem 3. *The stochastic broadcast control system consisting of N hysteric cellular units given by the Markov chain in Fig. 3.60 and a central controller that monitors aggregate output y_t and that broadcasts transition probabilities, p_{min} and q_{min}, based on the control law of (3.78), converges to a constant reference point $r \in [0, \eta N]$ with probability 1:*

$$e_t \to 1, \quad w.p.\ 1, \tag{3.81}$$

with the following exponential convergence properties:

$$E[e_t^2|\boldsymbol{X}_0] \leq \tilde{\alpha}^m e_0^2, \tag{3.82}$$

$$Pr\left(\sup_{\infty > i \geq t} e_i^2 \geq \mu\right) \leq \frac{\tilde{\alpha}^m}{\eta} e_0^2 \tag{3.83}$$

where $0 \leq \tilde{\alpha} \leq 1$, $\boldsymbol{X}_0 = x$ is an arbitrary initial state, and m is the floor $m = \lfloor t/\tau \rfloor$ with $\tau = \max(n_F, n_B)$. μ determines the open set $Q_\mu = \{e_t : e_t^2 < \mu\}$.

Proof. The control law (3.78) does not increase the mean squared error, i.e., $E[e_{t+1}^2|\boldsymbol{X}_t] \leq e_t^2$, for each e_t^2 in the sample space and in every time step. Taking expectation conditioned on $\boldsymbol{X}_0$ yields $E[e_{t+1}^2|\boldsymbol{X}_0] \leq E[e_t^2|\boldsymbol{X}_0]$. For the first τ time steps, we can write $e_0^2 \geq E[e_1^2|\boldsymbol{X}_0] \geq \cdots \geq E[e_\tau^2|\boldsymbol{X}_0]$. From Lemma 2, it follows that during this period there exists at least one time step $t' \in (0, \tau)$ such that $E[e_{t'}^2|\boldsymbol{X}_0] \leq \tilde{\alpha}' e_0^2$ with $\tilde{\alpha}'$ less than 1. Therefore, we can bound $E[e_t^2|\boldsymbol{X}_0] \leq \tilde{\alpha} e_0^2$ with $0 \leq \tilde{\alpha} < 1$ for all $t \in (\tau, 2\tau)$. For $(m-1)\tau \leq t \leq m\tau$, we can repeat this process m times since the fractional bound α is uniform in time. This yields (3.82).

From the above argument, $E[e_\tau^2|\boldsymbol{X}_0] \leq \tilde{\alpha} e_0^2$. Rewriting this yields $e_0^2 - E[e_\tau^2|\boldsymbol{X}_0] \geq (1-\tilde{\alpha})e_0^2$. Similarly for the next τ time steps, $E[e_\tau^2|\boldsymbol{X}_0] - E[e_{2\tau}^2|\boldsymbol{X}_0] \geq (1-\tilde{\alpha})E[e_\tau^2|\boldsymbol{X}_0]$. Repeating this process recursively for the following $(m-2)$ periods we can obtain a total of $m-1$ inequalities. Adding the left-hand sides of these $(m-1)$ inequalities yields $e_0^2 - E[e_{m\tau}^2|\boldsymbol{X}_0]$, which is larger or equal to the sum of the right-hand sides:

$$(1-\tilde{\alpha})\sum_{k=0}^{m-1} E[e_{k\tau}^2|\boldsymbol{X}_0] \le e_0^2 - E[e_{m\tau}^2|\boldsymbol{X}_0] \le e_0^2. \tag{3.84}$$

Since $1-\tilde{\alpha} > 0$,

$$\lim_{m\to\infty}\sum_{k=0}^{m-1} E[e_{k\tau}^2|\boldsymbol{X}_0] \le \frac{e_0^2}{1-\tilde{\alpha}} < \infty. \tag{3.85}$$

Namely, $\sum_{k=0}^{\infty} E[e_{k\tau}^2|\boldsymbol{X}_0]$ is bounded. Applying the Borel–Cantelli lemma [62], to this yields

$$E[e_{m\tau}^2|\boldsymbol{X}_0] \to 0, m \to \infty. \tag{3.86}$$

Since $E[e_{m\tau}^2|\boldsymbol{X}_0] \ge E[e_t^2|\boldsymbol{X}_0]$ for $\forall t \in [m\tau, (m+1)\tau]$,

$$E[e_t^2|\boldsymbol{X}_0] \to 0. \tag{3.87}$$

This implies (3.81); as t tends to infinity, e_t converges to 0 with probability 1. Furthermore, from the supermartingale inequality and the Chebyshev inequality [62] it follows that

$$\Pr\left(\sup_{\infty>i\ge t} e_i^2 \ge \mu\right) \le \frac{E[e_t^2|\boldsymbol{X}_0]}{\mu} \le \frac{\tilde{\alpha}^m}{\mu}e_0^2. \tag{3.88}$$

Thus, the theorem holds and the convergence properties are guaranteed. □

3.4.4.3 *Performance characterization*

While Theorem 3 is useful for proving convergence, it gives a very conservative bound on convergence rate, since in many cases, the fractional convergence will be much smaller than $\tilde{\alpha}$, and will be so much more frequently than once every τ time steps. A more useful notion of convergence rate comes from considering a fractional convergence *in the mean* at each time step, $\hat{\alpha}$, which leads to the following corollary.

Corollary 2 (Convergence Rate in the Mean). *Let the conditions of Theorem* 3 *hold. Then,* $0 \le \exists\hat{\alpha}_t \le 1$ *such that*

$$E[e_{t+1}^2|\boldsymbol{X}_0] \le \hat{\alpha}_t E[e_t^2|\boldsymbol{X}_0], \tag{3.89}$$

$$E[e_t^2|\boldsymbol{X}_0] \le \prod_{i=0}^{t-1}\hat{\alpha}_i e_0^2, \tag{3.90}$$

$$Pr\left(\sup_{\infty>i\ge t} e_i^2 \ge \mu\right) \le \frac{E[e_t^2|\boldsymbol{X}_0]}{\mu} \le \frac{\prod_{i=0}^{t}\hat{\alpha}_i}{\mu}e_0^2. \tag{3.91}$$

Proof. It is obvious from the proof of Theorem 3. Consider the incremental fractional convergence in the mean defined by

$$\hat{\alpha}_t^{min} = \frac{E[e_{t+1}^2|\boldsymbol{X}_0]}{E[e_t^2|\boldsymbol{X}_0]}, \tag{3.92}$$

where $E[e_t^2|\boldsymbol{X}_0] > 0$ and transition probabilities p_{min} and q_{min}, which minimize (3.73), are used. This incremental rate $\hat{\alpha}_t^{min}$ can be computed by evaluating the ensemble means $E[e_{t+1}^2|\boldsymbol{X}_0]$ and $E[e_t^2|\boldsymbol{X}_0]$ for each time t as

$$E[e_t^2|\boldsymbol{X}_0] = \sum_{State} e^2(\boldsymbol{X}_t)\Pr(\boldsymbol{X}_t|\boldsymbol{X}_0) \tag{3.93}$$

where $State$ is the state space defined in (3.68), $e^2(\boldsymbol{X}_t)$ is the squared error as a function of $\boldsymbol{X}_t$. The probability $\Pr(\boldsymbol{X}_t|\boldsymbol{X}_0)$ can be computed recursively starting from $t = 1$ (see Appendix A.8). □

Corollary 3 (Wide range stability). *Assume that all the conditions of Theorem 3 except the transition probabilities are held, and that in each time step the central controller broadcasts transition probabilities, p and q, within the following ranges*

$$\begin{aligned} &0 < p_1 \le p \le p_2 < \min[p_2', 1], q = 0; \quad e_t > 0, \\ &0 < q_1 \le q \le q_2 < \min[q_2', 1], p = 0; \quad e_t < 0, \\ &p = 0, q = 0; \quad e_t = 0, \end{aligned} \tag{3.94}$$

where

$$p_2' = \frac{2e_t - \eta}{\eta(N_t^{FR} - 1)}, \quad q_2' = \frac{-2e_t - \eta}{\eta(N_t^{BR} - 1)} \tag{3.95}$$

then the system converges with probability 1, and the same convergence properties as in the original Theorem 3 hold.

Proof. From Fig. 3.61 it is clear that the transition probabilities of (3.94) make $E[e_{t+1}^2|\boldsymbol{X}_t] - e_t^2$ negative. Therefore, from Lemma 2, the mean squared error decreases at least once in every $\tau = \max(n_F, n_B)$ time steps. The rest of the proof is the same as that of Theorem 3, and the corollary is proved. □

Corollary 3 guarantees that the cellular stochastic control is stable despite the lack of p_{min} and q_{min}, which may often occur in practical applications. Consider, for example, the total number of cells is not exactly known, or some fraction of the cells is not functioning, e.g., unable to transition. The cellular stochastic control system can still converge if the approximate probabilities are within the range guaranteed by Corollary 3. As will be discussed later, this cellular stochastic control is remarkably robust against cell failures. The above Corollary 3 supports the robustness of the cellular system.

3.4.5 Simulation

Basic properties and performance of the cellular stochastic control presented in this section are now evaluated through simulation. The section confirms first in a deterministic setting that the preloading scheme and refraction rule that constitute HLC can indeed improve the speed of response, and then demonstrate that the stochastic control can work nearly as well as the deterministic control. The incremental fractional convergence in the mean defined above will also be evaluated in order to examine convergence properties.

3.4.5.1 Deterministic control

Fig. 3.62 compares the deterministic HLC to the one without preloading and refractory control. The work in [263] (also shown in Section 3.2) briefly demonstrated that, if segmented SMA cells are individually preloaded and selected for immediate output transition, HLC does increase actuator bandwidth. To illustrate how effective HLC is, we consider a deterministic controller communicating with each cell individually, explicitly knowing the node of each cell and commanding each cell separately. Here, we simulate the behavior of $N = 100$ serially connected units with intermediate states such that $n_B = 4$ and $n_F = 2$. This could, for example, reflect the fact that Joule heated SMA heats very quickly, but requires longer to cool, meaning that more nodes are required in the cooling direction. For simplicity, all simulations take $\eta = 1$.

Our deterministic controller is algebraic, i.e., it computes how many cells need to be turned ON or OFF to reduce the current error to zero based only on the current error. Each cell is then commanded to transition accordingly. The deterministic controller without refractory HLC simply leaves a cell at S_0 or S_2 after its output has transitioned to ON or OFF, respectively, while the deterministic refractory HLC controller preloads the cell to S_1 or S_3. Fig. 3.62 shows both controllers responding to a series of step inputs, some with significant temporal separation and some very close to one another. Note that the ordinate is the number of ON cells. From the figure, the controller without HLC lags the reference significantly, even at the beginning, when the step inputs are significantly separated. This is because 4 time steps are required between when a command is given for cells to turn OFF and when their outputs reflect the command. In contrast, the controller using HLC is able to track the first 3 step inputs perfectly with a 1 time step lag (the lag results from it regulating using only the current time step error).

While HLC preloading is very helpful, it can only do so much. The perfect tracking with only 1 time step delay is possible only when enough cells have been preloaded and are ready for immediate transition. The second downward step input was given less than 4 time intervals after the previous upward step. This means that the cells which responded during the previous step input are not yet ready. Since the step input is larger in magnitude than the 60 units which were not used in the previous step, those 60 units can respond immediately, but the rest must wait for the 4 interval refractory period to expire, so that the remainder of the response can be achieved. The same behavior arises when the step commands come too quickly and there are insufficient

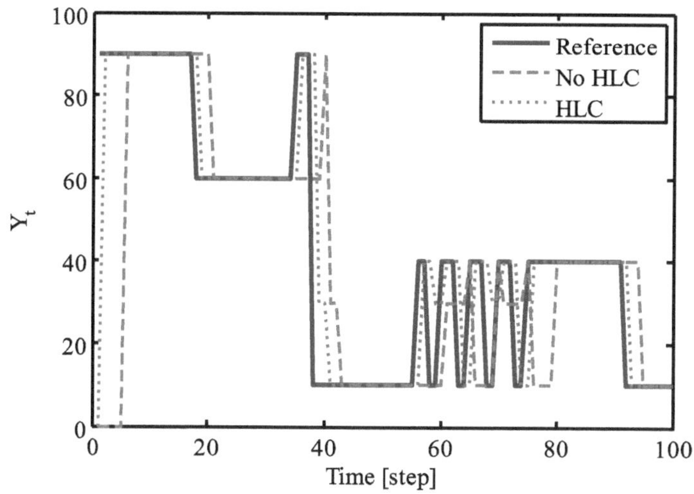

Figure 3.62 Deterministic regulator responding to a series of step inputs with and without preloading HLC.

units being preloaded to transition in one direction. The problem arises more often when the reference command is near one end of the actuator dynamic range.

Despite these physical limitations, HLC preloading is an effective method for coping with time delay associated with hysteric systems. As long as enough cells have been brought to ready states, the feedback system can respond to a command perfectly with only one time step delay.

3.4.5.2 *Probability broadcast control*

A probability broadcast feedback controller can perform nearly as well as its deterministic counterpart if it is properly tuned. Consider again the simulation parameters N, n_B, and n_F. This time, consider a reference comprising uniformly spaced step inputs sufficiently spaced so that enough units are always ready to respond to the command. Since the response of the probability broadcast control is different for each trial, we consider the ensemble mean of many trials.

Fig. 3.63 shows the root mean squared (RMS) error over 10,000 trials for the probability broadcast refractory HLC controller with the control law given by (3.78). For comparison, the deterministic HLC controller from the previous section is also shown in the figure. Both controllers see the same initial error of $e_0 = 65$. While the deterministic controller converges in one time step, the stochastic controller needs a few more steps to correct for overshoot or undershoot before convergence. The total error integrated along the time axis is 215.0 for the deterministic controller and 289.1 for the probability broadcast controller.

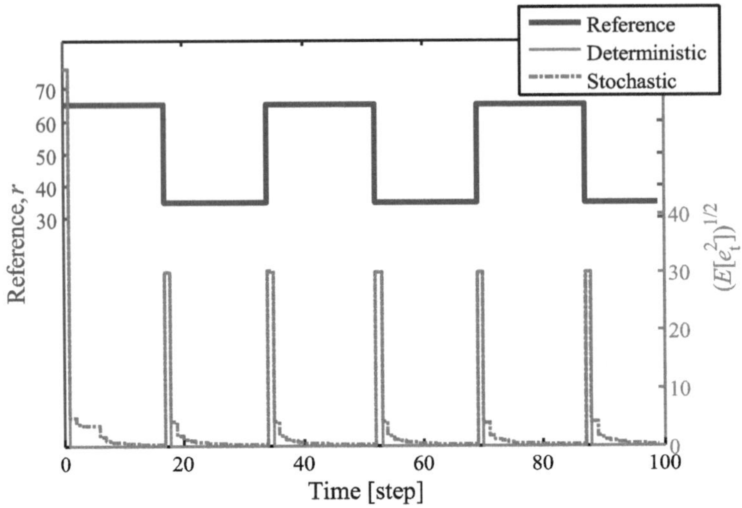

Figure 3.63 RMS error of deterministic HLC and probability broadcast HLC controllers responding to a series of step inputs.

3.4.5.3 Convergence properties

Although the above result of probability broadcast control is encouraging, the convergence process varies depending on the initial condition as well as the number of intermediate nodes, n_F and n_B. To investigate this we consider simulations of $N = 100$ cells, to be controlled to a constant reference, $r = 50$, and repeated 10,000 times for each of the following cases: (a) $n_B = 0, n_F = 0$, and $N_0^{FR} = N$; (b) $n_B = 4$, $n_F = 4$, and $N_0^{FR} = N$; (c) $n_B = 20$, $n_F = 4$, and $N_0^{FR} = N$; (d) $n_B = 4$, $n_F = 4$, and random initial conditions. Case (c) is an extreme asymmetric system with an extremely long backward preloading time. In case of (d), each cells has a 1/2 probability of starting ON or OFF with initial state distributed uniformly across the intermediate and ready states of each ON or OFF side. The ensemble means for each of the cases can be computed by averaging across the 10,000 trials.

Fig. 3.64 shows $E[e_t^2]$ and the estimate of $\hat{\alpha_2}_t^{min}$ computed from 10,000 trials for each case. Case (a) does not have any intermediate nodes and thus converges very quickly. Because cases (a), (b), and (c) have a lot of initial correction to make, $\hat{\alpha_2}_0^{min}$ is very close to zero. On the other hand, case (d) starts much closer to the desired reference on average and has a much larger $\hat{\alpha_2}_0^{min}$. Cases (b) and (c) from the figure both have many intermediate nodes, and these cases do not initially have any ON units. Thus, if the target is overshot, then it is not possible for any units to respond until the units which just turned to ON traverse all of the intermediate nodes. The system converges even for these extreme cases, as guaranteed by Theorem 3.

Since our stability criterion is to reduce the error distribution quickly, it is helpful to see the distribution of y_t from 10,0000 trials at various points in time. Fig. 3.65 shows the distribution after 1, 2, 5, 10, and 30 time steps for each of the cases (a), (b),

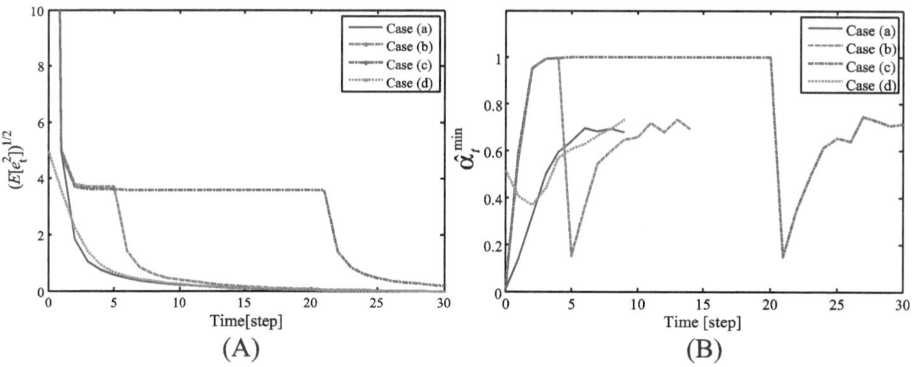

Figure 3.64 RMS error and best fractional convergence in the mean versus time for cases (a), (b), (c), and (d).

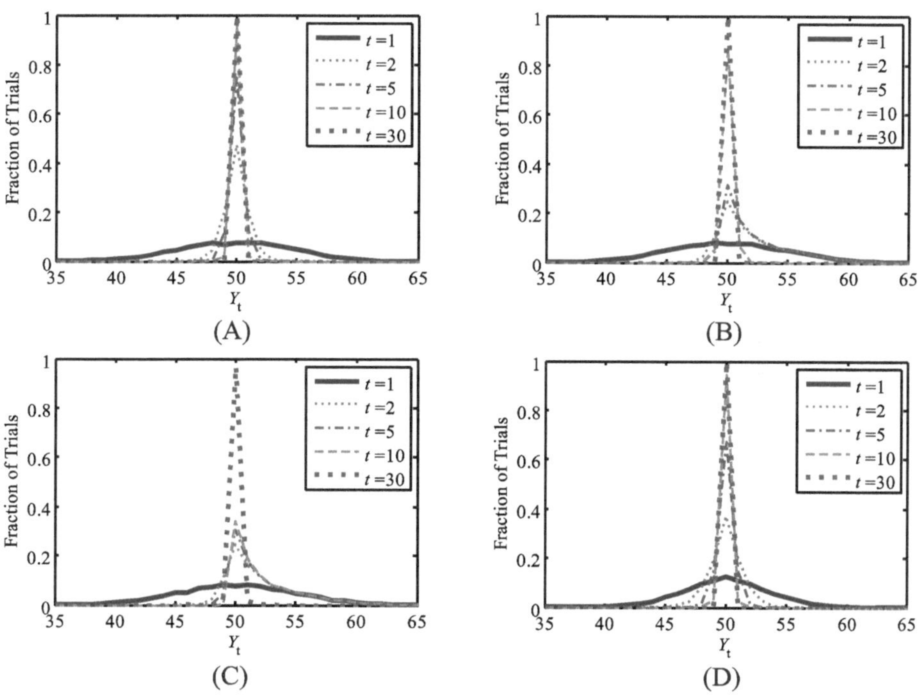

Figure 3.65 Fraction of 10,000 trials versus displacement for control to given each of the cases (a), (b), (c), and (d).

(c), and (d). Cases (a) and (d) show that the system converged with high probability after only a few time steps. Case (b) shows that 5 time steps are often not enough time to converge, which happens because $n_B = 4$, while case (c) shows that 10 time steps are not enough. Even so, each case is still ultimately stable and convergent.

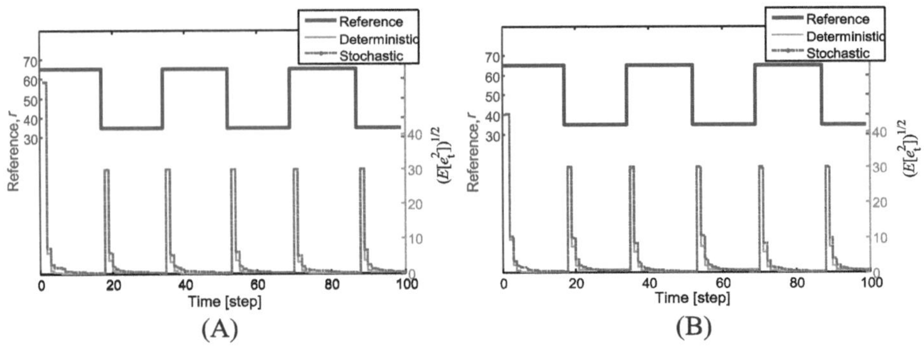

Figure 3.66 Fraction of 10,000 trials versus displacement for control to given each of the cases (a), (b), (c), and (d).

3.4.6 Robustness against cell failures

The cellular stochastic architecture is remarkably robust against failures of the cells. Fig. 3.66 shows responses where 20% and 40% of the 100 cells are not functional, respectively, leaving outputs of the cells either ON or OFF permanently. Dead cells are randomly distributed, with 50% of the dead cells in the on ready position and the remainder in the off ready position. The resultant average responses based on 10,000 simulation trials indicate slight increases of error, as the percentage of cell failures increases. For comparison, responses for the deterministic HLC control with the same percentage of cell failures are shown in the figure. Both stochastic and deterministic controllers do not know which cells and how many of them are non-functional. It is interesting to note that there is no significant difference between the deterministic and stochastic controllers when a significant percentage of cell failures occurred. Fig. 3.67 compares the deterministic and stochastic controllers in terms of the total time integral of the root mean squared error. Compared to the 100% healthy system (0% cell failure), the degradation of both controllers is relatively small. The stochastic controller always performs slightly worse than the deterministic controller, but its degradation is slower than that of the deterministic controller, as the dead cell percentage increases. The degradation is caused primarily by underestimation of the number of ready state units, leading to the slightly sluggish response. The probability broadcast control treats all the cells anonymously, and thereby the control performance does not depend on which cells are non-functional. As the number of cellular units increases, it is difficult to keep all the units perfectly healthy. Therefore, this robust performance against cell failures is an important feature.

3.4.7 Contribution of preloading and refraction rule

The preloading scheme and the refractory rule are the cornerstone enabling cell coordination and guaranteeing deterministic observability even though the cells all behave randomly. By prohibiting premature cells from output transition, the speed of response is improved and we can estimate the distribution of the cell states across the hystere-

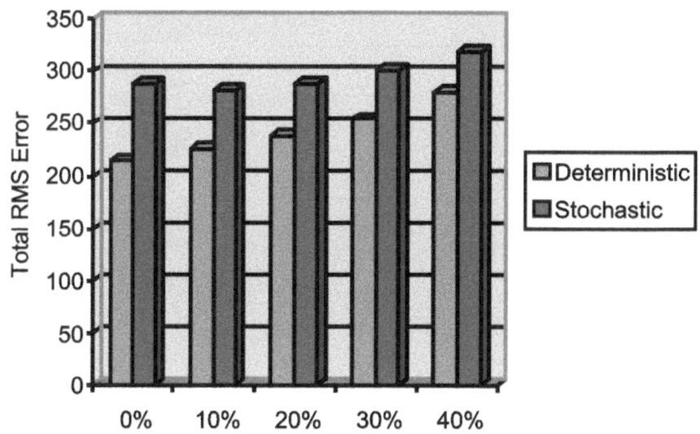

Figure 3.67 Total time integral of root mean squared error.

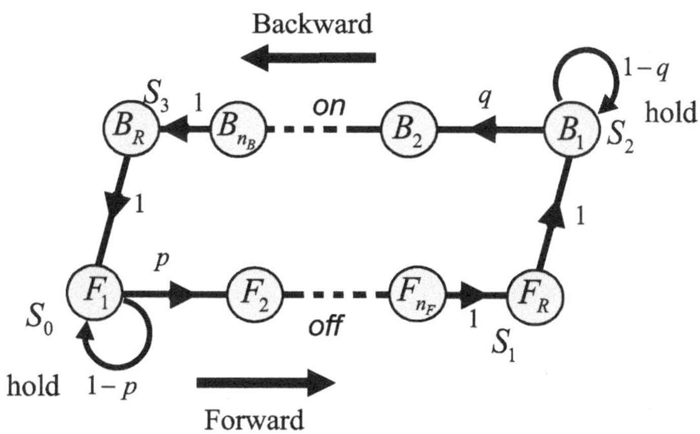

Figure 3.68 Local cell behavior without preloading and refraction.

sis loop from output observation. To better understand the necessity of these features, consider a local cell control system without preloading and refractory mechanisms. As shown in Fig. 3.68, the hold states in such a system are S_0 and S_2. The lack of preloading and refraction not only causes long latency times in transitions, but also makes the state observation very difficult. One portion of the problem is that, without preloading, the output of a cell does not change until sometime after it has decided to make the transition, meaning that the number of cells in each node cannot be determined. Further, the change to the output $y_t - y_{t-1}$ observed between times t and $t-1$ includes both forward and backward output transitions: those from F_R to B_1, which were caused by the cell controller's decision at time $t - n_F$, and those from B_R to F_1 caused by the decision at time $t - n_B$. It is infeasible to completely sort out the mixed output transitions into the two.

Fig. 3.69 shows simulations of the system without preloading and refraction, and with $n_B = 4$ and $n_F = 2$. Since the system is not observable due to the mixed transi-

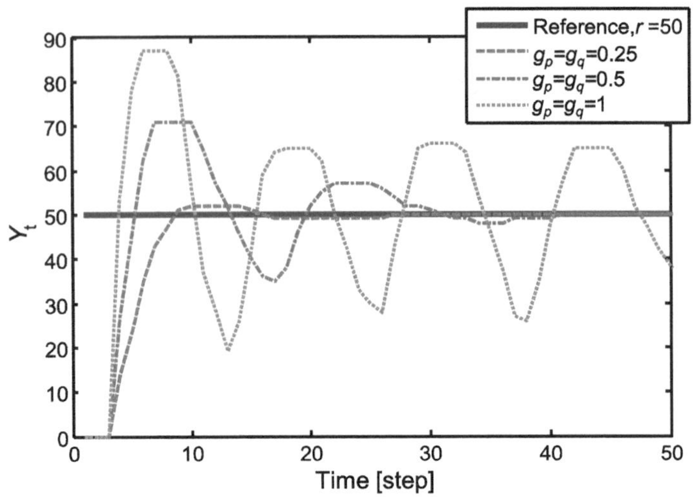

Figure 3.69 Control to $r = 50$ with different gains, g_p and g_q.

tions with time delays, the central controller uses a simple proportional controller,

$$p = \min(g_p e_t/N, 1), q = 0, \quad e_t \geq 0, \tag{3.96}$$

$$q = \min(g_q e_t/N, 1), p = 0, \quad e_t < 0, \tag{3.97}$$

for these simulations. The results are oscillatory unless the feedback gain is extremely low, $g_p = g_q = 0.5$, which makes the system response very slow.

Stabilizing a system with prominent hysteresis and improving the speed of response based on output feedback alone is a difficult problem. In contrast, the probability broadcast feedback control based on the preloading and refractory rule guarantees stability and improved speed of response in a straightforward manner, as demonstrated above.

3.4.8 Discussion

Feedback control of distributed hysteric systems remains a challenging problem. This section has developed a broadcast feedback method for controlling the collective behavior of an ensemble of stochastically behaving units. The control method guarantees convergence and allows for fast response despite distributed hysteric dynamics. The control system architecture is threefold:

- First, the bulk of the distributed system is segmented into a collection of cellular units having local controllers.
- Second, each local controller controls its own state following a hysteresis loop, and drives it to a ready state (preloading) so that it can make an immediate transition when called upon next time. Transition is prohibited unless the state has reached a ready state (the refraction rule).

- Third, a central controller does not dictate each cellular unit to take a specific control action, but allows them to make stochastic decisions with transition probabilities broadcasted anonymously to all of the cellular units. Limiting stochastic transitions at the individual cells to unilateral transitions, the central controller is able to construct, from an aggregate output, the distribution of state among the whole cellular units (observability).

Control algorithms that allow for rapid convergence of mean squared error have been developed for the cellular stochastic system, and the system stability has rigorously been proven. Simulations have shown that the presented probability broadcast feedback yields fast speed of response, is nearly as good as its deterministic counterpart, and is robust against failures. In contrast, the system tends to be oscillatory and difficult to control if the preloading and refraction rule are not adopted.

The presented control methodology has the potential to be a useful approach to broad application areas, which include both natural biological systems and engineered systems. The latter includes not only shape memory alloy actuators that we used as an exemplary case study for developing the control architecture and theory, but also any physical system exhibiting distributed hysteric dynamics whose aggregate output must be controlled stably and robustly. Segmenting a continuum into many cellular units allows us to apply this powerful, yet straightforward control method, which brings about fast, stable, and robust performance.

Such cellular architecture naturally resides in real biological systems, too. The authors are currently working on the control of biological cell populations, where collective behaviors of a multitude of cells must be controlled. The stochastic broadcast feedback approach is a promising methodology for those real biological cells, whose behaviors can be modeled as stochastic agents with prominent hysteresis, time delay, and refractory periods.

3.5 Signal-dependent variability of actuator arrays with floating-point quantization

3.5.1 Motor noise and cellular actuation

As discussed in section "Generation of natural movements", generation of human movement has been one of the most prevalent areas of inquiry in neuromuscular science for a long time. There are a number of hypotheses and computational models proposed to explain the principles in limb movement generation including the minimum-jerk (jerk is the time derivative of acceleration) model [76] and the minimum-torque-change model [277]. The concept of these computational models is that the central nervous system determines a limb trajectory by minimizing a cost function. Although some of the aforementioned computational models seem successful in reproducing a range of empirical data, there are still a number of unsolved problems from the engineering point of view. In other words, the reason is not clear why the central nervous system needs to minimize such a cost function and how the cost is

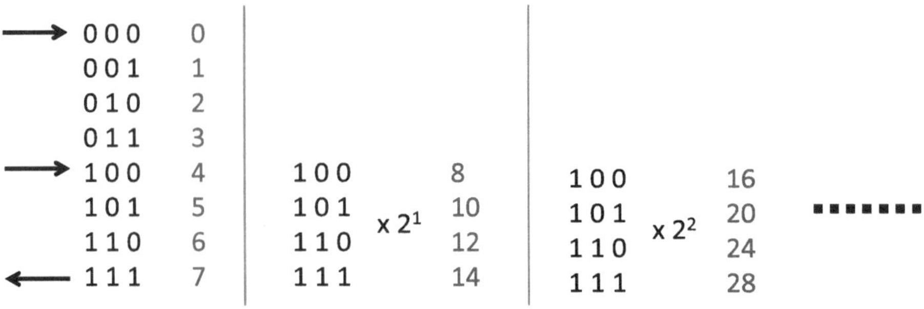

Figure 3.70 Floating-point quantization with 3-bit mantissa.

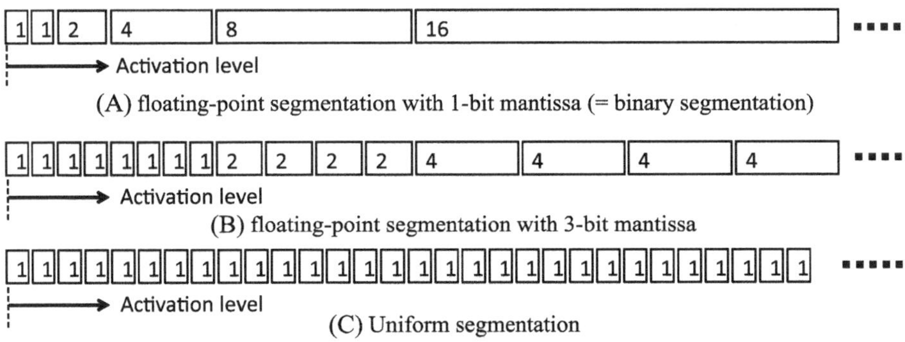

Figure 3.71 Floating-point segmentation of cellular actuator arrays.

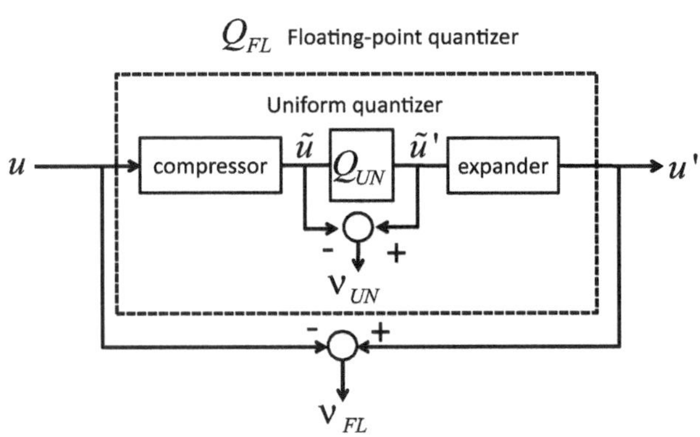

Figure 3.72 Equivalent block diagram representation of a floating-point quantizer.

quantizer is represented by

$$\nu_{FL} = \nu_{UN} \left(\frac{du'}{d\tilde{u}'} \right)_{\tilde{u}} \tag{3.99}$$

$$= \nu_{UN} \cdot 2^{Q_{UN}\left(\left(\log_2 \frac{|u|}{\Delta_{FL}}\right) + 0.5\right)} \tag{3.100}$$

where $Q_{UN}(\cdot)$ denotes a uniform quantizer, and ν_{UN} is its uniform quantization error. Theoretical lower and upper bounds of the expectation of ν_{FL}^2 are given as [287]

$$\frac{1}{12} \cdot 2^{-2\rho} \cdot E[u^2] \leq E[\nu_{FL}^2] \leq \frac{1}{3} \cdot 2^{-2\rho} \cdot E[u^2]. \tag{3.101}$$

In addition, if the so-called Pseudo Quantization Noise (PQN) conditions [287] are met (details omitted due to space limitations), quantization errors can be regarded as additive noise uncorrelated with the input. Hence, $E[\nu_{FL}^2]$ can be obtained more precisely as

$$\begin{aligned} E[\nu_{FL}^2] &= 2.16\ E[\nu_{UN}^2]\ E\left[\frac{u^2}{\Delta_{FL}^2}\right] \\ &= 0.18 \cdot 2^{-2\rho}\ E[u^2], \end{aligned} \tag{3.102}$$

where $E[\nu_{UN}] = \rho^2/12$ and $\rho/\Delta_{FL} = 2^{-\rho}$ were used [287]. If the range of u is sufficiently larger than ρ, we can assume that $E[\nu_{FL}] = 0$. Therefore,

$$Var[\nu_{FL}] = E[\nu_{FL}^2] - E[\nu_{FL}]^2 = E[\nu_{FL}^2]. \tag{3.103}$$

From (3.103), the standard deviation of the floating-point quantization error is give as

$$\sigma_{FL} = \sqrt{Var[\nu_{FL}]} = \sqrt{0.18 \cdot 2^{-2\rho}} \cdot \sqrt{E[u]^2}. \tag{3.104}$$

Refer to [287] for detailed statistical analysis of floating-point quantization.

3.5.2.2 Proportional signal-dependent noise

Harris and Wolpert modeled the variability in muscle forces as "signal-dependent" noise [98] with standard deviation *proportional* to the mean command x for isometric force generation. Let ν_{SDN} be the signal-dependent noise and $x + \nu_{SDN}$ be the resultant motor command. In [253], this was represented as

$$Var[u + \nu_{SDN}|u] = k_{SDN}^2 \cdot u^2. \tag{3.105}$$

where $k_{SDN} > 0$ is a constant. Note that we explicitly represent the above equation as a conditional variance since u is stationary for isometric contraction. Since $Var[u|u] = 0$, the noise component in motor commands can be represented as

$$Var[\nu_{SDN}|u] = k_{SDN}^2 \cdot u^2. \tag{3.106}$$

3.5.2.3 Equivalent floating-point quantization to signal-dependent noise: input-level matching law

This section's interest is to investigate the statistical property of ν_{FL} and ν_{SDN}. By using the variance decomposition formula, we have

$$Var[\nu_{SDN}] = E[Var[\nu_{SDN}|u]] + Var[E[\nu_{SDN}|u]]. \tag{3.107}$$

From (3.106) and since $E[\nu_{SDN}|u] = 0$, i.e., white noise,

$$\begin{aligned} Var[\nu_{SDN}] &= E[Var[\nu_{SDN}|u]] \\ &= E[k_{SDN}^2 u^2] = k_{SDN}^2 E[u^2]. \end{aligned} \tag{3.108}$$

Finally, the standard deviation of the proportional SDN is given as

$$\sigma_{SDN} = \sqrt{Var[\nu_{SDN}]} = k_{SDN} \cdot \sqrt{E[u^2]}. \tag{3.109}$$

Comparison between equations (3.104) and (3.109) clearly shows that the quantization error generated by a floating-point quantizer is statistically equivalent to that of Wolpert's SDN when

$$k_{SDN} = \sqrt{0.18 \cdot 2^{-2\rho}}. \tag{3.110}$$

In other words, a source of Gaussian noise, such as a random noise generator, is not necessary for implementation; simply quantizing the actuator array in a floating-point fashion is sufficient.

3.5.3 Numerical example

Note that the PQN conditions are not strictly met for floating-point quantization and the equivalence given by (3.110) is not trivial; however, as reported in [287], the conditions will be met very closely with mantissas of 4 bits (i.e., $\rho = 4$) or more. Consider an actuator array with N cells connected in series as shown in Fig. 3.71. Let u $(0 \le |u| \le 1)$ be a normalized command. For example, $u = 1$ means that all the N cells will be turned on. For mathematical convenience, negative u is allowed here. For this normalized u, $\Delta_{UN} = 1/N$.

Figs. 3.73 and 3.74 show a comparison between floating-point quantization and proportional signal-dependent noise for sinusoidal inputs with different mantissa values and numbers of units. The matching law given by (3.110) was applied. Fig. 3.75 shows standard deviations σ_{SDN} and σ_{FL} for a sinusoidal input and Gaussian input for α. The graphs also show the analytical upper and lower bounds given by (3.101). The statistical characteristics in terms of standard deviation between the FPQ and proportional SDN approach are not exactly the same; however, it is evident that the floating-point quantization behaves very much like the SDN.

3.5.4 Discussion

This section described a concept of floating-point quantization and activation of cellular actuator arrays, inspired by Henneman's size principle, for the realization of neuromotor variability. Results indicated that a source of Gaussian noise is not necessary for such implementation. Rather, quantization errors behave like SDN. The quantized control architecture could be an alternative to the conventional signal-dependent-noise approach for computational modeling of neuromotor mechanisms. Simulation results of the generation of point-to-point motion using a 2-DOF planar robot arm with FPQ can be found in [269].

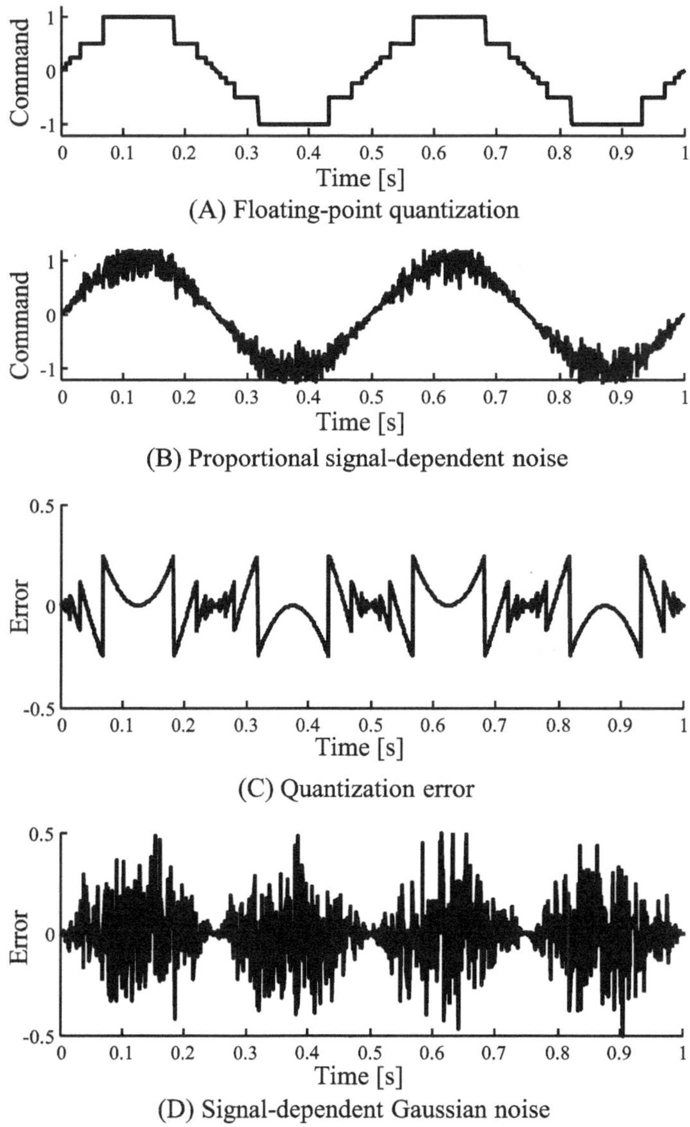

(A) Floating-point quantization

(B) Proportional signal-dependent noise

(C) Quantization error

(D) Signal-dependent Gaussian noise

Figure 3.73 Noise in the signal: sinusoidal input, $\rho = 1$, $N = 256$.

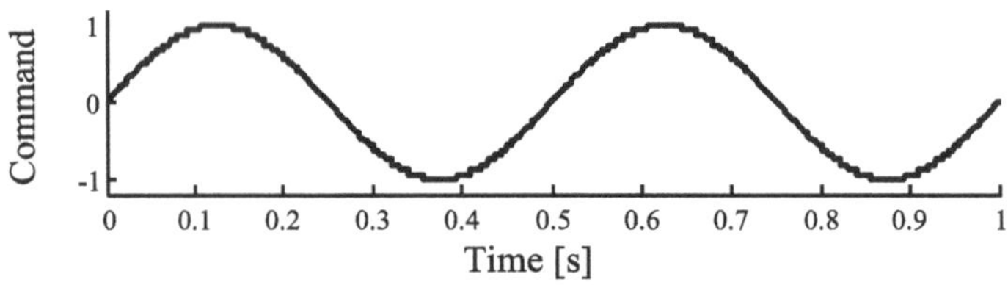

(A) Floating-point quantization

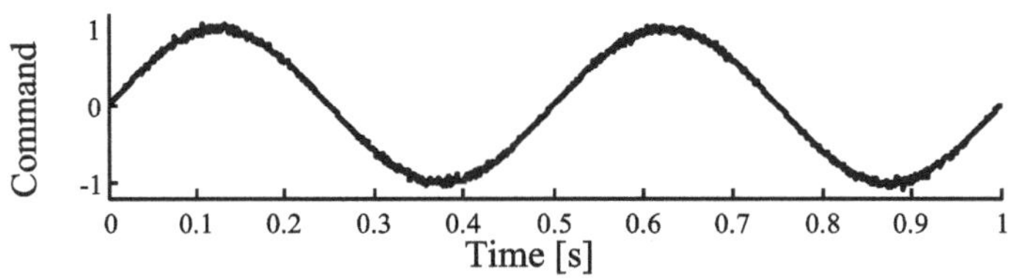

(B) Proportional signal-dependent noise

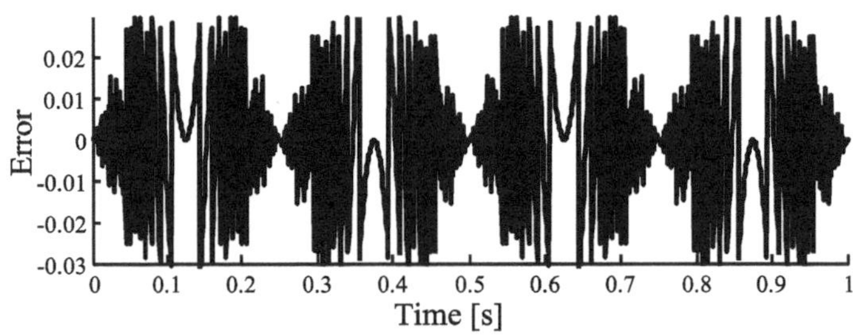

(C) Quantization error

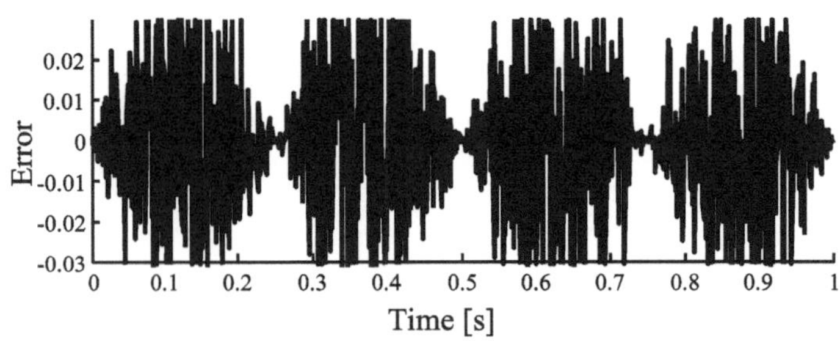

(D) Signal-dependent Gaussian noise

Figure 3.74 Noise in the signal: sinusoidal input, $\rho = 4$, $N = 2^{16} = 16,384$.

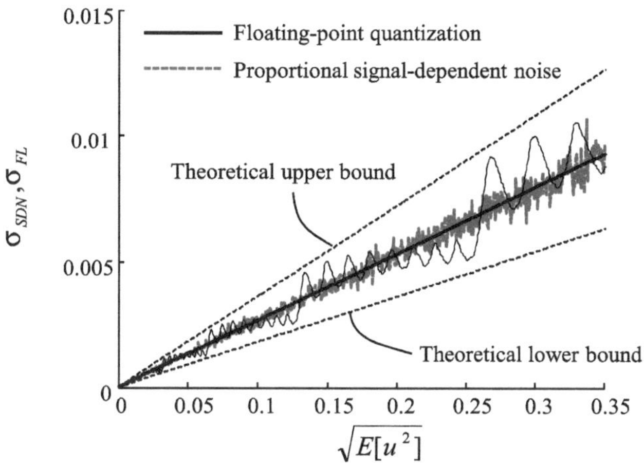

(A) Sinusoidal input (4Hz)

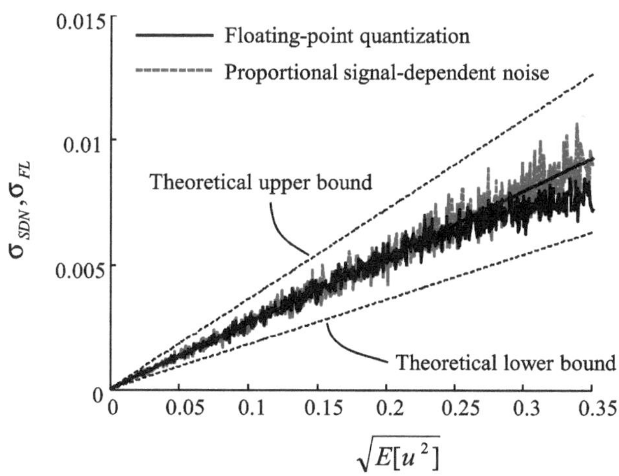

(B) Gaussian input

Figure 3.75 Comparison of standard deviation between floating-point quantization and proportional signal-dependent noise: $\rho = 4$, $N = 2^{16} = 16,384$, with the matching law in Eq. (3.110).

Application of cellular actuators

4

4.1 Variable stiffness cellular actuators

4.1.1 Variable stiffness actuators

Variable stiffness actuators have the capability to meet the increased demand for safety, robust locomotion, and dynamic manipulation. In robots interacting with humans, variable stiffness actuators allow maximum forces to be held to acceptable levels [213]. In mobile robots, joint compliance absorbs impulsive forces from the ground and allows a vehicle to better negotiate with rough terrain. Actuator compliance is also essential in manipulating objects in unstructured environments, where the properties of the manipulated object are not known *a priori*. Although feedback control is one way of varying task space stiffness [104], the bandwidth of the control loop is usually insufficient for rapid and highly dynamic physical interaction. Therefore, it is desirable to have variable mechanical compliance inherent in the actuator's physical construction.

Several groups have investigated both constant stiffness and variable stiffness actuators. In the early work on the elastic hand [95,96], elastic elements were inserted in an actuator drive train for stable grasping and dexterous manipulation. The seminal work of Pratt [200] demonstrated the advantages of constant series elasticity to increase shock tolerance, reduce reflected inertia, and improve force control. These concepts are still employed in current systems such as advanced protheses that utilize both series and parallel elasticity to enhance force bandwidth [13]. Although constant stiffness in a drive train is beneficial, additional advantages are obtained when the stiffness is allowed to vary. A majority of the work in variable stiffness actuators has been directed at developing new variable stiffness rotary joints. For example, in [39], the stiffness of a joint is varied by changing the overlapping area of two permanent magnets. In [213], a combined spring and belt drive system achieves variable stiffness. Contemporary designs using similar ideas were proposed by [278] and [7]. A theme among these actuator designs is the requirement of two actuators for a single degree-of-freedom. Moreover, the designs all employ standard electromechanical actuators: one actuator allows for stiffness tunability and another for angular motion.

Apart from variable stiffness actuators is the use of resonance in robotics. Resonance is a condition exhibited by linear (and weakly nonlinear) systems and is a manifestation of maximal potential and maximal kinetic energy oscillating 180° out of phase. For multi degree-of-freedom systems, this condition can occur at multiple frequencies. The frequencies where resonance occurs depend upon the distribution of mass and stiffness within the oscillatory system. In robotics, resonance has been recognized as an important phenomenon that can be used to increase power transmission to a load, reduce the effort of actuators, and achieve a large amplitude motion for cyclic tasks such as running (e.g., [204,107]), flapping (e.g., [114,295]), or fin-based swimming (e.g., [293]).

Cellular Actuators, DOI: 10.1016/B978-0-12-803687-7.00004-1
Copyright © 2017 Elsevier Inc. All rights reserved.

Variable stiffness and resonance can be intimately connected because the ability to vary actuator stiffness provides the ability to tune a robotic system's resonant frequencies. Indeed, some recent work has simulated tunable stiffness as a means for tuning resonant-like frequencies in multi-DOF manipulators [276]. Other work in actuator design has considered the constant resonant frequencies in certain material systems such as piezoelectric beam actuators [153] and electrostrictive polymers [135]. Because of the great importance of both variable stiffness and tunable resonance, there exists a need for an actuator that can provide both functions.

This section presents the design, analysis, and testing of a linear artificial muscle actuator that incorporates both variable stiffness and tunable resonant frequencies by using a modular, cellular architecture. Artificial muscle actuator design is considered because of its importance to many areas of biorobotics that require smooth, natural motions that cannot be achieved with either linear or rotary electromechanical actuators. The basic principle of the actuator design is to connect a plurality of actuator cells in series or parallel (as in natural muscle) and then control stiffness and resonance by turning units ON to increase compliance and OFF to reduce compliance. The distribution of stiff versus compliant units within the actuator also determines the mass distribution and thereby allows the resonance properties to be tuned.

This section first presents the conceptual design and method for varying stiffness in a cellular actuator based on discrete PZT actuator units. The effect of distributed mass in the actuator is taken into account, which leads to tunable resonant frequencies. The properties and ranges for tuning resonant frequencies are rigorously established using an idealized analytical model. For experimental comparison and design, the idealized model is extended to include realistic parasitic effects such as damping. Finally, the actuator is tested experimentally.

4.1.2 Design of variable stiffness cellular architecture

This section summarizes basic principles of variable stiffness cellular actuators (VSCAs). Any VSCA implementation must possess two fundamental characteristics: (i) cellular units having discrete stiffness states that can be selectively switched ON or OFF, and (ii) cellular units that are connected to form an aggregate output. Within this framework many implementations are possible. For clarity, a particular PZT-based design will be described, yet the basic principles of VSCAs are applicable to a large class of cellular actuators using other smart materials and structures.

4.1.2.1 Variable stiffness mechanism

Fig. 1.37 in Section 1.3.4 shows the design concept for a nested-flexure PZT cellular actuator. Fig. 4.1 shows a modification to this design to achieve variable stiffness. Outside the second layer flexure is a rigid structure that limits the stroke of the output displacement in the Y direction. When the PZT is not activated, the output node of the second layer flexure rests on the stroke limiting beam. The output node movement is also limited when an excessive tensile force acts on the output node. As the applied

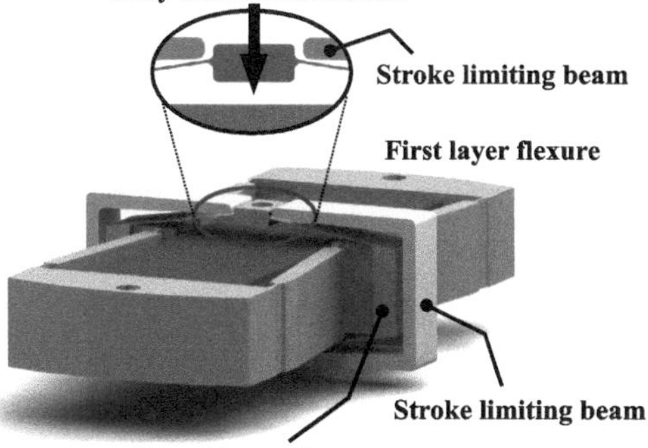

Figure 4.1 Design of a variable stiffness PZT-based cell. The system consists of two strain amplification layers. The second layer flexure incorporates a stroke limiting beam.

PZT voltage increases, the output node is pulled inward and is detached from the stroke limiting beam.

Fig. 4.2 shows the displacement versus compliance characteristics for a single cell and corresponding schematic representations of a cell at various points along the operating curve. In the schematic representations (A), (B), and (C), the outermost thin lines indicate the stroke limiter, while the vertical thick line represents the output node of the flexure. The schematics show all of the stiffness within the cell lumped into a single element with a constant value $\tilde{k}$ (Note that, according to the convention given in Section 1.3, the lumped stiffness of a two-layer mechanism is given by $\tilde{k}_2$. In this section, the subscript 2 will not be shown for simplicity.) Similarly, all of the mass of a cell is lumped into a single mass element $\tilde{m}$ and damping effects are neglected to simplify the initial analysis.

As the PZT-induced force f_{pzt} increases, inward displacement also increases and the thick vertical line detaches from the limiter. For the ideal cell, the equivalent compliance seen at the output node is $1/\tilde{k}$. The nonlinearity in the actual compliance curve occurs because of the geometric nonlinearity of the flexure that becomes more pronounced as the cell contracts. The nonlinearity is often negligible during typical operation of the PZT-based system.

Note that it is also possible to design the stroke limiting system so that the stroke is limited only after the maximum contraction is achieved. This would provide a compliance characteristic that drops to zero upon achieving maximum contraction. Such a cell would be actively OFF and passively ON rather than actively ON and passively OFF. For brevity, this section only consider cells that are actively ON and passively OFF.

The above cellular units can be connected in series, parallel, or antagonistic configurations, creating diverse stiffness characteristics as a collective sum. The most fundamental VSCA is a serial connection of N units. This arrangement is referred to as a strand. For simplicity, assume that the stiffness of an OFF state unit is infinitely large, while stiffness in the ON state is a constant $\tilde{k}$. If N^{ON} units are ON and the

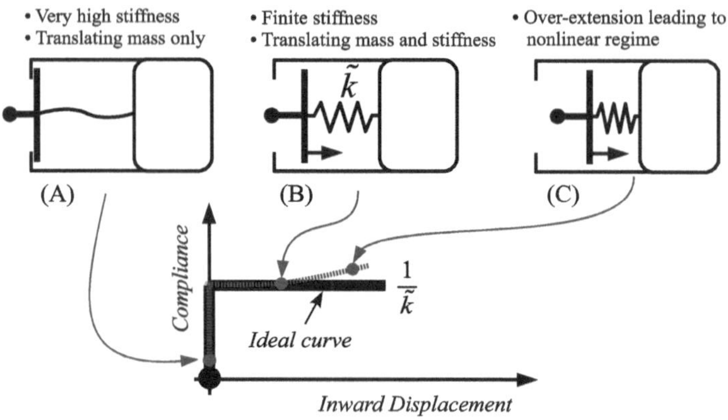

Figure 4.2 Compliance versus displacement characteristics and schematic representations for a variable stiffness, PZT-actuated cell: (A) cell in the OFF state, (B) cell in the ON state and linear regime, and (C) cell in the ON state and nonlinear regime.

others are OFF, then the resultant stiffness of an N-unit strand reduces to

$$\tilde{k}_S = \frac{\tilde{k}}{N^{ON}} \quad 1 \leq N^{ON} \leq N. \tag{4.1}$$

4.1.3 Tunable resonant frequencies

4.1.3.1 Principle

As demonstrated for varying stiffness, the resonant frequencies of a collection of cellular units can also be varied by exploiting the cellular architecture. The key behavior is that turning on a specific number of units to achieve a desired static stiffness still allows for numerous ON–OFF unit combinations, each of which exhibits different vibration modes with different resonant frequencies. Although the total number of ON state units remains the same, the resonant frequencies may vary significantly depending upon the location of the ON units within the actuator.

To illustrate the basic concept, recall the simplified dynamic model shown in Fig. 4.2. Consider three of these units connected serially. Suppose that two out of the three units are turned ON so that the static stiffness of the serial connection is $\tilde{k}_S = \tilde{k}/2$. There are three unique ways of selecting two units to turn ON, and they are shown in the upper portion of Fig. 4.3. Depending upon which two units are ON, the assembly dynamic behavior is different while the static behavior is the same.

Since all of the actuator configurations in Fig. 4.3 have two degrees of freedom, each configuration has two distinct vibration modes with distinct resonant frequencies. Table 4.1 shows the normalized resonant frequencies for each of the three ON–OFF states having $N^{ON} = 2$. The second mode of configuration (1), OFF–ON–ON, provides the highest resonant frequency overall ($1.62\sqrt{\tilde{k}/\tilde{m}}$), while the lowest resonant

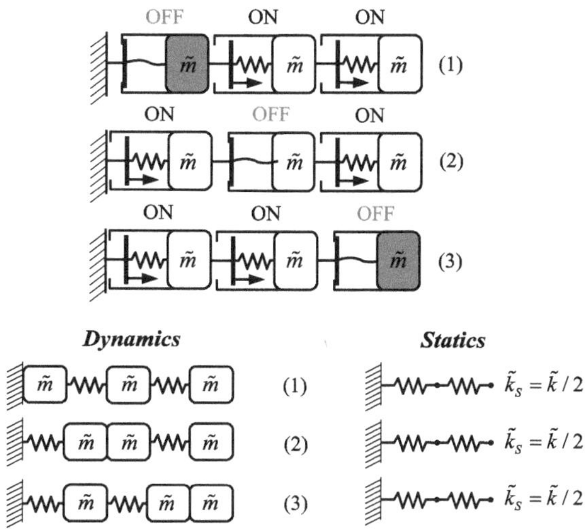

Figure 4.3 Comparison of static and dynamic behavior for all possible ON–OFF distributions for three serially connected units having two units activated.

Table 4.1 First and second natural frequencies of each of the three configurations.

Configuration	$\frac{\omega_1}{\sqrt{\tilde{k}/\tilde{m}}}$	$\frac{\omega_2}{\sqrt{\tilde{k}/\tilde{m}}}$
1. OFF–ON–ON	0.62	1.62
2. ON–OFF–ON	0.54	1.31
3. ON–ON–OFF	0.47	1.51

frequency ($0.47\sqrt{\tilde{k}/\tilde{m}}$) occurs for the first mode of the ON–ON–OFF configuration. In each case, the OFF unit mass creates a different dynamic system and thereby changes the resonant frequencies.

Since all three cases have the same static stiffness $\tilde{k}_S$ but different resonance properties, this example demonstrates that actuator stiffness and resonant frequencies can be changed independently. Although the resonant frequencies do not vary continuously, multiple choices are available for different task requirements and, as the number of cellular units increases, the number of possible stiffness levels and resonant frequencies also increases. The next subsection establishes the model for determining resonant frequency bounds of N serially connected units.

4.1.3.2 Basic model of serial connection dynamics

Consider a single serially connected strand of N cellular units. Suppose that N^{ON} units are in ON state so that N^{ON} springs are detached from the stroke limiters. This creates an N^{ON} degree-of-freedom system with each degree-of-freedom correspond-

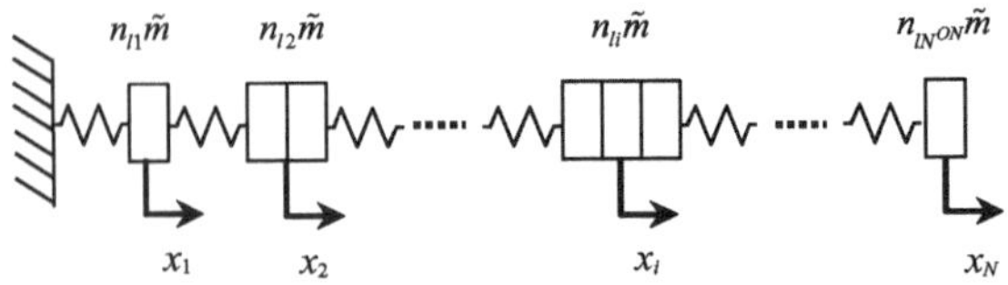

Figure 4.4 Idealized dynamic model for an N-unit strand of cells having N^{ON} units activated.

ing to a single lumped mass that moves independently along a single axis. This means that $(N - N^{ON})$ OFF-state units are rigidly connected to other units as illustrated in Fig. 4.4. Let x_i be the position of the ith combined masses and $\boldsymbol{x}$ be a vector collectively representing the N^{ON} lumped mass displacements, $\boldsymbol{x} = [x_1, x_2, \dots, x_N]^T$.

The ith lumped mass consists of n_{li} cellular units with total mass of $n_{li}\tilde{m}$ with $n_{li} \in \mathbb{N}$. From Fig. 4.4, the equations of unforced motion are obtained as

$$n_{li}\tilde{m}\ddot{x}_i = -2\tilde{k}x_i + \tilde{k}x_{i-1} + \tilde{k}x_{i+1}, \quad 2 \le i \le N^{ON} - 1, \tag{4.2}$$

and for $i = 1$ and $i = N^{ON}$,

$$n_{l1}\tilde{m}\ddot{x}_1 = -2\tilde{k}x_1 + \tilde{k}x_2, \quad n_{lN^{ON}}\tilde{m}\ddot{x}_n = -\tilde{k}x_{N^{ON}} + \tilde{k}x_{N^{ON}-1}. \tag{4.3}$$

These equations can be arranged into vector–matrix form

$$\boldsymbol{M}\ddot{\boldsymbol{x}} + \boldsymbol{K}\boldsymbol{x} = \boldsymbol{0}, \tag{4.4}$$

where $\boldsymbol{M} = \tilde{m} \cdot \text{diag}(n_{l1}, n_{l2}, \dots, n_{lN}) \triangleq \tilde{m}\boldsymbol{L}$ and

$$\begin{aligned} \boldsymbol{K} &= \tilde{k} \begin{pmatrix} 2 & -1 & & 0 \\ -1 & 2 & \ddots & \\ & \ddots & 2 & -1 \\ 0 & & -1 & 1 \end{pmatrix} \\ &\triangleq \tilde{k} \cdot \boldsymbol{C}. \end{aligned} \tag{4.5}$$

One important assumption of the model is that the longitudinal vibrations in a strand are the dominant dynamic behavior for the actuator. This assumption will be addressed in the practical system through an application of preload tension as discussed in Section 4.1.5.

4.1.3.3 Maximum and minimum resonant frequencies for a specified static stiffness

The model developed in the previous subsection can now be utilized to determine how widely the resonant frequencies can be tuned while satisfying the static stiffness requirement, i.e., a specified number of ON units N^{ON}. For the model in (4.4), the squared resonant frequencies of the N^{ON} degree-of-freedom mass–spring system are given by the eigenvalues of matrix product $\boldsymbol{M}^{-1}\boldsymbol{K}$,

$$\omega_i^2 = \lambda_i(\boldsymbol{M}^{-1}\boldsymbol{K}) = \frac{\tilde{k}}{\tilde{m}}\lambda_i(\boldsymbol{L}^{-1}\boldsymbol{C}), \quad 1 \le i \le N^{ON}. \tag{4.6}$$

The N^{ON} eigenvalues of the matrix are ordered from the minimum to the maximum and written as

$$\lambda_{\min}(\boldsymbol{M}^{-1}\boldsymbol{K}) \le \lambda_2(\boldsymbol{M}^{-1}\boldsymbol{K}) \le \cdots \le \lambda_{\max}(\boldsymbol{M}^{-1}\boldsymbol{K}). \tag{4.7}$$

For a fixed number of ON-state units N^{ON}, the stiffness matrix, $\boldsymbol{K}$, remains the same regardless of the arrangement of ON–OFF units. The mass matrix, $\boldsymbol{M}$, on the other hand, varies depending on the ON–OFF arrangement within the strand. Therefore, the above $\lambda_{min}(\lambda_{max})$ can be further decreased (increased) by changing the ON–OFF unit configurations and thereby changing the mass matrix.

Given a specific static stiffness $\tilde{k}_S$, which will uniquely determine N^{ON}, consider the set of all mass matrices associated with the possible distributions of N cells into N^{ON} clusters of masses:

$$\mathfrak{M}_{N^{ON}} = \left\{ \boldsymbol{M} = \tilde{m} \cdot \mathrm{diag}(n_{l1}, \ldots, n_{lN^{ON}}) \middle| \sum_{i=1}^{N^{ON}} n_{li} = N - N^{OFF,base}, \right.$$
$$\left. N^{OFF,base} = 0, \ldots, N - N^{ON};\ n_{li} \in \mathbb{N} \right\}, \quad N^{ON} = 1, 2, \ldots, N. \tag{4.8}$$

The number of possible arrangements of N^{ON} ON units within a strand is equivalent to the cardinality of $\mathfrak{M}_{N^{ON}}$ and is $\binom{N}{N^{ON}}$. If the first $N^{OFF,base}$ units are turned OFF, then they become fixed to the base structure and do not participate in the dynamics. With $N^{OFF,base}$ units OFF at the base, the total number of ON units is then $(N - N^{OFF,base})$ where $N^{OFF,base}$ can take on values $1, 2, \ldots, (N - N^{ON})$. Thus, the subset of $\mathfrak{M}_{N^{ON}}$ defined by $N^{OFF,base} = 0$ in (4.8) is the set of all possible mass matrices assuming that the first unit in the strand is in the ON state so that all cell masses are participating dynamically.

Let $\sigma_{\min}(N^{ON})$ be the minimum of the positive square root of $\lambda_{\min}(\boldsymbol{M}^{-1}\boldsymbol{K})$ with respect to all $\boldsymbol{M} \in \mathfrak{M}_{N^{ON}}$,

$$\sigma_{\min}^2(N^{ON}) = \min_{\boldsymbol{M} \in \mathfrak{M}_{N^{ON}}} \lambda_{\min}(\boldsymbol{M}^{-1}\boldsymbol{K}). \tag{4.9}$$

$\sigma_{\min}(N^{ON})$ provides the lowest resonant frequency among all the ON–OFF unit distributions having N^{ON} ON-state units. Recall that minimum eigenvalue of $\boldsymbol{M}^{-1}\boldsymbol{K}$ is given by the minimum of the Rayleigh quotient,

$$\lambda_{\min}(\boldsymbol{M}^{-1}\boldsymbol{K}) = \min_{\boldsymbol{x} \in \mathbb{R}^{N^{ON}}} \frac{\boldsymbol{x}^T\boldsymbol{K}\boldsymbol{x}}{\boldsymbol{x}^T\boldsymbol{M}\boldsymbol{x}}. \tag{4.10}$$

Similarly, the highest resonant frequency is given by

$$\begin{aligned}\sigma_{\max}^2(N^{ON}) &= \max_{\boldsymbol{M}\in\mathfrak{M}_{N^{ON}}} \lambda_{\max}(\boldsymbol{M}^{-1}\boldsymbol{K}) \\ &= \max_{\boldsymbol{M}\in\mathfrak{M}_{N^{ON}}} \left(\max_{\boldsymbol{x}\in\mathbb{R}^{N^{ON}}} \frac{\boldsymbol{x}^T\boldsymbol{K}\boldsymbol{x}}{\boldsymbol{x}^T\boldsymbol{M}\boldsymbol{x}} \right). \end{aligned} \tag{4.11}$$

Using (4.11), one can prove the following proposition concerning the highest resonant frequency $\sigma_{\max}(N^{ON})$ and the ON–OFF unit distribution that provides $\sigma_{\max}(N^{ON})$.

Proposition 1. *Let* $\boldsymbol{M} = \tilde{m}\boldsymbol{L}$ *and* $\boldsymbol{K} = \tilde{k}\boldsymbol{C}$ *be, respectively, the* $N^{ON} \times N^{ON}$ *mass and stiffness matrices of N serially-connected cellular units among which* N^{ON} *units are active (turned ON). The highest resonant frequency occurs when the first* $N - N^{ON}$ *units are turned OFF. This maximum frequency is given by*

$$\sigma_{\max}(N^{ON}) = \sqrt{\frac{\tilde{k}}{\tilde{m}}\lambda_{\max}(\boldsymbol{C})}. \tag{4.12}$$

Proof. Since $\boldsymbol{x}^T\boldsymbol{M}\boldsymbol{x} \geq \tilde{m}\boldsymbol{x}^T\boldsymbol{I}\boldsymbol{x}\ \forall \boldsymbol{M} \in \mathfrak{M}_{N^{ON}}$ and $\forall \boldsymbol{x} \in \mathbb{R}^{N^{ON}}$

$$\begin{aligned}\lambda_{\max}(\boldsymbol{M}^{-1}\boldsymbol{K}) &= \max_{\boldsymbol{x}\in\mathbb{R}^n} \frac{\boldsymbol{x}^T\boldsymbol{K}\boldsymbol{x}}{\boldsymbol{x}^T\boldsymbol{M}\boldsymbol{x}} \\ &\leq \max_{\boldsymbol{x}\in\mathbb{R}^{N^{ON}}} \frac{\tilde{k}\boldsymbol{x}^T\boldsymbol{C}\boldsymbol{x}}{\tilde{m}\,\boldsymbol{x}^T\boldsymbol{I}\boldsymbol{x}} \\ &= \frac{\tilde{k}}{\tilde{m}}\lambda_{\max}(\boldsymbol{C}). \end{aligned}$$

Therefore,

$$\arg\max_{\boldsymbol{M}\in\mathfrak{M}_{N^{ON}}} \lambda_{\max}(\boldsymbol{M}^{-1}\boldsymbol{K}) = \mathrm{diag}(\tilde{m}, \tilde{m}, \ldots, \tilde{m}).$$

Hence, the arrangement where the first $(N - N^{ON})$ units are turned OFF (i.e., $N^{OFF,base} = (N - N^{ON})$ in (4.8)) yields $\sigma_{\max}(N^{ON})$. □

To address the lowest possible resonant frequency configuration, first consider the following two lemmas:

Lemma 1. *The fundamental mode of a fix-free vibrating system as shown in Fig. 4.4 has nodal displacements that satisfy* $0 \leq x_1 \leq x_2 \leq \cdots \leq x_{N^{ON}}$.

Lemma 2. *Consider the scalar function defined by*

$$J(n_{l1}, n_{l2}, \ldots, n_{lN^{ON}}) \triangleq \boldsymbol{x}^T \mathit{diag}(n_{l1}, n_{l2}, \ldots, n_{lN^{ON}})\boldsymbol{x}, \tag{4.13}$$

where the n_{li} satisfy the conditions in (4.8) *and the vector* $\boldsymbol{x}$ *has elements that are monotonically increasing*

$$0 \le x_1 \le x_2 \le \cdots \le x_{N^{ON}}. \tag{4.14}$$

The scalar function J is maximized when $n_{l1} = 1$, $n_{l2} = 1$, ..., $n_{lN^{ON}-1} = 1$, $n_{lN^{ON}} = N - N^{ON} + 1$.

A proof of Lemma 1 can be found in [86] while a proof of Lemma 2 is given in Appendix A.9; both may be used in the following proposition for $\sigma_{\min}(N^{ON})$.

Proposition 2. *The arrangement having all the inactive units placed at the unconstrained end of the serial connection gives the lowest undamped resonant frequency $\sigma_{\min}(N^{ON})$.*

Proof. By Lemma 1, the first mode shape can be considered positive and monotonically increasing. Therefore,

$$\begin{aligned}
\min_{M \in \mathfrak{M}_{N^{ON}}} \frac{\tilde{k}\boldsymbol{x}^T \boldsymbol{C}\boldsymbol{x}}{\boldsymbol{x}^T \boldsymbol{M}\boldsymbol{x}} &= \min_{\substack{M \in \mathfrak{M}_{N^{ON}} \\ 0 \le x_1 \le \cdots \le x_{N^{ON}}}} \frac{\tilde{k}\boldsymbol{x}^T \boldsymbol{C}\boldsymbol{x}}{\boldsymbol{x}^T \boldsymbol{M}\boldsymbol{x}} \\
&= \min_{0 \le x_1 \le \cdots \le x_{N^{ON}}} \frac{\tilde{k}\boldsymbol{x}^T \boldsymbol{C}\boldsymbol{x}}{\boldsymbol{x}^T \boldsymbol{M}_0\boldsymbol{x}},
\end{aligned}$$

where $\boldsymbol{M}_0 \triangleq \tilde{m} \cdot \mathrm{diag}(1,\ 1, \ldots,\ N - N^{ON} + 1)$ by Lemma 2. This represents the case where all of the inactive units are placed at the unconstrained end. □

In many practical situations, the first (i.e., fundamental) mode is the most readily excited. Therefore, to establish a range for the fundamental frequency tunability, the highest value for the first resonant frequency will be given the notation $\sigma_{\max,\text{first}}(N^{ON})$ and defined as

$$\begin{aligned}
\sigma^2_{\max,\text{first}}(N^{ON}) &= \max_{M \in \mathfrak{M}_{N^{ON}}} \lambda_{\min}(\boldsymbol{M}^{-1}\boldsymbol{K}) \\
&= \max_{M \in \mathfrak{M}_{N^{ON}}} \left(\min_{\boldsymbol{x} \in \mathbb{R}^{N^{ON}}} \frac{\boldsymbol{x}^T \boldsymbol{K}\boldsymbol{x}}{\boldsymbol{x}^T \boldsymbol{M}\boldsymbol{x}} \right).
\end{aligned} \tag{4.15}$$

Corollary 1 (of Proposition 2). *The arrangement having the first $(N - N^{ON})$ units turned OFF gives the highest possible value for the first undamped resonant frequency $\sigma_{\max,\text{first}}(N^{ON})$.*

Proof. The proof follows from the proof of Proposition 1 since $\forall \boldsymbol{M} \in \mathfrak{M}_{N^{ON}}$ and $\forall \boldsymbol{x} \in \mathbb{R}^{N^{ON}}$, $\boldsymbol{x}^T \boldsymbol{M} \boldsymbol{x} \geq \tilde{m} \boldsymbol{x}^T \boldsymbol{I} \boldsymbol{x}$. Hence,

$$\begin{aligned}
\lambda_{\min}(\boldsymbol{M}^{-1}\boldsymbol{K}) &= \min_{\boldsymbol{x}\in\mathbb{R}^{N^{ON}}} \frac{\boldsymbol{x}^T \boldsymbol{K} \boldsymbol{x}}{\boldsymbol{x}^T \boldsymbol{M} \boldsymbol{x}} \\
&\leq \min_{\boldsymbol{x}\in\mathbb{R}^{N^{ON}}} \frac{\tilde{k}\boldsymbol{x}^T \boldsymbol{C} \boldsymbol{x}}{\tilde{m}\,\boldsymbol{x}^T \boldsymbol{I} \boldsymbol{x}} \\
&= \frac{\tilde{k}}{\tilde{m}} \lambda_{\min}(\boldsymbol{C}).
\end{aligned}$$

Therefore, the same configuration as Proposition 1 gives the highest resonant frequency of the first mode. □

4.1.3.4 Global maximum and minimum resonant frequencies

The above analysis provides the highest and lowest resonant frequencies for a fixed N^{ON}, i.e., a given static stiffness. If instead the synthesis objective is to achieve the highest or the lowest resonant frequency regardless of the static stiffness, then the highest and lowest bounds can be further extended by considering all possible values of N^{ON} varying from 1 to N.

The global minimum $\Sigma_{\min}(N)$ is then defined with an additional minimization over N^{ON} as

$$\begin{aligned}
\Sigma^2_{\min}(N) &= \min_{1\leq N^{ON}\leq N} \left\{ \min_{\boldsymbol{M}\in\mathfrak{M}_{N^{ON}}} \left\{ \lambda_{\min}(\boldsymbol{M}^{-1}\boldsymbol{K}) \right\} \right\} \\
&= \min_{1\leq N^{ON}\leq N} \left\{ \sigma^2_{\min}(N^{ON}) \right\}.
\end{aligned} \tag{4.16}$$

A similar expression is obtained for the global maximum $\Sigma_{\max}(N)$. For the maximum, the following definition and lemma are required.

Definition 1. *For the eigenvalue problem*

$$\boldsymbol{K}\boldsymbol{x} = \lambda \boldsymbol{M} \boldsymbol{x} \tag{4.17}$$

with $\boldsymbol{M}, \boldsymbol{K} \in \mathbb{R}^{N\times N}$, *the associated nth reduced eigenvalue problem is*

$$\boldsymbol{K}^{(n)}\boldsymbol{x}^{(n)} = \lambda^{(n)} \boldsymbol{M}^{(n)} \boldsymbol{x}^{(n)}, \quad n = 0, 1, \ldots, N-1, \tag{4.18}$$

where $\boldsymbol{M}^{(n)}$ *and* $\boldsymbol{K}^{(n)}$ *are obtained by deleting the first n rows and first n columns in* $\boldsymbol{M}$ *and* $\boldsymbol{K}$*, respectively.*

Lemma 3. *The eigenvalues of the nth reduced eigenvalue problem* $\lambda_i^{(n)}$, $i = 1, 2, \ldots, (N-n)$ *and the eigenvalues of the* $(n+1)$*th reduced eigenvalue problem*

$\lambda_j^{(n+1)}$, $j = 1, 2, \ldots, (N - n - 1)$ *satisfy the following relation:*

$$\lambda_1^{(n)} \leq \lambda_1^{(n+1)} \leq \lambda_2^{(n)} \leq \lambda_2^{(n+1)} \leq \cdots \\ \leq \lambda_{N-n-1}^{(n)} \leq \lambda_{N-n-1}^{(n+1)} \leq \lambda_{N-n}^{(n)}. \tag{4.19}$$

Lemma 3 can be proved using the technique found in [21], while a proof for the more general case of Jacobi matrices can be found in [81]. Now the following proposition may be proved.

Proposition 3. *A serial strand of N units takes the globally maximum resonant frequency $\Sigma_{\max}(N)$ at a configuration where all the N units are turned ON.*

Proof. From Proposition 1, the maximum resonant frequency for a given N^{ON}, $\sigma_{\max}(N^{ON})$, occurs when the first $(N - N^{ON})$ units are turned OFF, i.e., fixed to the base structure. The associated mass matrix is given by deleting the first $(N - N^{ON})$ rows and the first $(N - N^{ON})$ columns of the $N \times N$ identity matrix multiplied by $\tilde{m}$, $\tilde{m}\boldsymbol{I}_N$. The stiffness matrix is similarly obtained by deleting the first $(N - N^{ON})$ rows and the first $(N - N^{ON})$ columns of the $N \times N$ stiffness matrix given in (4.5). Therefore, the global maximum of the resonant frequency can be found by comparing the largest eigenvalues of all the reduced eigenvalue problems in (4.18) with $n = 1, 2, \ldots, N - 1$.

Setting $n = N - 1$ in Lemma 3 gives $\lambda_1^{(N-1)} \leq \lambda_2^{(N-2)}$. Repeatedly applying Lemma 3 for $n = N - 2, N - 3, \ldots, 1$ yields

$$\lambda_1^{(N-1)} \leq \lambda_2^{(N-2)} \leq \cdots \leq \lambda_{N-1}^{(1)} \leq \lambda_N^{(0)}. \tag{4.20}$$

The final eigenvalue in (4.20), $\lambda_N^{(0)}$, is the maximum eigenvalue of $\boldsymbol{M}^{-1}\boldsymbol{K}$ for the original $N \times N$ matrices $\boldsymbol{K} = \tilde{k}\boldsymbol{C}$ and $\boldsymbol{M} = \tilde{m}\boldsymbol{I}_N$. Therefore,

$$\begin{aligned} \Sigma_{max}^2(N) &= \lambda_{\max}\left[(\tilde{m}\boldsymbol{I}_N)^{-1}\boldsymbol{K}\right] \\ &= \frac{\tilde{k}}{\tilde{m}}\lambda_{\max}(\boldsymbol{C}) \\ &= \lambda_N^{(0)} \\ &\geq \lambda_{N-n}^{(n)}, \quad n = 1, 2, \ldots, N, \end{aligned} \tag{4.21}$$

which concludes the proof. □

Corollary 2 (of Proposition 3). *As N increases, the globally maximum resonant frequency of an N unit strand increases monotonically, namely*

$$\Sigma_{\max}(N) \leq \Sigma_{\max}(N+1). \tag{4.22}$$

Proof. Considering $\boldsymbol{M}$ and $\boldsymbol{K}$ of sufficiently large dimension and applying Lemma 3 immediately proves the corollary. □

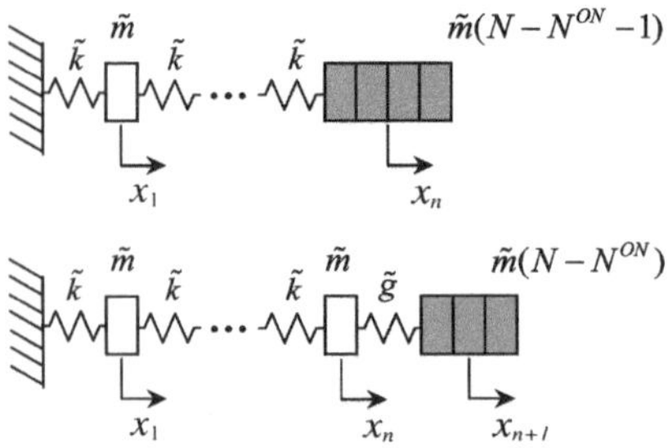

Figure 4.5 Comparison of the minimum resonant frequency for the case with N^{ON} units ON (top) and with $(N^{ON}+1)$ units ON (bottom).

To obtain the global minimum $\Sigma_{\min}(N)$, recall Proposition 2. The lowest resonant frequency for a given N^{ON} occurs when the last $(N-N^{ON})$ units are turned OFF, i.e., creating a lumped mass of $\tilde{m}(N-N^{ON}+1)$ at the free end. Therefore, the question is how many units must be lumped together at the free end to minimize the resonant frequency. This can be solved by comparing the lowest eigenvalue of a strand with $(N-N^{ON})$ OFF units at the free end to the case with virtually $(N-N^{ON}-1)$ OFF units. This situation is depicted in Fig. 4.5.

The lower portion of Fig. 4.5 shows the case with $(n+1)$ units ON with a stiffness of $\tilde{g}$ connecting the N^{ON}th and the $(N^{ON}+1)$th unit. The stiffness of $\tilde{g}$ is assumed to have a lower bound value of $\tilde{k}$ so that $\tilde{k} \leq \tilde{g} < \infty$. As $\tilde{g} \to \infty$, the vibrating system behaves as if N^{ON} units are ON and as $\tilde{g} \to \tilde{k}$, the vibrating system behaves as if $N^{ON}+1$ units are ON. Using this definition of the stiffness $\tilde{g}$, it can be shown that $\sigma_{\min}(N^{ON}) \geq \sigma_{\min}(N^{ON}+1)$, which leads to the following proposition.

Proposition 4. *A serial strand of N units takes the global minimum resonant frequency $\Sigma_{\min}(N)$ at a configuration where all the N units are turned ON.*

Proof. For the $(N^{ON}+1)$th order system, define the mass matrix $\boldsymbol{M} = \tilde{m} \cdot \mathrm{diag}(1, 1, \ldots, 1, N-N^{ON})$ and define the stiffness matrix $\boldsymbol{K}'$ as

$$\mathbf{K}' = \left(\begin{array}{ccc|cc} 2\tilde{k} & -\tilde{k} & & & 0 \\ -\tilde{k} & \ddots & \ddots & & \\ & \ddots & 2\tilde{k} & -k & \\ \hline & & -\tilde{k} & \tilde{k}+\tilde{g} & -\tilde{g} \\ 0 & & & -\tilde{g} & \tilde{g} \end{array}\right). \tag{4.23}$$

The quadratic form $\boldsymbol{x}^T \boldsymbol{K}' \boldsymbol{x}$ can be written as two separate scalar functions,

$$\boldsymbol{x}^T \boldsymbol{K}' \boldsymbol{x} = \Phi(x_1, x_2, \ldots, x_{N^{ON}}, \tilde{k}) + \tilde{g}(x_{N^{ON}+1} - x_{N^{ON}})^2, \tag{4.24}$$

where $\Phi(x_1, x_2, \ldots, x_{N^{ON}}, \tilde{k})$ is a positive function for any nonzero x_i with $i = 1, 2, \ldots, N^{ON}$. Note that $\tilde{g}$ is factored out in the second scalar function. Hence,

$\forall \boldsymbol{x} \in \mathbb{R}^{N^{ON}+1}$,

$$
\begin{aligned}
\frac{\boldsymbol{x}^T \boldsymbol{K}' \boldsymbol{x}}{\boldsymbol{x}^T \boldsymbol{M} \boldsymbol{x}} &= \frac{\Phi(x_1, x_2, \ldots, x_{N^{ON}}, \tilde{k}) + \tilde{g}(x_{N^{ON}+1} - x_{N^{ON}})^2}{\boldsymbol{x}^T \boldsymbol{M} \boldsymbol{x}} \\
&\geq \frac{\Phi(x_1, x_2, \ldots, x_{N^{ON}}, \tilde{k}) + \tilde{k}(x_{N^{ON}+1} - x_{N^{ON}})^2}{\boldsymbol{x}^T \boldsymbol{M} \boldsymbol{x}} \\
&= \frac{\boldsymbol{x}^T \boldsymbol{K} \boldsymbol{x}}{\boldsymbol{x}^T \boldsymbol{M} \boldsymbol{x}}. \qquad (4.25)
\end{aligned}
$$

Now let $\boldsymbol{x}^*$ denote the fundamental mode shape for the system having a stiffness matrix $\boldsymbol{K}'$. Then, using (4.25) and the properties of the Rayleigh quotient, it follows that

$$
\begin{aligned}
\lambda_{\min}(\boldsymbol{M}^{-1}\boldsymbol{K}') &= \frac{\boldsymbol{x}^{*T} \boldsymbol{K}' \boldsymbol{x}^*}{\boldsymbol{x}^{*T} \boldsymbol{M} \boldsymbol{x}^*} \\
&\geq \frac{\boldsymbol{x}^{*T} \boldsymbol{K} \boldsymbol{x}^*}{\boldsymbol{x}^{*T} \boldsymbol{M} \boldsymbol{x}^*} \\
&\geq \min_{\boldsymbol{x} \in \mathbb{R}^{N^{ON}+1}} \frac{\boldsymbol{x}^T \boldsymbol{K} \boldsymbol{x}}{\boldsymbol{x}^T \boldsymbol{M} \boldsymbol{x}} \\
&= \lambda_{\min}(\boldsymbol{M}^{-1}\boldsymbol{K}). \qquad (4.26)
\end{aligned}
$$

The above inequality shows that the case having only N^{ON} units ON will have a higher fundamental frequency than the case with $n + 1$ units ON because the case having only N^{ON} units ON is a special case of the system with the stiffness matrix $\boldsymbol{K}'$ when $\tilde{g} \to \infty$. Therefore, $\sigma_{\min}(N^{ON}) \geq \sigma_{\min}(N^{ON} + 1)$ for $N^{ON} = 1, \ldots, N - 1$, from which it immediately follows that

$$
\sigma_{\min}(1) \geq \sigma_{\min}(2) \geq \cdots \geq \sigma_{\min}(N) = \Sigma_{\min}(N). \qquad (4.27)
$$

This result shows that turning all units ON leads to the global minimum resonant frequency. □

A numerical illustration of Propositions 1 to 4 is offered in Fig. 4.6 for $N = 5$. Fig. 4.6 shows the locations of all possible resonant frequencies as triangles. The maximum and minimum bounds indicated by the uppermost and lowermost solid lines are the resonant frequencies based on the configurations in Proposition 1 and 2, respectively. That is, the uppermost points on the solid lines are achieved by placing all OFF units next to the base structure while the lowermost points are achieved by placing all OFF units at the free end. Note that as N^{ON} increases toward the maximum value of $N = 5$, the upper and lower resonant frequency bounds monotonically approach the limits specified by Propositions 3 and 4 where all units are ON. As N^{ON} increases, the reduction in the absolute slope of the overall upper and lower resonance bounds implies that including more than $N \approx 5$ units in a strand does not significantly improve the range of attainable resonant frequencies although it would increase the static stiffness tuning range.

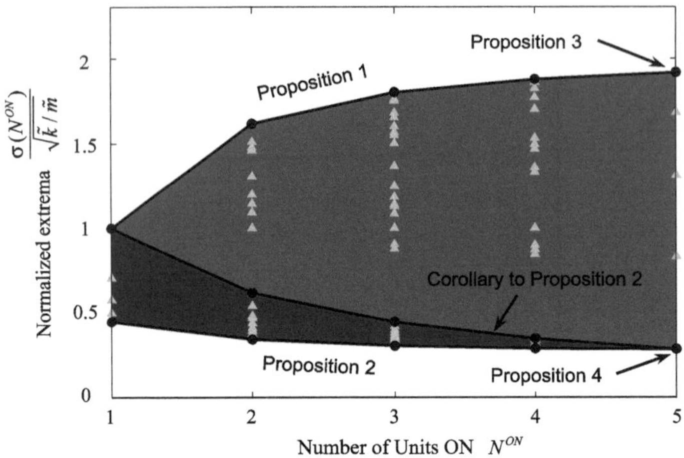

Figure 4.6 An illustration of the theoretical propositions for $N = 5$. All possible resonant frequencies are shown as well as numerical bounds on the minimum and maximum achievable resonant frequencies for $N = 5$.

The darker region in Fig. 4.6 indicates the tunable range of the first resonance only. This region's upper bound is dictated by Corollary 1 to Proposition 2. As in Proposition 1, the configurations for highest first resonance have all OFF units placed at the base. The general trend of this first mode region is a downward and narrowing behavior.

A further note regarding the theoretical resonance tunability is that there is a gap between the first mode and higher modes for $N^{ON} \geq 2$. For a given N^{ON}, this spacing creates a band of frequencies over which no resonance can occur. However, if all values of N^{ON} are considered, the overall modal density is fairly uniform between the global limits $\Sigma_{\min}(N)$ and $\Sigma_{\max}(N)$.

4.1.3.5 Maximal tuning for long strands

The previous propositions establish the conditions under which maximum and minimum resonant frequencies are achieved for a finite N. In this section, the case where $N \to \infty$ is considered to establish the maximal tunability using the proposed actuation method. Given an N, Propositions 3 and 4 indicate that $\Sigma_{min}(N)$ and $\Sigma_{max}(N)$ are given by $\sqrt{\frac{\tilde{k}}{\tilde{m}}}$ multiplied by the positive square root of the maximum and minimum eigenvalues of $\boldsymbol{C} \in \mathbb{R}^{N \times N}$ having the form defined in (4.5). As described in [86], the eigenvalues for $\boldsymbol{C}$ can be written in closed form:

$$\lambda_i(\boldsymbol{C}) = 4\sin^2\left(\frac{(2i-1)\pi}{2(2N+1)}\right), \quad i = 1, 2, \ldots, N. \tag{4.28}$$

Therefore, the following two propositions bound $\Sigma_{\max}(N)$ and $\Sigma_{\min}(N)$ for all N.

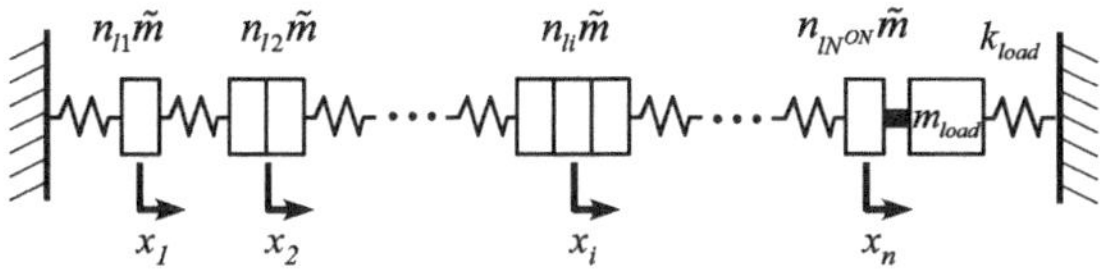

Figure 4.7 Idealized dynamic model for an N-unit strand of cells rigidly connected to a spring–mass load and having N^{ON} units activated.

Proposition 5. *The maximum undamped resonant frequency* $\Sigma_{\max}(N)$ *has the least upper bound of* $2\sqrt{\frac{\tilde{k}}{\tilde{m}}}$.

Proof. Applying (4.28) with $i = N$, yields

$$\Sigma_{\max}(N) = 2\sqrt{\frac{\tilde{k}}{\tilde{m}}}\sin\left(\frac{1-\frac{1}{2N}}{2+\frac{1}{N}}\pi\right). \tag{4.29}$$

Letting $N \to \infty$ gives an upper bound of $2\sqrt{\frac{\tilde{k}}{\tilde{m}}}$. □

Proposition 6. *The minimum undamped resonant frequency* $\Sigma_{\min}(N)$ *has the greatest lower bound of* 0.

Proof. Applying (4.28) with $i = 1$, yields

$$\Sigma_{\min}(N) = 2\sqrt{\frac{\tilde{k}}{\tilde{m}}}\sin\left(\frac{1}{2(2N+1)}\pi\right). \tag{4.30}$$

Letting $N \to \infty$ gives a lower bound of 0. □

4.1.3.6 Loading effects

While the system shown in Fig. 4.4 possesses fixed-free boundary conditions, practical applications entail a load end condition. A simple extension of the model in Fig. 4.4 is provided in Fig. 4.7, which shows the addition of a mass and spring load connected rigidly to the N^{ON}th node. The load mass and stiffness will dictate the following dimensionless ratios:

$$\gamma_m \triangleq \frac{m_{load}}{\tilde{m}}, \tag{4.31}$$

and

$$\gamma_k \triangleq \frac{k_{load}}{\tilde{k}}. \tag{4.32}$$

As $\gamma_m \to 0$ and $\gamma_k \to 0$, the system in Fig. 4.7 approaches the system of Fig. 4.4, and the theoretical conclusions of the previous sections apply directly. In this section, the effects of the load ratios on the static and dynamic tuning properties are considered.

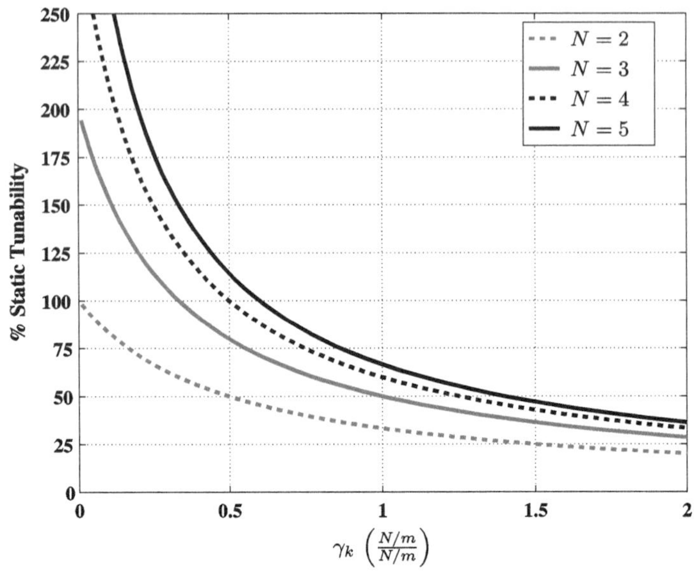

Figure 4.8 Illustration of static tunability as a function of the load stiffness ratios for $N = 2$, $N = 3$, $N = 4$, and $N = 5$.

The static behavior of a loaded cell strand can be classified by the tunability of the output stiffness, which will be defined as

$$\begin{aligned}\text{Static Tunability} &= \left(\frac{\max(\tilde{c}_S) - \min(\tilde{c}_S)}{\min(\tilde{c}_S)}\right) \times 100 \\ &= \left(\frac{N-1}{1+\gamma_k N}\right) \times 100, \end{aligned} \tag{4.33}$$

where $\tilde{c}_S$ is the equivalent compliance as viewed by forces applied at the N^{ON}th node. A plot of the static tunability as a function of N and γ_k is shown in Fig. 4.8. Beyond $\gamma_k \approx 1.5$, the tunability drops below 50% even for the larger N values. In general, the static tunability is reduced greatly if the load stiffness exceeds approximately $0.5\tilde{k}$, where the curvature of the tunability curves nears its maximum.

The dynamic behavior of a strand of cells is characterized by the ability to tune resonant frequencies. The new dynamic model is the same as described previously but with $\boldsymbol{M} = \tilde{m} \cdot \text{diag}(n_{l1}, n_{l2}, \ldots, n_{lN^{ON}} + \gamma_m)$ and the (N^{ON}, N^{ON}) entry of $\boldsymbol{K}$ given by $\tilde{k}(1 + \gamma_k)$. In practical applications, the first (i.e. fundamental) mode is the most readily excited. Therefore, the tunability of the dynamic system's fundamental frequency will be defined as

$$\text{Dynamic Tunability} = \left(\frac{\sigma_{\max,\text{first}}(1) - \Sigma_{\min}}{\Sigma_{\min}}\right) \times 100. \tag{4.34}$$

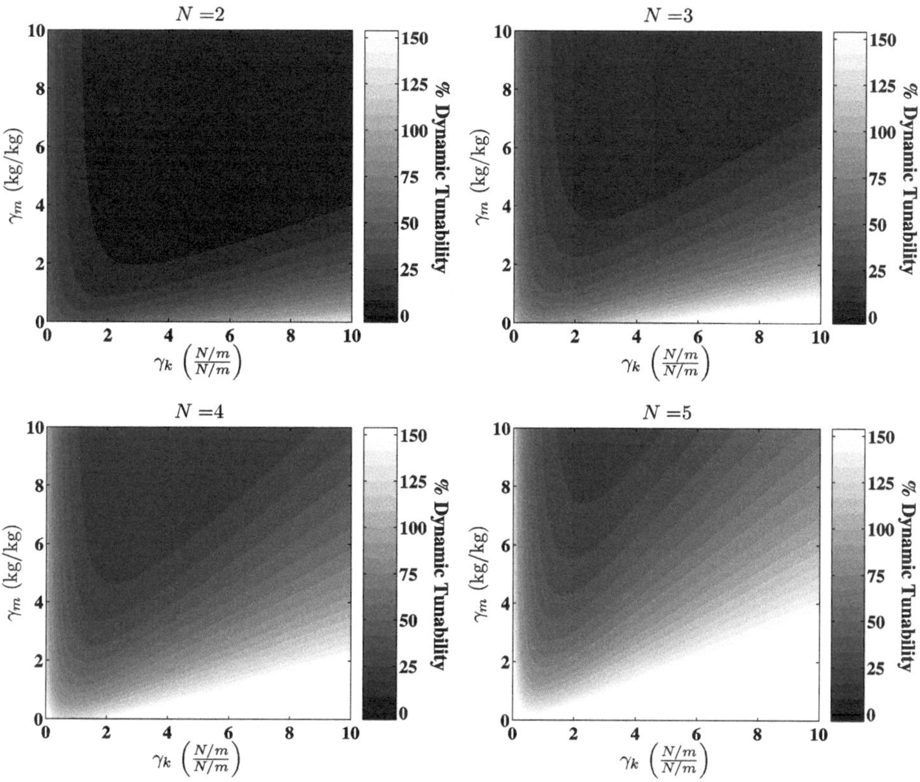

Figure 4.9 Illustration of fundamental frequency tunability as a function of the load stiffness and mass ratios for $N = 2$, $N = 3$, $N = 4$, and $N = 5$.

A plot of (4.34) is provided in Fig. 4.9 for $N = 2$ to $N = 5$. Fig. 4.9 shows that the tunability decreases as γ_m and γ_k increase. However, for low values of γ_k, increasing γ_m has a very weak effect on tunability. Likewise, for low γ_m, the tunability is only a weak function of γ_k. The dynamic tunability also increases as N is increases. Overall, the frequency spacing between possible fundamental modes decreases as the stiffness and mass of the load increase or as N decreases.

Based on numerical evaluation, the propositions described previously remain valid except under the condition when $N = 2$ and $\gamma_k = \gamma_m + 1$. In this case, the configuration having $N^{ON} = 1$ unit ON with the first unit remaining OFF (i.e., grounded) and the case having $N^{ON} = N = 2$ (i.e., all units ON) both achieve a minimum fundamental frequency of $\Sigma_{\text{min}} = \sqrt{\frac{\tilde{k}}{\tilde{m}}}$. Overall, the loading effects illustrate that this cellular tuning method is most effective when N is large and the actuator is scaled such that γ_k and γ_m remain within the high tunability regions of Fig. 4.9.

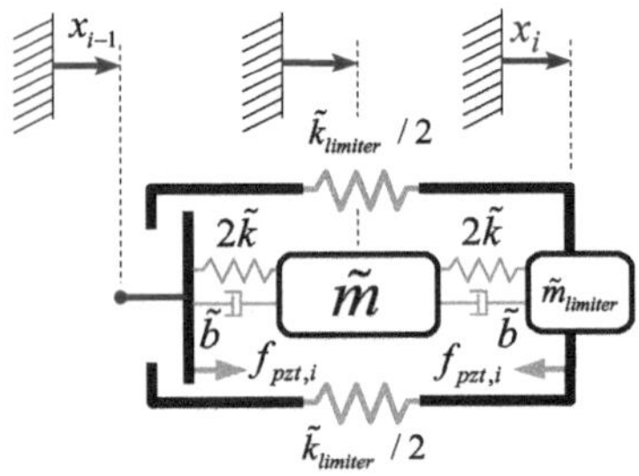

Figure 4.10 Single cell model that includes the parasitic effects of the mass $\tilde{m}_{limiter}$ and stiffness $\tilde{k}_{limiter}$ in the stroke limiting beam, as well as flexure damping $\tilde{b}$.

4.1.4 Implementation of a PZT-based VSCA

The analysis in the previous section provides useful insight into the mechanism of variable resonant frequencies and the associated tunable range under idealized conditions. This section will consider a specific implementation of the VSCA and extend the previous analysis to account for parasitic dynamics including additional mass, damping, and compliance of the stroke limiter for the PZT-based cell design. An image of the PZT-based cell (without the stroke limiter) is given in Fig. 1.37B. Each cell has a 20 N blocking force and 1.5 mm free displacement.

4.1.4.1 Single cell dynamic model

A detailed lumped parameter model for the specific PZT-based VSCA cell is shown in Fig. 4.10 for the ith cell in a strand. This model is a necessary extension of the idealized model because it includes the parasitic effects of stroke limiter mass, stroke limiter stiffness, and flexure damping that are present in the implemented system. Moreover, the new detailed model allows for a straightforward translation from the actual cell construction to the lumped parameter representation. The stroke limiter mass is lumped in to the element $\tilde{m}_{limiter}$ while the finite stroke limiter stiffness is given by $\tilde{k}_{limiter}$. The flexure stiffness is spatially divided into two springs of equal value $2\tilde{k}$ on each side of the large suspended mass $\tilde{m}$. The stiffness $2\tilde{k}$ does vary slightly with displacement, but this variation is assumed to be small for the purpose of model development. The damping effects of the flexures are assumed to be viscous in nature and modeled using a dashpot with a constant value of $\tilde{b}$. The damping of the flexures, as well as any damping in the load, will shift the resonant frequencies away from their undamped counterparts. However, this shift is small provided that $\tilde{b}$ is relatively small. The PZT-generated force at the output of the second layer is now denoted by $f_{pzt,i}$.

Each cell requires three generalized coordinates to describe its configuration with respect to an inertial reference frame. The first coordinate, x_i, determines the position of the stroke limiter mass. The second coordinate determines the position of the second layer unit suspended mass. The third coordinate determines the location of the output node junction. Note that if another cell is connected to the output node, then the third coordinate is x_{i-1}.

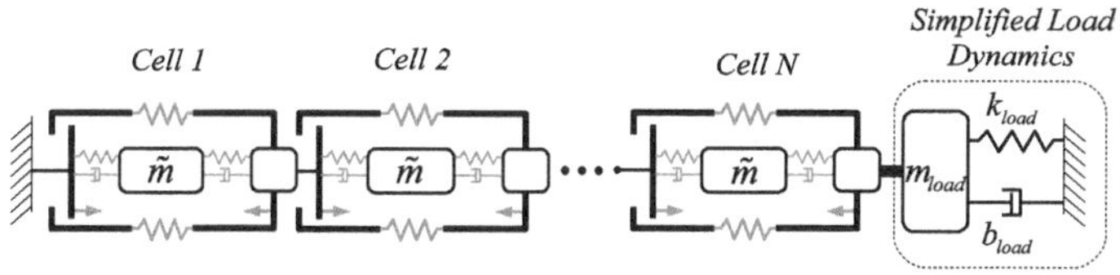

Figure 4.11 Model of an assembled strand of PZT-driven cellular units connected to a general spring–mass–damper load.

4.1.4.2 Assembled system behavior

Consider an arbitrary strand of cells having length N as shown in Fig. 4.11. The strand is connected to a load mass, m_{load}, which is further connected to ground through a parallel arrangement of a spring and dashpot having values of k_{load} and b_{load}. The arrangement depicted in Fig. 4.11 represents the usage of the actuator described in Section 4.1.5.

Within the actuator, each cellular unit behaves according to the lumped model in Fig. 4.10. For the low frequency or static behavior, mass and damping in the dynamic model may be neglected yielding a series compliance of

$$\tilde{c}_S(N^{ON}, N) = \frac{N^{ON}}{\tilde{k}} + \frac{(N - N^{ON})}{\tilde{k} + \tilde{k}_{limiter}}, \tag{4.35}$$

where the first term is consistent with the idealized model in (4.1) and the second term accounts for the finite stiffness of the stroke limiting beams.

If the effects of mass and damping are included, then the dynamic model for a serial chain of units, as well as the dynamic models presented in [221], can be written in the standard vibratory form as

$$\boldsymbol{M}\ddot{\boldsymbol{q}} + \boldsymbol{B}\dot{\boldsymbol{q}} + \boldsymbol{K}(t)\boldsymbol{q} = \boldsymbol{F}(t). \tag{4.36}$$

The vector $\boldsymbol{q}$ contains the generalized coordinates of the strand while the vector $\boldsymbol{F}(t)$ contains the piezoelectrically generated forces, $f_{pzt,i}(t)$, acting within each cell. For simplicity, these forces are taken as the voltage applied to the cell $V_{pzt,i}$, scaled by an constant electromechanical transduction factor α.

The parameters of the physical system used in the theoretical model are given in Table 4.2. The stiffness parameters were taken from finite element simulation. As is standard practice in vibration models, the damping $\tilde{b}$ was used as a free parameter. Therefore, the damping constant $\tilde{b}$ was tuned to match the experimentally observed resonant peak amplitudes.

4.1.5 Experimental results

4.1.5.1 Static stiffness tunability

To demonstrate that the static tunability obeys (4.35), the stiffness of a serial chain having $N = 5$ was measured using the experimental apparatus shown in Fig. 4.12.

Table 4.2 **Model parameters.**

Parameter	Value	Units
$\tilde{k}$	16.2	N/mm
$\tilde{k}_{limiter}$	80.1	N/mm
$\tilde{b}$	0.62	N · s/mm
$\tilde{m}_{limiter}$	2.5	g
$\tilde{m}$	26.0	g
m_{load}	21.9	g
k_{load}	0.23	N/mm
b_{load}	≈ 0.62	N · s/mm

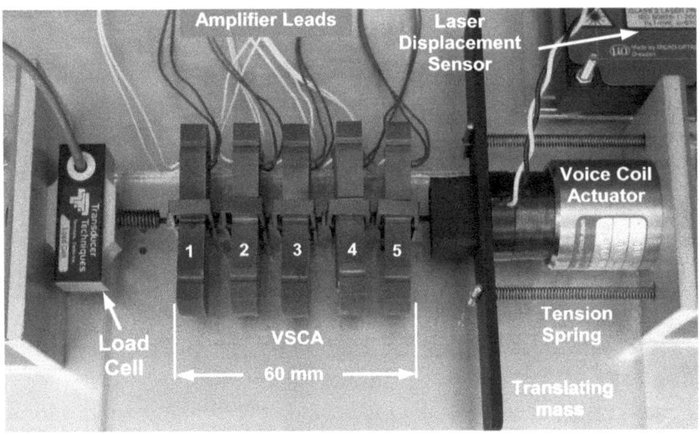

Figure 4.12 Experimental apparatus for measuring static and dynamic properties of a 5-cell system.

With $N = 5$, there are $2^5 = 32$ possible arrangements of ON–OFF units. As shown in Fig. 4.12, the serial chain was connected to a voice coil actuator and preload tension springs. Units were turned ON or OFF by applying voltages of 150 and 0 V, respectively. Voltages were generated using a Cedrat CA-45 amplifier. The ON–OFF switching was performed with a manual switchboard. With the voice coil actuator, a 1 N peak–peak sinusoidal force was applied at 0.5 Hz about the preload force of 8 N. In general, some preload force is necessary to ensure that the units remain OFF when they receive a zero voltage input and that non-longitudinal vibration modes remain at high frequencies. Forces were measured using a Transducer Techniques load cell while strand displacements were measured using a MicroEpsilon laser displacement sensor. All measured signals were sampled at 1 kHz with a National Instruments data acquisition board.

The results for the static experiment are shown in Fig. 4.13. For each N^{ON} there were $\binom{N}{N^{ON}}$ measured data points. Each compliance data point was generated from a least squares fit to the force versus displacement data. According to (4.35), the equivalent series compliance is constant for a fixed value of N^{ON}. However, in the real

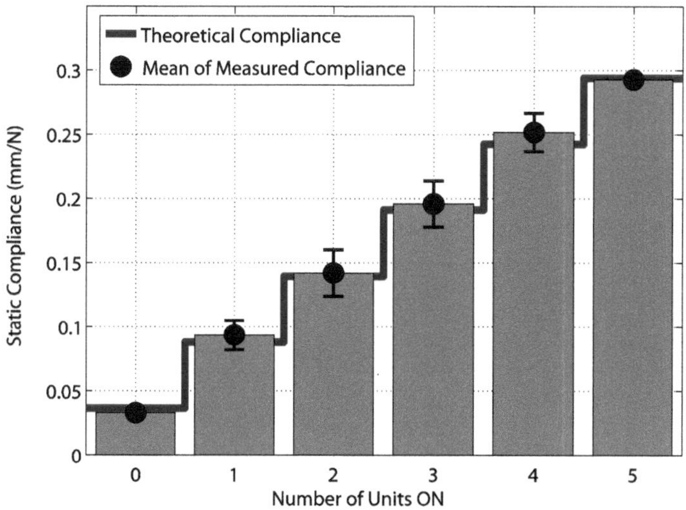

Figure 4.13 Experimentally measured compliance compared to the predicted values based on theoretical stiffness.

system, there are variations in the series compliance for a given N^{ON} due to individual cell differences. The mean and standard deviations of the measured data were computed for each value of N^{ON}. In Fig. 4.13, the mean is indicated by a circle while the vertical bars represent one standard deviation about the mean. The solid staircase line is the predicted compliance based on (4.35). The main cell–cell difference occurs in the slight dimensional variation of the second layer flexure angle, which can have a notable influence on the stiffness $\tilde{k}$. Other causes of stiffness variation are the slight geometric nonlinearity of the flexure and the variation in ON versus OFF stiffness of the PZT stacks. Despite these sources of variation, the model and the data agree well over the entire range of ON–OFF cases and both illustrate the large tunable range for static compliance. For the experimentally considered case of $N = 5$, the compliance tunability computed from (4.33) is over 350%.

4.1.5.2 Resonant frequency tunability

This experimental section reinforces and illustrates the theoretical concepts developed in Sections 4.1.3 and 4.1.4. Frequency response tests were conducted for a VSCA with $N = 3$. The case with $N = 3$ units provides $2^3 - 1 = 7$ possible cases that can be clearly illustrated on frequency response plots. The experimental apparatus is the same as that shown in Fig. 4.12 only oriented so that gravity is acting along the strand.

A chirp voltage input was used to obtain the frequency response characteristics of the actuator. The chirp signal ranged from 5 to 150 Hz and contained a DC offset of 145 V. The peak-to-peak voltage amplitude was 10 V.

The theoretical model and experimental model are compared in Fig. 4.14 for all of the 7 possible cases. Each case is given a binary number where the most significant

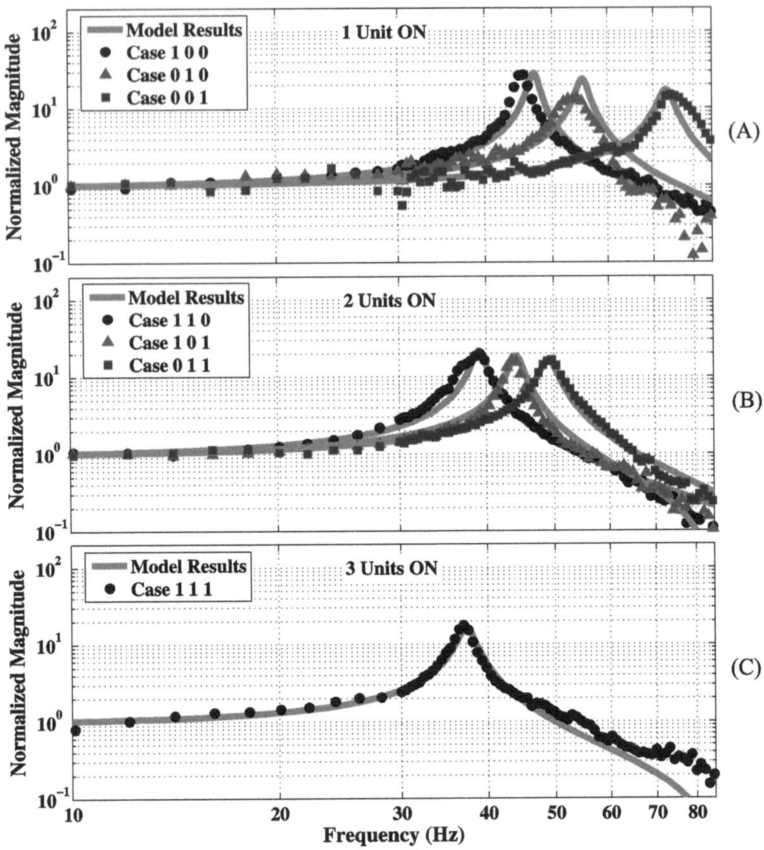

Figure 4.14 Experimental results demonstrating the variable resonance concept for three serially connected units: (A) cases with one unit ON, (B) cases with two units ON, and (C) all three units ON.

bit denotes the unit connected to ground and the least significant bit denotes the unit connected to the load. The gain in the frequency response is normalized with respect to the DC displacement. Therefore, the vertical axes in Fig. 4.14 represent the frequency dependent amplification of displacement. The average measured amplification at the fundamental resonance was 17.8 with a standard deviation of 4.4. When comparing the fundamental frequency location between the model and the experimental results, the coefficient of determination is very high ($R^2 = 0.993$) for the entire tuning range of nearly 38 Hz.

For the loaded strand, the load mass and stiffness ratios can be computed from the parameters in Table 4.2 as $\gamma_m = 21.9/26.0 = 0.84$ and $\gamma_k = 0.23/16.2 = 0.014$. From Fig. 4.9 for the case with $N = 3$, the tunability as predicted by the idealized model is over 100%. Furthermore, as indicated by theoretical propositions, the lowest fundamental frequency ($\Sigma_{\min} = 37.5$ Hz) was obtained in the [1 1 1] case while the highest fundamental resonant frequency ($\sigma_{\max,\text{first}}(1) = 75.2$ Hz) was obtained was in the [0 0 1] case. Computing the dynamic tunability from (4.34) yields a value of over 100%, which agrees well with the idealized model.

4.1.6 Discussion

This section presented two main contributions: (i) quantification of the essential theoretical properties of VSCAs, and (ii) the implementation and testing of a novel PZT-based VSCA design. The change in both stiffness and resonant frequencies is achieved by selectively turning variable stiffness units ON or OFF within a serial strand. Experiments and models illustrate the shift in resonant frequency based on a simple change in ON–OFF configuration. The experimental results illustrate three key features of the VSCA approach.

1. The VSCA approach allows for a wide tunability of both the static stiffness and the first resonant peak in an assembled system. For the experimental system having only $N = 3$ units, the tunability exceeds 100%. The tunability is also increased as N becomes larger and as higher modes are excited.
2. The results show the high degree of dynamic displacement amplification (≥ 5) when there is low damping in the system. Hence, the pre-existing large static amplification produced by the nested flexures is augmented in the VSCA approach such that resonant peaks occur over a wide range of frequencies useful for robotic locomotion and manipulation.
3. The experimental results and the model prediction are in close agreement for nearly all cases tested. The small model residuals, especially near resonance, validate linearity assumptions and demonstrate a sound model for the dominant energy storage and dissipation modes in the driven system.

The PZT-based actuators considered in this section may be valuable in many biorobotics applications because they achieve static strain that is commensurate with skeletal muscle (10–20%) and their strain is further amplified under resonance conditions. Furthermore, VSCA designs based on PZT actuator cells are durable, low power, high bandwidth devices. One particular advantage of PZT is its near-zero power consumption when statically holding a load.

The tunability of the piezoelectric actuator also holds particular promise as an energy harvesting device. Harvesting devices that vary resonant frequency do exist (e.g., [146,37]), but such systems require an additional actuator to tune the resonant frequency. The cellular PZT system described provides a means for easily tuning resonance over a wide range in real time without the use of additional actuators. This may be very advantageous for mobile robots capable of passively gathering energy from their environment (e.g., fish robots within a flow). Further extensions of the technology include using cells with differing stiffnesses to achieve more uniform modal density across the tuning range. The theoretical treatment may also be extended to include parallel and antagonistic strands.

4.2 Bipolar buckling actuators

4.2.1 Strain amplification by structural buckling

Buckling is a pronounced nonlinear effect that occurs at a structurally singular point. A small piezoelectric displacement on the order of 10 μm results in a large buckling displacement on the order of mm. Furthermore, the usable stroke is doubled if both sides of the singular point can be reached resulting in bipolar motion. Despite the large gain, buckling is an erratic, singular phenomenon; the side on which deflection will occur is unpredictable. In this section multiple design concepts are presented for regulating the buckling direction as well as for extending its usable stroke to bipolar motion.

The key idea is to exploit "buckling", a pronounced nonlinearity of structural mechanics. The presented method produces a displacement amplification ratio of 100 without the addition of much structure to the piezoelectric elements. Consider a tall PZT stack constrained between two rigid walls. It may buckle as it tends to expand. Suppose now that the tall PZT stack is evenly divided into two stacks connected in series with a rotational joint in the center, with additional rotational joints connecting the other ends of the PZT stacks to the walls, as shown in Fig. 4.15. Then, the buckling direction is constrained within a plane, and a large displacement occurs at the central rotational joint between the PZT stacks, called a keystone. The keystone displacement is typically 100 times larger than the displacement of the individual PZT stacks.

While this nonlinear and singular phenomenon can produce an order-of-magnitude larger displacement amplification than typical flexure based methods, buckling is an erratic, unpredictable phenomenon; although constrained in a plane, the direction in which buckling will occur is unpredictable. The objective of the current work is to regulate the buckling direction and extend the usable stroke to both sides. The resultant effective stroke of bipolar motion can be more than 200 times larger than the displacement of the individual PZT stacks.

Two kinematic methods for regulating buckling direction and doubling the usable stroke are discussed. The first kinematic method utilizes a stiff mechanical element to redirect the PZT stack force [283]. The present work expands upon the previous work by introducing a refined model validated by a newly presented functional prototype incorporating a redirecting stiff element. The second kinematic method utilizes a plurality of buckling actuators [224]. The present work contributes to previous dual buckling actuator work by expanding the model to incorporate the kinematic change that occurs when a PZT stack no longer bears a compressive load. Additionally this new model is validated by careful experimentation with new precise results.

Although force redirecting mechanical elements [201,10] and multiple flextensional units have been used for balancing kinematic properties [80], to the authors' knowledge, neither method has been used to control displacement through a singular configuration.

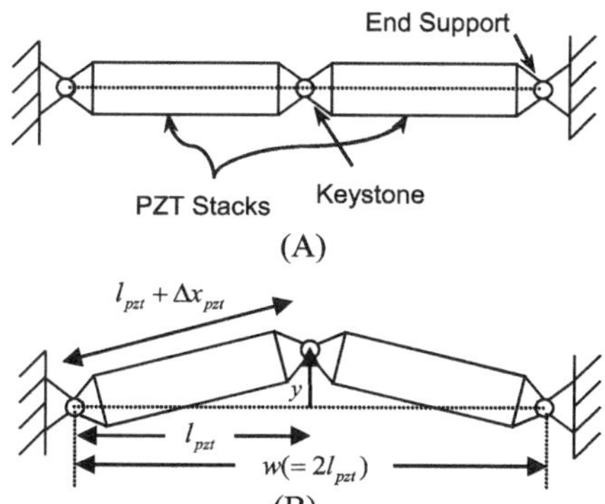

Figure 4.15 Kinematics of a single buckling unit: (A) unactivated, (B) activated. © 2014 IEEE, reprinted with permission

4.2.2 Buckling for large displacement amplification

Figs. 4.15A–B show the schematic of a PZT buckling actuator, consisting of a pair of PZT stacks, a keystone connecting the two PZT stacks in the middle, and end supports that secure the PZT stacks to two rigid walls at both ends. The keystone and the end supports are ideally rotational joints that rotate within a plane.

When the PZT stacks are activated, they tend to elongate, generating a large stress along the longitudinal direction. When the two PZT stacks are completely aligned, the longitudinal forces cancel out, creating an unstable equilibrium. With any disturbance, the two PZT stacks tend to rotate, i.e., "buckle", as shown in Fig. 4.15B. Let Δx_{pzt} be the elongation of each PZT stack and y be the vertical position of the keystone output node. Since $y = \sqrt{(l_{pzt} + \Delta x_{pzt})^2 - l_{pzt}^2}$, the displacement amplification ratio, $a = \Delta y / \Delta x_{pzt}$ can be computed as

$$a = \frac{\partial y}{\partial \Delta x_{pzt}} = \frac{l_{pzt} + \Delta x_{pzt}}{\sqrt{(l_{pzt} + \Delta x_{pzt})^2 - l_{pzt}^2}} = \frac{l_{pzt} + \Delta x_{pzt}}{y} \approx \frac{l_{pzt}}{y} \to \infty, \text{ as } y \to 0. \tag{4.37}$$

Note that the amplification gain a tends to infinity as y approaches 0. This is a type of kinematic singularity. Even for a finite piezoelectric displacement, the amplification gain, a, is significantly large.

Fig. 4.16 plots the amplification gain, a, as a function of the output displacement, y. The force–displacement characteristics of the PZT buckling actuators are highly nonlinear. Despite the large displacement, the maximum force that can be produced is still large due to the high efficiency of the energy transfer.

It is not feasible to quasi-statically bring the output keystone from one side to the other across the middle singular point. Once the keystone goes upwards, it tends to stay there, and vice versa. This is in a sense "mono-polar" activation, in which the stroke of the output keystone is half of the total possible displacement. Therefore,

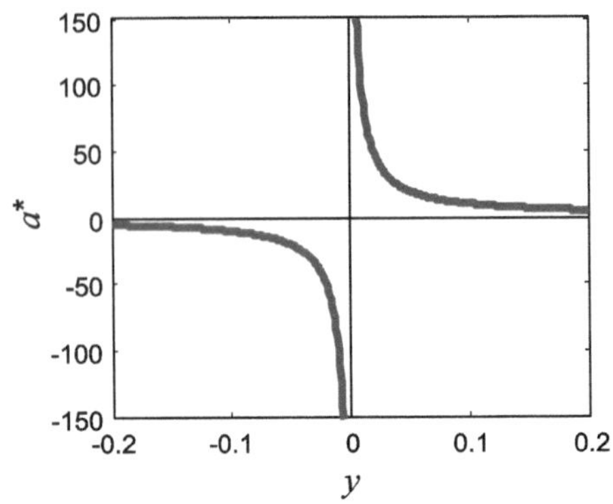

Figure 4.16 Displacement gain of a single buckling unit a^* (normalized for $l_{pzt} = 1$) as a function of output displacement y. © 2014 IEEE, reprinted with permission

it is desirable to both control the buckling direction and have the capability to pass through the singularity point to the other side once buckling has occurred. This bipolar activation can double the displacement stroke.

4.2.3 Redirecting stiffness

The buckling direction may be controlled if asynchronous activation is used for the two PZT stacks and the mechanical structure is modified accordingly. As shown in Fig. 4.17A, an additional stiffness element indicated by a spring is placed in an oblique orientation. If the left PZT stack is first activated, then a positive displacement of Δx is generated in the output keystone as shown in Fig. 4.17B. Because of the redirecting stiffness, this displacement in the x-direction generates a force in the y-direction and forces the buckling unit to buckle upwards, Fig. 4.17C. After buckling, the right PZT stack may be activated to further push the output keystone upwards. Similarly, if the right PZT stack were to be activated first, a negative x-displacement would be generated, causing a force from the redirecting stiffness to cause the buckling unit to buckle downwards.

The key to this controlled buckling via lateral activation is the two-dimensional stiffness in which the output keystone is suspended. Let $\boldsymbol{K}$ be a 2×2 matrix relating the restoring force vector $(F_x \ F_y)^T$ acting on the keystone to the displacement vector of the keystone $(\Delta x \ \Delta y)^T$ as

$$\begin{pmatrix} F_x \\ F_y \end{pmatrix} = -\boldsymbol{K} \begin{pmatrix} \Delta x \\ \Delta y \end{pmatrix}, \boldsymbol{K} = \begin{pmatrix} K_{xx} & K_{xy} \\ K_{yx} & K_{yy} \end{pmatrix}. \tag{4.38}$$

If the off-diagonal element is less than zero ($K_{xy} < 0$), then the y-directional force becomes positive ($F_y > 0$) in response to positive Δx, initially generated by the left PZT stack. Note that $\boldsymbol{K}$ is a positive definite, symmetric matrix with two principal axes associated with two positive eigenvalues, $\lambda_{max} > \lambda_{min} > 0$. As long as $K_{xy} \neq 0$, the two eigenvalues are distinct, and the directions of the principal axes are not aligned

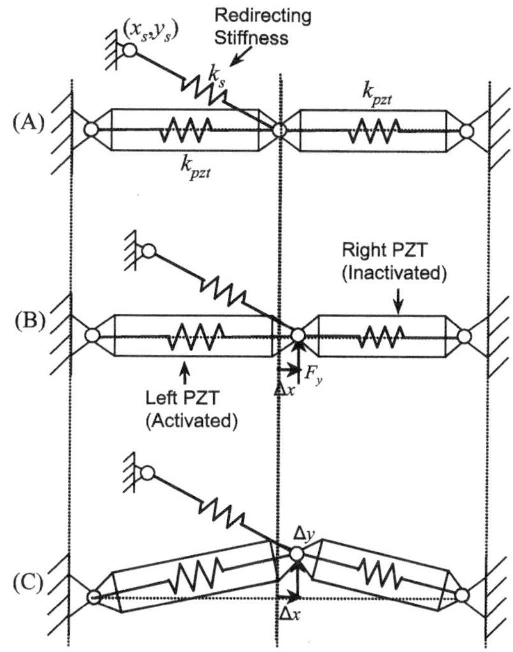

Figure 4.17 Simplified static model of PZT buckling mechanism with redirecting stiffness. Force in the y-direction is generated by displacement in the x-direction. © 2014 IEEE, reprinted with permission

with the x- and y-axes, the longitudinal and transverse directions of the dual PZT stacks. This condition can be realized with numerous structure designs.

4.2.4 Dual buckling unit mechanism

Consider two sets of buckling actuator units arranged next to each other with output axes aligned. As shown in Fig. 4.18, the two keystone output nodes, B and E, are connected with a rigid coupling that is shorter than the distance between the two sets of buckling actuators, $\overline{BE} < \overline{AD} = \overline{CF}$. In this configuration, if the bottom buckling unit DEF is activated first, the keystones move upwards, and the keystone of the top inactivated unit B may pass through its singular point as shown in Fig. 4.18A at time t_2. Then, if the top buckling unit ABC is activated, the output nodes will move further upwards as shown in Fig. 4.18A at time t_3. At this time both units work together to generate force and displacement in the upwards direction. Conversely, if the top buckling unit is activated first followed by the bottom, then the output will first move down (as in Fig. 4.18B at time t_2) and then move down further (as in Fig. 4.18B at time t_3). In both cases, the active unit forces the inactive unit through its singularity position. Once the output nodes of both units are on the same side of their respective singular points, activation of both units causes further displacement in the same initial direction as shown in Figs. 4.18A–B at time t_3. This actuation method requires the

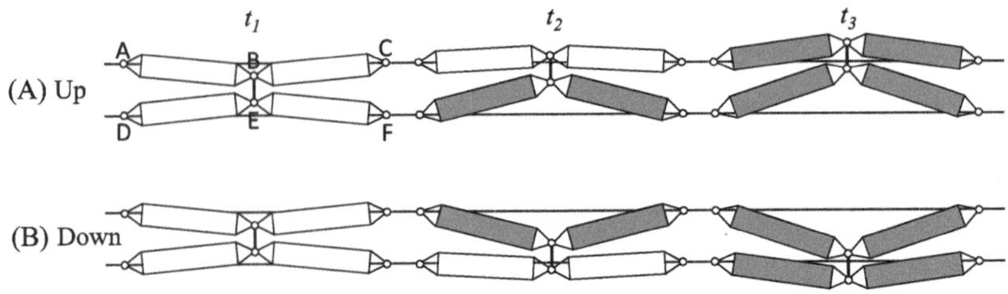

Figure 4.18 Asynchronous activation time sequence (t_1, t_2, t_3) of a dual buckling unit phase-shifted actuator, showing (A) upwards free displacement, and (B) downwards free displacement. White and gray represent inactive and active PZT stack actuators, respectively. © 2014 IEEE, reprinted with permission

top and bottom buckling actuator units to be activated asynchronously or temporally out of phase. This phased array actuator utilizes both being out of phase spatially and being activated out of phase temporally.

4.2.5 Force–displacement analysis

4.2.5.1 Basic properties of buckling mechanism

The buckling actuator can be modeled as a series of two springs in parallel between two grounded nodes, as shown in Fig. 4.19. The activation of the PZT stacks is modeled as effectively changing the rest length of the springs from l_{pzt} to $l_{pzt} + \Delta x_{pzt}$. For the ideal actuators, the length of each spring at the singularity point shown in Fig. 4.19A would be equal to the rest length of each spring. However, this is not physically realizable due to manufacturing tolerances, nor necessarily desirable due to preload benefits (as described in Section 4.2.7). A preload is achieved by positioning the grounded ends of the buckling actuator a preload displacement length closer to the center than the rest length of the spring, as shown in Fig. 4.19B.

With this model, the potential energy U stored in a spring can be calculated as a function of free displacement of the PZT stack at a given voltage input, Δx_{pzt}, and output displacement y. From Fig. 4.18B,

$$U = \frac{1}{2}\tilde{k}\left((l_{pzt} + \Delta x_{pzt}) - \sqrt{y^2 + \left(l_{pzt} - l_{preload}\right)^2}\right)^2. \tag{4.39}$$

The negative derivative of U with respect to y yields the force along the output axis exerted by the buckling actuator,

$$F = -\frac{dU}{dy} = -\tilde{k}y + \frac{(l_{pzt} + \Delta x_{pzt})\tilde{k}y}{\sqrt{y^2 + (l_{pzt} - l_{preload})^2}}. \tag{4.40}$$

A plot of the force vs. output displacement is shown in Fig. 4.20 for two activation levels, $\Delta x_{pzt} = 0$ (OFF) and $\Delta x_{pzt} = \Delta x_{pzt,max}$(ON). The parameter values for these plots are based on prototypes using 40 mm PZT stack actuators with a maximum free

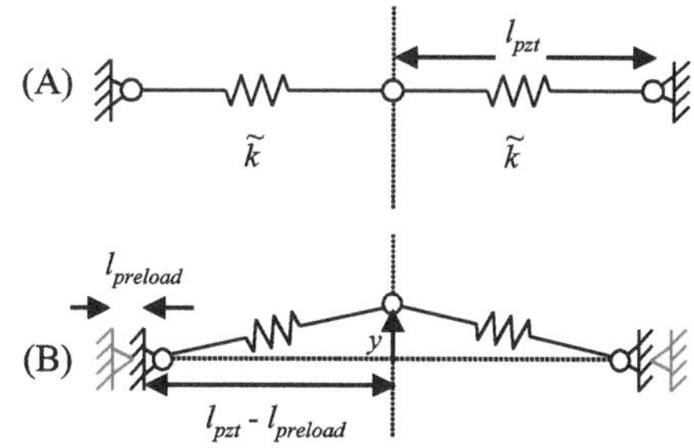

Figure 4.19 Simplified static model of PZT buckling mechanism at (A) singularity point ($y = 0$) and (B) at finite displacement ($y \neq 0$). © 2014 IEEE, reprinted with permission

displacement of $\Delta x_{pzt,max} = 42$ μm, (NEC TOKIN AE0505D44H40). The spring constant $\tilde{k}$ is calculated based on the geometry and material properties of the steel flexure joints as well as the stiffness of the PZT stacks [180]. The bending stiffness of the flexures was calculated and accounted for.

The negative slope of the force-displacement curve, i.e., the stiffness, is also plotted with broken lines in Fig. 4.20. Note that the stiffness is negative in the middle, which implies that instability occurs in the vicinity of $y = 0$. This singular point is unstable even for $\Delta x_{pzt} = 0$ (OFF), as a preloading displacement is applied. The preload displacement $l_{preload}$ depends on manufacturing tolerances and the necessity of a compressive load in the PZT stacks. In the following analysis, $l_{preload}$ is conservatively modeled as equal to the maximum free displacement $\Delta x_{pzt,max}$.

There are a few important features to note about a single buckling actuator that can be seen in Fig. 4.20.

Remark 1. The significant displacement amplification of the actuator. The displacement in a single direction is greater than 1.5 mm at full activation, i.e., 150 V. Utilizing bipolar actuation, the total displacement is greater than 3.0 mm. Compared to the maximum free displacement of the PZT stack (42 mm) at 150 V, this buckling actuator produces 71 times larger displacement.

Remark 2. Nonlinear force–displacement characteristics. The maximum force is not generated at zero-displacement ($y = 0$) as is the case for solenoids and typical PZT stack actuators, but at a point in the middle of the stroke. In this ideal model, no force is generated at the singular point at $y = 0$, and in realized prototypes this force is very small compared to the maximum achievable force.

Remark 3. Varying stiffness. Near the singular point, the stiffness is zero or negative, whereas it increases sharply as displacements get larger in both directions.

The feature from Remark 2 above significantly differs from the inherent PZT stack properties and the properties of conventional displacement amplification mechanisms, in which the peak force, i.e., blocking force, is created when no displacement is made; the output force decreases monotonically as displacement increases. In contrast, the buckling actuator produces its peak force mid-stroke. This nonlinear force–displacement relationship is useful, as we exploit in designing multi-unit actuators.

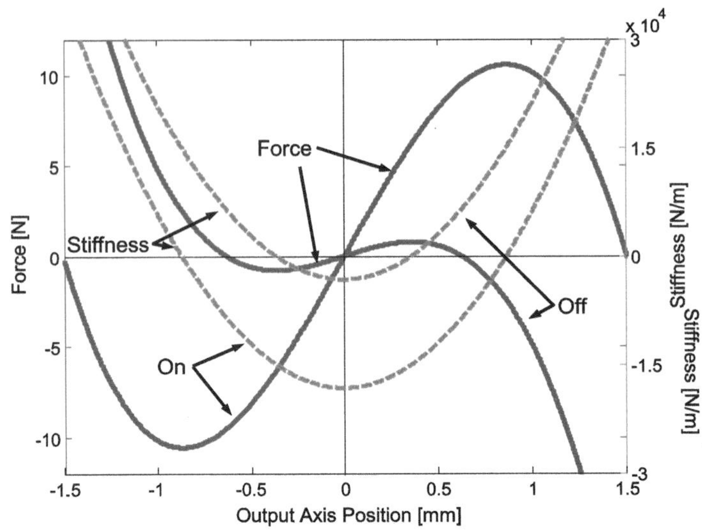

Figure 4.20 Stiffness and force of the actuator output node along the output axis as a function of output node position for two activation levels, 0 and full activation. © 2014 IEEE, reprinted with permission

From Remark 3, the buckling actuator exhibits a unique stiffness characteristic; stiffness becomes zero, or even negative with nonzero activation level, in the vicinity of the singular point. This allows us to control the buckling direction of a single unit with the redirecting stiffness mechanism. Additionally, the stiffness at the singularity is also used for arranging multiple units in an array, as with the dual buckling unit design. When one unit moves in the vicinity of the singular point, it is effectively "disengaged" from other unit(s), so that it may not be a "load" for the other units producing forces.

4.2.5.2 Single buckling actuator with the redirecting stiffness

Consider the simplified static model of a buckling unit with a redirecting stiffness shown in Fig. 4.17. By incorporating the potential energy associated with the redirecting stiffness, U_{red}, into the potential energy equation, (4.39), we can compute the force–displacement relationship with two independent activation levels of the PZT stacks. Because of the redirecting stiffnesses effect on symmetry, the output force and displacement are now two dimensional, having both x- and y- components. Fig. 4.21 shows the iso-potential curves of a few demonstrative combinations of right and left PZT activation levels: (A) both not activated, (B) only left activated, and (C) both activated. Points in the potential energy field where any movement along either axis results in additional potential energy, or "valleys", represent unforced, steady state positions of the actuator. The output keystone is forced towards these valleys.

For the simulation, we model PZT stacks with length $l_{pzt} = 40$ mm, maximum free displacement $\Delta x_{pzt,max} = 42$ µm, and stiffness 2.0×10^7 N/m. Flexure joint compressive and bending stiffness values were modeled as steel beams as described

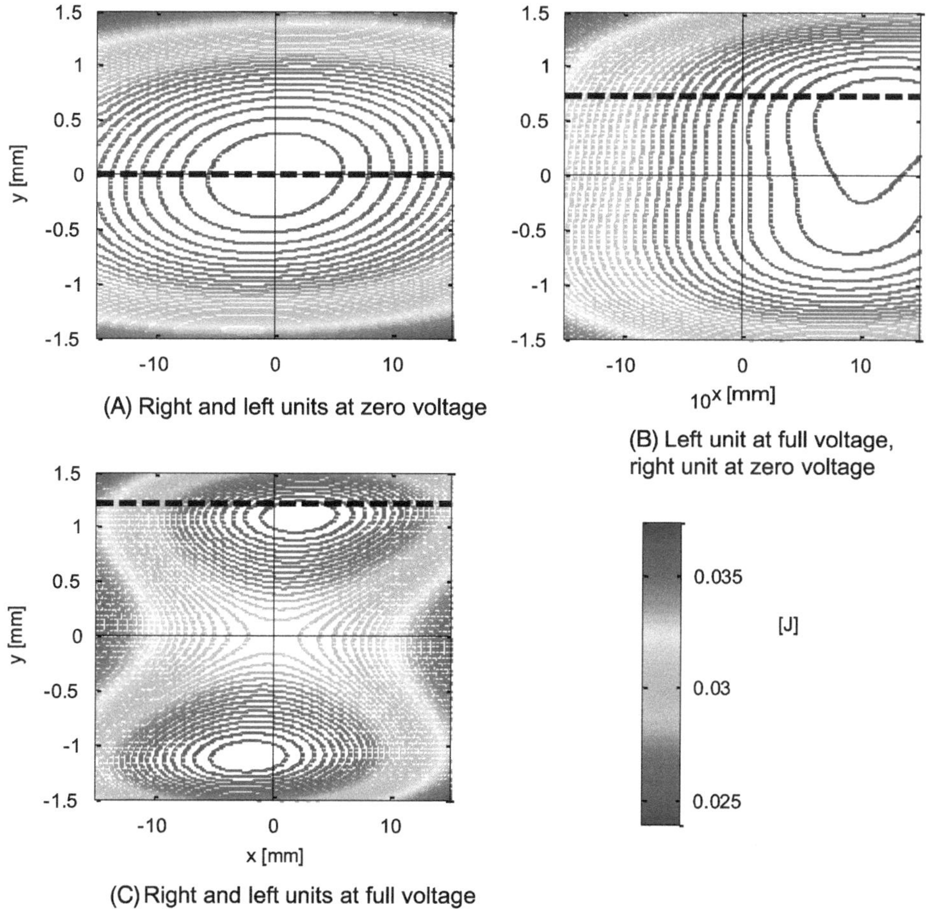

Figure 4.21 Isopotential energy curves in joules as a function of output axis position, y, and lateral position, x, of a buckling actuator with redirecting stiffness. The dashed line in each plot represents realized prototype output displacement performance. © 2014 IEEE, reprinted with permission

in Section 4.2.7. The ratio of $-x_s$ to y_s is $y/(2l_{pzt})$ where y is the pre-computed simple single buckling unit free displacement in the output direction. The grounding point of the redirecting spring is in the second quadrant and its stiffness, k_s, is designed to be approximately equal to the stiffness of the series stiffness of the PZT stacks and flexure joints. This rule-of-thumb value provides a good balance of providing a high K_{xy} vs. increasing K_{yy} which degrades performance, but this value may be adjusted depending on required controllability and performance requirements.

When both units are inactive, there is a single valley at the origin (Fig. 4.21A) and the output is forced toward the origin from any other point. When a single PZT unit is active, there is a nonzero force at the origin, and the output node is directed towards a valley away from the origin (Fig. 4.21B). When both units are fully activated, there are two valleys on either side of the buckling position with a saddle point at the origin, showing that if the output has been positioned on a specific side, it will stay on that side.

There are two subtle but important points. First, the axes of symmetry are aligned with the eigenvectors of the stiffness matrix, $\boldsymbol{K}$, of (4.38). Second, it can be shown that there is a finite force at the origin when the activation levels of the input stacks are different for finite values of x_s, y_s, and k_s. There may be two valleys, and thus two stable positions that exist for just one unit fully activated depending on the relative stiffness, k_s/k_p, and orientation of the redirecting element, $-x_s/y_s$. However, the origin cannot be a saddle point; there will always be a finite force towards only one of those valleys. Therefore, the buckling direction is always controllable for a finite range of external force.

4.2.5.3 Dual buckling actuator

As with the simulation for the single unit buckling actuator, the series stiffnesses of the piezoelectric actuators and axial stiffness of the joints were modeled as springs, while the activation levels of the piezoelectric actuators were modeled as effectively changing the rest length of those springs. A significant change, however, is when the actuators are modeled in tension. In implementation, unless a local preload or coupling is used, the PZT stacks begin to lose firm contact with the structure as they go into tension. Therefore, this highly nonlinear phenomenon must be modeled. It is simply modeled as a change in axial stiffness by a factor of 0.2 when in tension. Additionally, the bending stiffness of the flexures was modeled as rotational springs.

Fig. 4.22A shows the potential energy curves when both units are inactive. Depending on the rest angle, θ_0, shown in the figure, the potential curve shows a different pattern. If θ_0 is large enough, as is 2 degrees, there exists a global minimum, and the output is forced towards the origin at every displacement value. Even if the output displacement of one of the buckling units is extended past its singularity point, the actuator still provides a restoring force to the zero displacement position. In contrast, the origin is unstable for smaller rest angles. This is because the preloading displacement causes instability in the vicinity of a singular point. Notice that for $\theta_0 = 1$ [deg], there are 3 local minima.

When a single unit is active, as shown in Fig. 4.22B, the slope of the potential curve at the origin becomes nonzero for $\theta_0 > 0$. A nonzero slope means that at the rest position, the output nodes will be forced in one direction. Notice that in Fig. 4.22B as θ_0 increases, the local maximum on the left-hand side gets higher, providing a higher energy barrier to prevent the actuator from moving in the unwanted side. This gives more control over the output motion, but at the cost of efficiency because a greater amount of energy from the input actuators is converted to strain energy within the actuator when both buckling units are activated. This can be seen by the less deep wells for $\theta_0 = 2$ [deg] in Fig. 4.22C. When both units are active, two symmetric unforced equilibrium points exist at nonzero displacement. These equilibrium displacements are greater in magnitude than the equilibrium displacements from activating just one unit. Thus, the simulation shows that activating one unit, followed by the other after the output is beyond the zero displacement point, is a method of controlling the buckling direction of the actuator.

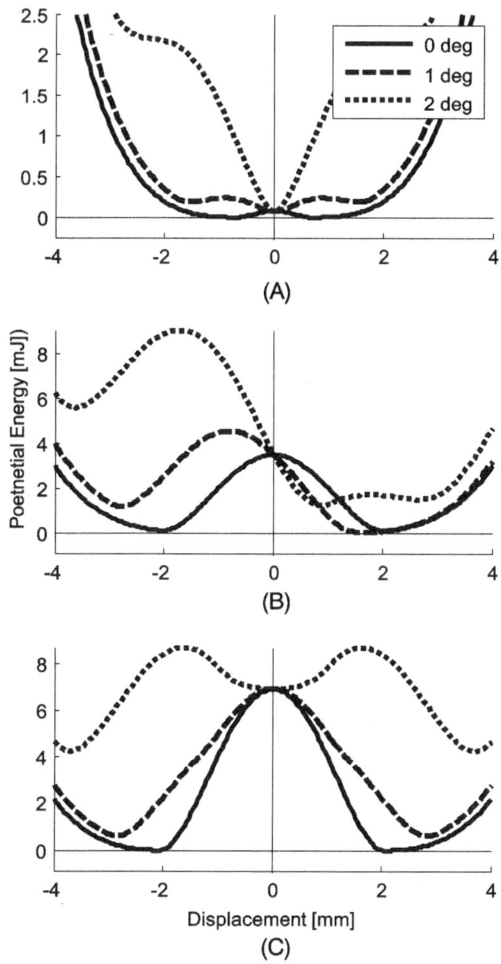

Figure 4.22 Potential energy vs. displacement simulation plots of dual-unit out-of-phase actuator when (A) both units inactive, (B) left unit active, and (C) both units active. © 2014 IEEE, reprinted with permission

4.2.6 Dynamic bipolar motion

4.2.6.1 Nonlinear dynamic model

Due to the prominent nonlinearity of the buckling mechanism, the dynamics of the buckling actuators are highly complex. This section analyzes dynamic behaviors of the buckling actuator coupled to a simple mass–damper load. The analysis will focus on bipolar motion, which doubles the stroke.

This analysis is intended to directly evaluate and clearly describe insightful phenomenon about the unique mechanism presented. Linearization about the equilibrium points clearly shows stable oscillations about two points, and a saddle point at the zero displacement point. Further analysis will show that trajectories that lay outside the two critical trajectories (separatrices) that meet at the saddle point will generate bipolar oscillations. Derivation of equations of motion for a similar system can be found in Nayfeh and Mook [179]. More in depth analysis for a bistable piezoelectric generator with similar dynamics can also be found in [244].

Let $\tilde{m}$ be the total effective mass of the load and the buckling actuator reflected to the output node, and b the total damping coefficient at the output node. From (4.40) we obtain the following nonlinear dynamic equation:

$$\tilde{m}\ddot{y} + \tilde{b}\dot{y} + \tilde{k}y = \frac{(l_{pzt} + \Delta x_{pzt})\tilde{k}y}{\sqrt{y^2 + (l_{pzt} - l_{preload})^2}}. \tag{4.41}$$

The load may also be modeled as a mass–spring–damper using this model if the stiffness of this load is simply added to the stiffness, $\tilde{k}$, of (4.41). Note that this is a simplified model. While the mass is distributed across the buckling mechanism, it is lumped to a single effective mass at the output node. Furthermore, the mass is assumed to be constant despite the fact that the PZT stacks rotate and their effective mass changes accordingly. These assumptions are justifiable since the angular displacement is very small, and small rotational inertia of the stacks combine to form a negligible effect on the effective mass of the system.

Fig. 4.23 illustrates representative phase plane plots of undamped ($\tilde{b} = 0$) and damped system governed by (4.41), where the input was set to zero, $\Delta x_{pzt}(t) = 0$, and a preloading offset length, $l_{preload}$, was equal to the maximum free displacement. Note that in Fig. 4.23A there exists an undamped trajectory that passes through the origin ($y = 0$, $\dot{y} = 0$). This is a critical trajectory separating bipolar motion from mono-polar motion for the undamped system. Any state of position and velocity that lies within this trajectory is part of a trajectory staying only on one side of the singularity, meaning mono-polar motion. Conversely, any state that is outside of the critical trajectory is part of a trajectory that passes through the singular point and sweeps both sides of the singularity, yielding bipolar motion. For the damped system, shown in Fig. 4.23B, the trajectory starting off at a point outside the critical trajectory exhibits bipolar motion, going across the singular point for the first few cycles, but ends up with mono-polar oscillation as it enters the inside of the critical trajectory.

4.2.6.2 Equilibrium points and phase plane analysis

There are three equilibrium points in the system of (4.41); two are stable, and one is unstable. In the nonlinear dynamics literature, an equilibrium point is also called a singularity point. To differentiate it from the kinematic singularity discussed in this section, those dynamic singularity points are called equilibrium points. We first look at the properties of the unstable equilibrium point ($y = 0$, $\dot{y} = 0$) and show that the point is a saddle point for both damped and undamped systems. This can be shown by linearizing the system at an equilibrium point and examining the poles of the linearized system.

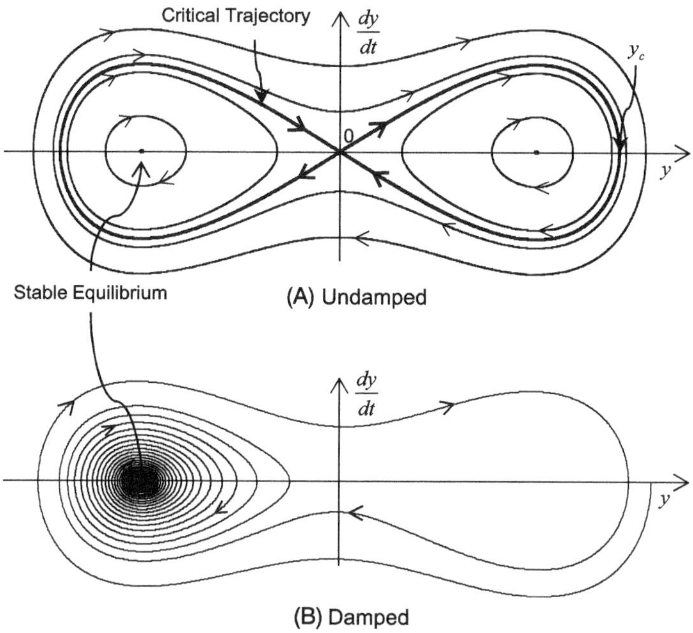

Figure 4.23 Phase portraits of undamped (A) and damped (B) single buckling unit dynamics with unit parameter values. © 2014 IEEE, reprinted with permission

Linearizing (4.41) about $y = 0$ yields

$$\tilde{m}\ddot{y} + \tilde{b}\dot{y} - \tilde{c}y = 0 \tag{4.42}$$

where

$$\tilde{c} = \tilde{k}\left(\frac{l_{pzt} - \Delta x_{pzt}}{l_{pzt} - l_{preload}} - 1\right). \tag{4.43}$$

Notice that $\tilde{c} > 1$ for all $\Delta x_{pzt} \geq 0$ because $l_{preload} > 0$. The state equations can be written as

$$\frac{d}{dt}\begin{bmatrix} y \\ \dot{y} \end{bmatrix} = \begin{bmatrix} 0 & 1 \\ \frac{\tilde{c}}{\tilde{m}} & -\frac{\tilde{b}}{\tilde{m}} \end{bmatrix}\begin{bmatrix} y \\ \dot{y} \end{bmatrix}. \tag{4.44}$$

The poles of the linearized system are given by

$$\lambda_{1,2} = \frac{-\tilde{b} \pm \sqrt{\tilde{b}^2 + 4\tilde{c}\tilde{m}}}{2\tilde{m}}. \tag{4.45}$$

For the undamped system, these values are simply

$$\lambda_{nodamping1,2} = \pm\sqrt{\frac{\tilde{c}}{\tilde{m}}}. \tag{4.46}$$

The eigenvalues of the damped and undamped systems are always opposite in sign, therefore, the point (0, 0) is a saddle point. This is indicative of the unstable, singular nature of the point. The trajectories corresponding to the positive value can be interpreted as instability. Fig. 4.23 presents several key points. The eigenvector associated with the positive value is in the direction of the line between the two arrows pointing away from the origin on the critical trajectory. The trajectories associated with the negative value can be interpreted as the system's ability to draw trajectories towards the origin. The eigenvector associated with the negative eigenvalue is in the direction of the line between the two arrows pointing towards the origin on the critical trajectory. The saddle point is clearly observed at the origin.

Similar general analysis is now shown for the stable equilibrium points. The equilibrium points are

$$\pm y_0 = \pm\sqrt{(l_{pzt} + \Delta x_{pzt})^2 - (l_{pzt} - l_{preload})^2},\ \dot{y} = 0. \tag{4.47}$$

The elastic forcing terms about these points can be expressed as

$$\begin{aligned} F(y) &\approx \tilde{c}_0 - \tilde{c}_1 y, \\ \tilde{c}_0 &= y_0\tilde{k}\left[1 + \frac{(l_{pzt} + \Delta x_{pzt})y_0^2}{\left(y_0^2 + (l_{pzt} - l_{preload})^2\right)^{3/2}}\right], \\ \tilde{c}_1 &= \tilde{k}\left[1 + \frac{(l_{pzt} + \Delta x_{pzt})y_0^2}{\left(y_0^2 + (l_{pzt} - l_{preload})^2\right)^{3/2}} - \frac{l_{pzt} + \Delta x_{pzt}}{\sqrt{y_0^2 + (l_{pzt} - l_{preload})^2}}\right]. \end{aligned} \tag{4.48}$$

Note that $\tilde{c}_1 > 0$. The linearized state equations can be written as

$$\frac{d}{dt}\begin{bmatrix} y \\ \dot{y} \end{bmatrix} = \begin{bmatrix} 0 & 1 \\ -\frac{\tilde{c}_1}{\tilde{m}} & -\frac{\tilde{b}}{\tilde{m}} \end{bmatrix}\begin{bmatrix} y \\ \dot{y} \end{bmatrix} + \begin{bmatrix} 0 \\ \tilde{c}_0 \end{bmatrix}. \tag{4.49}$$

The eigenvalues with and without damping of the Jacobian are

$$\lambda_{1,2} = \frac{-\tilde{b} \pm j\sqrt{4\tilde{c}_1\tilde{m} - b^2}}{2\tilde{m}}, \tag{4.50}$$

$$\lambda_{nodamping1,2} = \pm j\sqrt{\frac{\tilde{c}_1}{\tilde{m}}}. \tag{4.51}$$

Assuming the system is not critically or over damped, the poles are complex conjugates of each other with negative real parts. This means these points are stable foci with trajectories orbiting and approaching them as shown in Fig. 4.23B. In the undamped case, there is no real part, thus local trajectories constantly cycle around but never approach the point as can be seen in Fig. 4.23A. Linearized equations of motion around these stable points behave as simple mass–spring–damper systems.

The critical trajectory passing through the saddle point of the undamped system provides the minimum amplitude to achieve bipolar oscillations. In Fig. 4.23A, this minimum amplitude is the y-coordinate of the critical trajectory when crossing the y axis ($\dot{y} = 0$). The critical value y_c can be derived by using conservation of energy. Since the kinetic energy is zero on the y-axis, the potential energy in the system at is a constant value for a constant Δx_{pzt} in the undamped system. Therefore, the potential energy is equal at $y = 0$ and $y \pm y_c$. Derived from expressions for potential energy at these positions, y_c can be expressed as

$$y_c = \sqrt{4(l_{pzt} l_{preload} + l_{pzt} \Delta x_{pzt} + l_{preload} \Delta x_{pzt} + \Delta x_{pzt}^2)} \simeq \sqrt{l_{pzt}(l_{preload} + \Delta x_{pzt})}. \tag{4.52}$$

Dynamic bipolar motion is limited to a class of large amplitude oscillations. Small amplitude bipolar oscillations less than the critical value y_c do not exist. The period of bipolar oscillation can be obtained by using the standard technique of phase plane analysis. This integral diverges for the critical trajectory, since it take infinite time to reach/leave the origin.

4.2.7 Prototyping buckling actuators

Four prototypes have been built and tested to validate the bipolar buckling actuator performance. High yield stress alloy steel was machined via wire EDM to produce the keystones, flexures and support frames for the single unit, redirecting stiffness unit, and dual unit buckling actuators shown in Figs. 4.24, 4.25, and 4.26, respectively. The PZT stacks in these prototypes are 40 mm NEC Tokin AE0505D44H40 stacks, have a free displacement of 42 μm and a blocking force of 850 N.

To prevent damaging the brittle ceramic structure of the PZT stacks, (i) care was taken to minimize the bending load imparted by the flexures, (ii) the PZT stacks were loaded through their centers by collinearly aligning the axes of the flexures and stacks, and (iii) the load was distributed evenly on the ends of the stacks via the triangular end-caps that interface the ends of the stacks with the flexures. Through rigorous testing, no PZT stack was damaged through use in any of the presented prototypes.

The dynamic actuator prototype shown in Fig. 4.27 uses 18 mm NEC Tokin AE0505D18F PZT stacks with a total of 15 μm stroke. To reduce weight and cost, a plastic material (acrylonitrile butadiene styrene) was used for this prototype. The structure was 3D printed by using a Dimensions 3D printer, model SST 1200ES. Thin compliant beams connect the keystone to the rigid frame to simplify assembly. These beams add negligible stiffness to the system. In each case, the distance between paired PZT interfaces was made slightly less than the length of the PZT, i.e., to provide a preloading initial displacement. Preload is beneficial because it eliminates overly compliant initial contact stress, which causes nonlinear contact effects. It is preferable to not adhesively connect the PZT stacks to the grounded nodes for two reasons. First, adhesive or other coupling methods would introduce additional compliance in

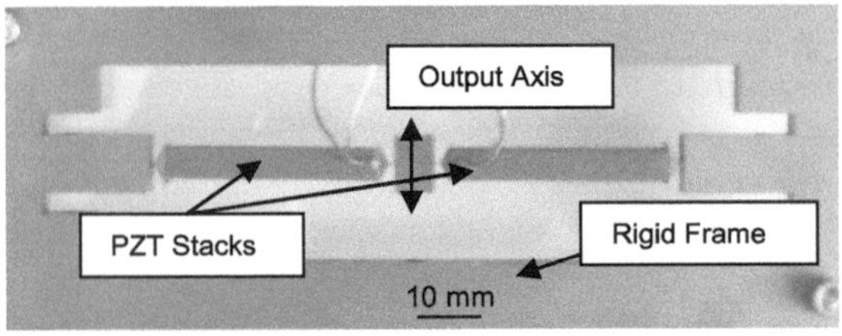

Figure 4.24 Single unit PZT buckling actuator prototype with a pair of 40 mm PZT stacks. © 2014 IEEE, reprinted with permission

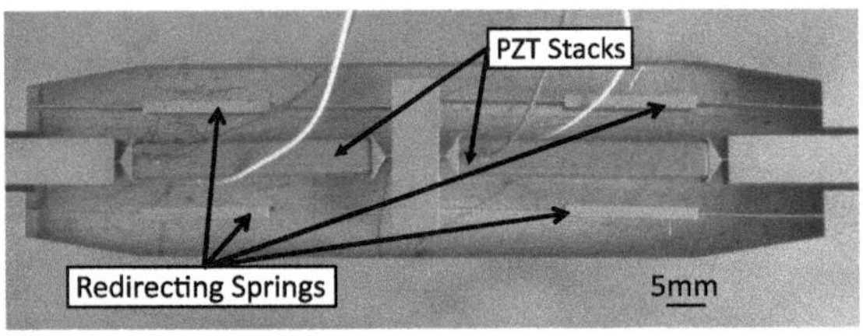

Figure 4.25 Buckling actuator prototype with stiffness redirecting elements. © 2014 IEEE, reprinted with permission

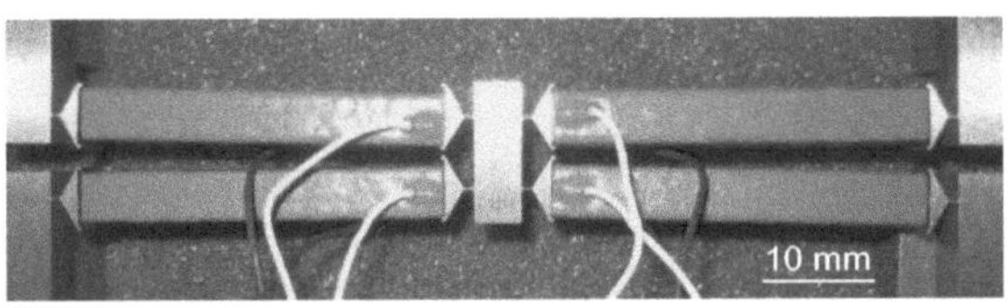

Figure 4.26 Dual-unit phase-shifted buckling actuator prototype.

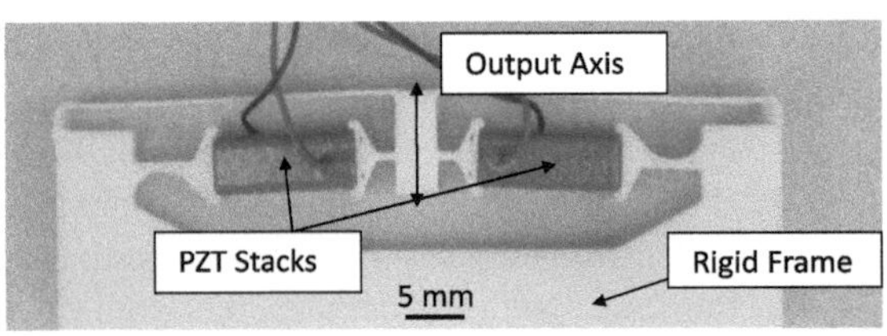

Figure 4.27 Lightweight dynamic bipolar PZT actuator. © 2014 IEEE, reprinted with permission

series with the PZT stacks, and second, the coupling should not be used in tension because the PZT stacks of ceramic layers could become delaminated due to cyclic tensile stresses. Compliance from any source that is in series with the PZT stacks degrades output performance by storing strain energy that could be otherwise applied to the load upon activation. With sufficient preload, the PZT stacks stay in contact with interfacing surfaces. Note that hysteresis is a significant issue for PZT-based applications; however, the present work presents concepts that are independent of the hysteresis issue. The experimental data presented are for single voltage actuations in order to clearly present the novel concepts.

Displacements were measured with a laser micrometer from Micro-Epsilon, model NCDT 1401 with displacement resolution of 1 µm. Force was measured with a load cell from Transducer Techniques, model MLP-10 with resolution of 0.04 N.

4.2.8 Static performance

The single unit prototype was used to validate the nonlinear force–displacement relationship shown in Fig. 4.28. The voltage applied was 88% of the maximum voltage. A free displacement of 1.74 mm was achieved for each side of the singular point. Comparing this to the free displacement of the PZT itself, approximately 50 times larger displacement was obtained. Over 10 N of force was generated in the midrange of the stroke, $0.75\,\text{mm} < y < 1.22\,\text{mm}$. The maximum force was reached in the middle of the stroke. The slight variations from the model are due largely to machining tolerances from production.

The redirecting stiffness prototype is shown in Fig. 4.25. Four redirecting springs redirect the stiffness of the keystone. These steel springs are axially stiff and provide miniscule bending stiffness. Their arrangement in this configuration provides symmetric performance and two four bar linkage structures that prevent parasitic rotation of the keystone. They are oriented approximately 1 degree clockwise from the neutral horizontal position of the PZT stacks. This gives the springs a negative slope of roughly $y_{max}/(2l_{pzt})$ as used for simulation. Because of the connectivity provided by the redirecting springs, all steel components can be made from a single monolithic piece of steel, simplifying manufacturing and assembly. When one PZT stack is actuated at 100 V and the other is at 0 V, the measured displacement is 0.72 mm. The direction of displacement depends on which stack is actuated with 100 V. When both stacks are actuated at 100 V, the unforced displacement is 1.22 mm, with the direction depending on which stack was actuated first. The prototype output free displacement values are represented as dashed lines for the conditions simulated in Fig. 4.21. Thus, the total bipolar displacement range is 2.44 mm. The unforced displacement of the prototype is slightly greater in magnitude than the modeled energy wells shown in Fig. 4.21 most likely because the EDM process results in sections of the flexure being slightly less than the designed thickness. In simulation, a decrease in flexure thickness of 10% results in nearly identical prototype and simulation free displacements.

The dual-unit phase-shifted prototype is shown in Fig. 4.26. The measured force–displacement curves were integrated to generate the potential energy curves of Fig. 4.29. As expected, when both buckling units are grounded (inactive), there is a single stable well at zero displacement. When a single buckling unit is active, there is a nonzero slope at zero displacement forcing the output towards a dominant energy well on the other side. Finally, when both units are active, there are two dominant, symmetric wells on either side of the zero displacement point. The distance between the two wells, 6.2 mm, gives the peak-to-peak free displacement of the dual unit buckling actuators. The slope of the tangent to the potential curves provides the force generated by the buckling actuator. A maximum force of 9.6 N was achieved at approximately $y = \pm 1$ mm on both sides of the singular point.

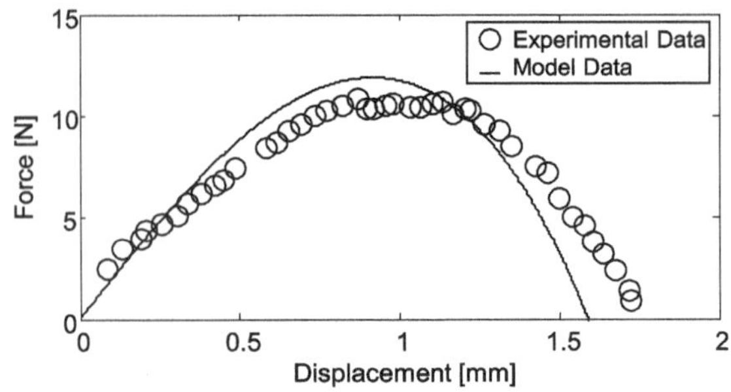

Figure 4.28 Theoretical and experimentally measured force–displacement data using an activation voltage of 88%. © 2014 IEEE, reprinted with permission

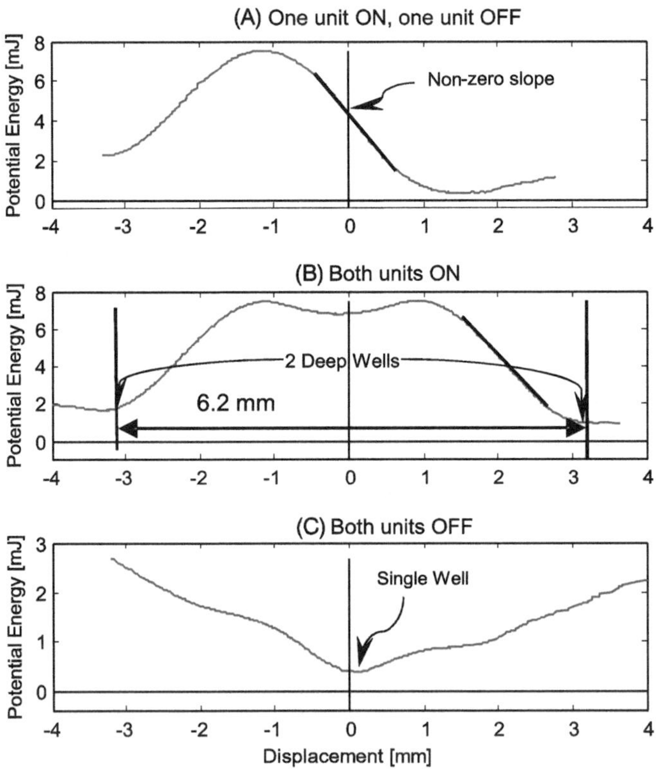

Figure 4.29 Potential energy vs. displacement profiles of dual unit phase shifted prototype for various states of activation: (A) one unit ON, one unit OFF, (B) both units OFF, and (C) displacement [mm]. © 2014 IEEE, reprinted with permission

This dual unit buckling design allows for robust buckling direction control, as depicted by the steep slope of the potential curve in Fig. 4.29A. The output node, the keystone, is pushed with a large force (> 9 N) towards a desired direction. This design was tuned to generate a steep potential curve when one unit is active. Based on the potential energy analysis in Fig. 4.22B, the rest angle θ_0 was set to 1.0 [deg]. Comparing Fig. 4.29 to Fig. 4.22, we notice that the distances between energy wells, when both stacks are on are within 10%, but, more interestingly, we notice numerous qualitative similarities: (i) for both units off, there is an energy well at zero displacement, (ii) for a single unit on, there is a steep slope at zero displacement forcing the output away, and (iii) for both units on there are two deep wells nearly symmetric about zero. If the rest angle θ_0 were to be decreased, output performance in terms of blocking force and displacement would increase but at the expense of limited fidelity of directional control. Note that the asymmetry in Fig. 4.29 is mostly due to slight misalignment of the stack actuators shown in Fig. 4.26, and does not reflect the positioning repeatability.

4.2.9 Dynamic performance

The existence of dynamic bipolar motion was verified experimentally with use of the third prototype shown in Fig. 4.27. Fig. 4.30 shows the phase plane plot of experimental data. The large phase plane portrait outside the theoretical critical trajectories indicates continuous bipolar oscillations sweeping both sides of the singular point. To generate the continuous mono-polar and bipolar oscillation, the PZT stacks were activated with periodic 0 to 100 V square wave voltage inputs at various frequencies. The PZT stacks cycle ON and OFF at twice the frequency of the output motion of the actuator therefore the output frequency is half of the activation frequency. In the phase plane for Fig. 4.30, two undamped model trajectories are shown, one with $\Delta x_{pzt} = 0$ and the other with $\Delta x_{pzt} = \Delta x_{pzt,max}$. The experimental portrait data lies primarily between these two curves as experimentally varies between 0 and $\Delta x_{pzt,max}$. Slight deviations outside of these modeled curves occur due to energy exchanging as the input PZT actuators are cycled. Notice that at higher frequencies, mono polar activation forms the expected closed loops around the stable equilibrium points defined by Eq. (4.47).

4.2.10 Discussion

This section presented a nonlinear piezoelectric buckling actuator. With the buckling configuration, the capability for large displacement has been presented with little added structural volume/mass, and few compliant parasitic energy storage elements. To emphasize this point, let's compare the present work to two typical dual layer flextensional amplification structures that achieve on the order of 100X amplification. In Kim et al. [130], an energy density per stroke by volume of 91.5 J/m^3 was achieved, whereas the simple buckling actuator shown in Fig. 4.24 achieves 2600 J/m^3. Similarly, Ueda et al. [267] achieved an energy density per stroke by mass of 0.14 J/kg whereas the simple buckling actuator achieves 0.74 J/kg.

To control the direction of buckling and generate bipolar motion spanning both sides of the singular point, we have utilized three separate methods. By incorporating a redirecting stiffness, the asynchronous activation of each individual input PZT stack can control direction, as can utilizing a plurality of spatially phased buckling units in parallel. Furthermore, activating a single buckling unit dynamically can sweep the out-

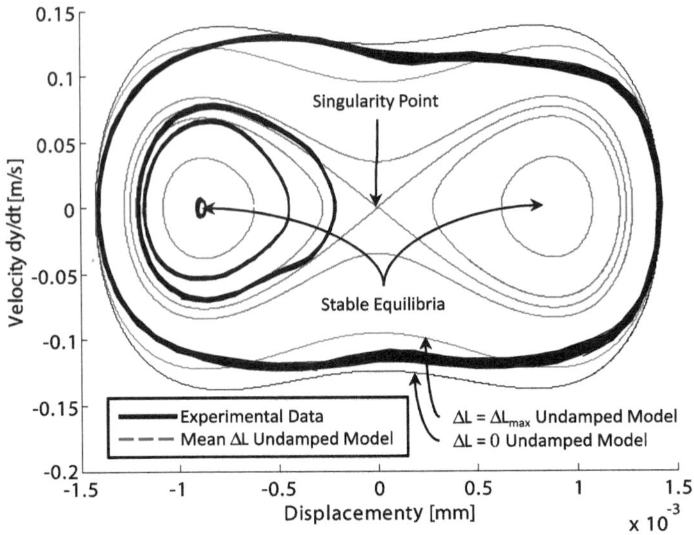

Figure 4.30 Input square wave signal and output response for actuator dynamically passing through the singularity point. Labeled frequencies are input activation frequencies of the PZT stacks. © 2014 IEEE, reprinted with permission

put to both sides of the singular point, doubling the stroke of mono-polar motion. The dynamic bipolar motion, however, occurs only at amplitudes greater than a defined critical value.

The most notable limiting factor for this actuator is the nonlinear force–displacement profile. Additionally, a rigid frame is necessary to mechanically ground the ends of the buckling units. Current work includes preloading the individual stack actuators within the buckling unit to allow substantial tensile force generation while limiting inefficient strain energy storage [19,258]. This actuator technology may be applicable to braking systems that utilize large displacement and zero-current holding force of PZT actuators, and a rotational motor with high holding torque which requires millimeter scale displacements for gear-like operation while still requiring high force. Unlike traditional friction drive piezoelectric motors, the use of gear teeth opens up applications for driving larger loads and higher torques. Practical gear teeth would require displacements on the order of multiple millimeters. Additionally, both of these applications make use of the fact that the maximum force occurs within the middle of the stroke.

4.3 Self-sensing piezoelectric grasper

4.3.1 Self-sensing of amplified PZT actuators

Self-sensing actuation, or simultaneous actuation and sensing, is a technique that takes advantage of the unique properties of piezoelectric materials not only to use them as sensors, but to also simultaneously produced forces and displacements. This is achieved by using a specially designed circuit. This concept was originally proposed by Dosch et al. [63]. A bridge circuit was developed by placing capacitances in series and parallel with the piezoelectric actuator that produced a voltage output proportional to velocity or force. The method was subsequently refined and applied by other researchers [83,141]. The three main drawbacks of the method are (i) that the operation of the bridge circuit requires close matching of the piezo capacitance, (ii) only dynamic (i.e., vibration) measurements are possible, and (iii) the hysteresis of the material is neglected. New models have also been developed to take hysteresis into account [87,85]. In [87], hysteresis is observed between voltage and charge but not charge and displacement, which suggests that the hysteresis occurs in the electrical domain. This means that for quasistatic operation charge is linearly related to displacement if there is zero external force on the actuator, or vice versa, so the linear model can appear correct. If one attempts to extend this model to a self-sensing scheme with unknown loading conditions, it is quickly seen that a model considering hysteresis is needed.

Badel et al. use the hysteretic model of [87] to implement force control based on self-sensed measurements [15]. In this work, calibration of the hysteretic model required two extreme loading conditions, namely, fully blocked and fully free. These loading conditions are easily be realizable in piezoelectric actuators when used "as-is," but cannot be replicated in "built-in" actuators including amplified PZT actuators. The proposal will later show that intricate coupling between parameters in the mathematical model makes it difficult to calibrate the hysteresis model. Kurita et al. investigated this problem for a tweezer-style end effector, and proposed using one actuator out of serially-connected actuators solely as a sensor [142]. Although the presented results were promising, one of the PZT units was used only for sensing in their implementation. In addition, the hysteresis was not modeled and the loading condition at the tip needed be known a priori. Consequently, there is no existing method that provides means to calibrate the electromechanical hysteretic of amplified PZT actuators for simultaneous actuation and sensing of amplified PZT actuators.

In this section, an on-site calibration procedure is presented that does not require specific loading conditions on the piezo actuator. A self-sensing technique is attractive because it does not sacrifice actuation capability. This section develops a calibration procedure based on a mathematical model with a hysteretic piezoelectric model and a two-port network model of the compliant amplified PZT actuators. This formalism reveals mathematical insights that lead to an effective calibration procedure. The methods will be implemented to the tweezer-style actuator introduced in Section 1.3.5.

4.3.2 Force magnification for tweezer-style piezoelectric end-effector

A prototype of a tweezer-style piezoelectric end-effector with nested-rhombus multilayer stain amplification mechanisms has been briefly introduced in Section 1.3.5. This prototype end-effector produces 1.1 N of pinching force when fully blocked or

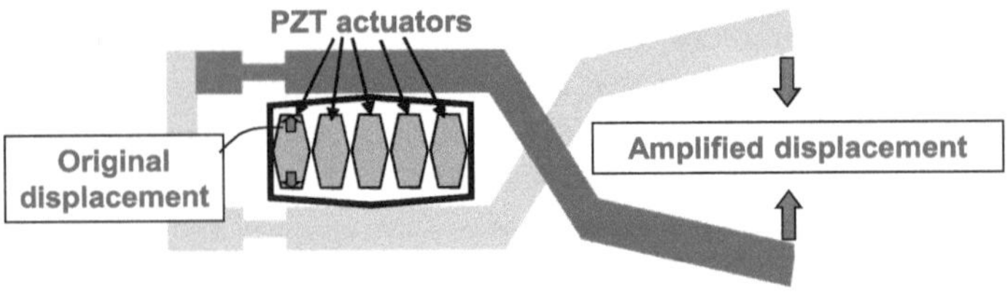

(A) Amplification of displacement: nested actuator to end-point

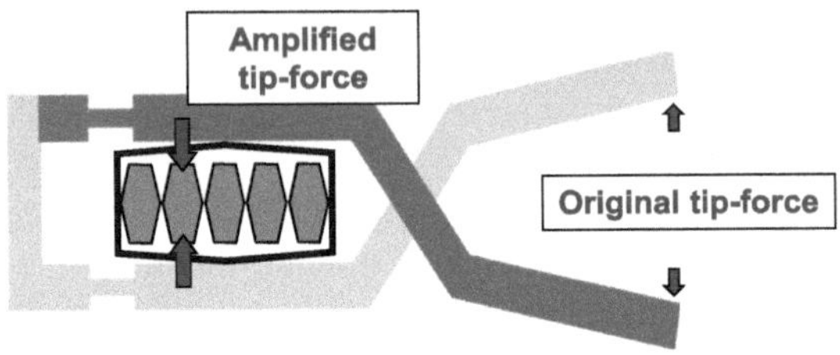

(B) Amplification of force: end-point to nested actuator

Figure 4.31 Concept of high-accuracy force sensing through strain amplification mechanisms.

8.8 mm of displacement at the tip when fully free, and requires a supply voltage of 0 to 150 V. A bandwidth of actuation up to 35 Hz is confirmed (see Fig. 1.47C), which is much wider than existing tendon-driven robotic graspers.

Sensing of both displacement and force at the tip is of interest. It is quickly seen that attaching additional sensors, e.g., strain gauges on the tweezer arm, is not a solution because (i) it requires additional components, (ii) tip-force is expected to be small (< 2 N) which would not induce a large strain in the arm structure to be precisely measured by strain gauges, and (iii) adding one type of sensor provides only displacement or force, not both.

As shown in Fig. 4.31A, a large displacement at the tip is achieved by amplifying the original small displacement of the PZT actuators by the multi-layer rhomboidal amplification structure. The salient feature of this architecture is that the strain amplifier is advantageous not only for actuation but also for sensing. In general, the high rigidity of PZT ceramics (20–30 GPa) is not suitable for sensing small forces. However, as shown in Fig. 4.31B, small tip-force can be measured by measuring induced electrical potential across one of the nested PZT actuators. Since the force acting on the PZT actuator is amplified tip-force on the order of hundreds through the strain amplification mechanism, small tip-force is magnified and accurately measured. Kurita et al. used 4 PZT units out of 5 for actuation and one for sensing [142]. McPherson et al. extended this idea and realized a full self-sensing device [170]. The high-sensitivity of force measurement is extremely important for medical applications that deal with delicate tissues and organs. When this technology is used for a medical tele-robot, measured end-point force is sent back to the operator's side for improved haptic feedback.

4.3.3 Mechanical modeling

Under quasistatic operation, a rhomboidal strain amplifier can be represented by a two-port network model of the form as described in Section 2.1, namely,

$$\begin{bmatrix} f_{pzt} \\ f_1 \end{bmatrix} = \begin{bmatrix} s_1 & s_3 \\ s_3 & s_2 \end{bmatrix} \begin{bmatrix} \Delta x_{pzt} \\ \Delta x_1 \end{bmatrix}. \tag{4.53}$$

Forces and displacements are defined in Fig. 1.24. The parameters of the model can be interpreted intuitively as follows: s_1 is the stiffness at the input when the output is blocked, s_2 is the stiffness at the output when the input is blocked, and finally, s_3 is the ratio of force produced at the blocked output to a given input displacement. The second and third amplification layers can be similarly represented by two-port networks. Schematic diagrams are shown in Figs. 4.32 and 4.33.

$$\begin{bmatrix} f_1 \\ f_2 \end{bmatrix} = \begin{bmatrix} s_4 & s_6 \\ s_6 & s_5 \end{bmatrix} \begin{bmatrix} \Delta x_1 \\ \Delta x_2 \end{bmatrix} \tag{4.54}$$

More detailed modeling is found in Sections 2.1 and 2.3. The lever action of the tweezer arms provides the final level of strain amplification. Assuming quasistatic operation, another two-port network can be written describing the third layer:

$$\begin{bmatrix} \Delta x_2 \\ \Delta x_{tip} \end{bmatrix} = \begin{bmatrix} Q_1 & Q_3 \\ Q_3 & Q_2 \end{bmatrix} \begin{bmatrix} f_2 \\ f_{tip} \end{bmatrix}, \tag{4.55}$$

$$Q_1 = \left(\frac{C_{A2}}{2EI_1} + \frac{C_{A4}}{2EI_3} \right), \tag{4.56}$$

$$Q_3 = \left(\frac{C_{A1}}{2EI_1} + \frac{C_{A3}}{2EI_3} \right), \tag{4.57}$$

$$Q_2 = \left(\frac{C_{B1}}{2EI_1} + \frac{C_{B3}}{2EI_2} + \frac{C_{B4}}{2EI_3} \right), \tag{4.58}$$

where the coefficients Q_1, Q_2, and Q_3 are obtained using Castigliano's Theorem and the Bernoulli–Euler beam model and are given in Appendix A.2.

The input–output behavior of nested rhomboidal strain amplifiers can be represented by single, combined two-port network if every layer is represented by a two-port network [267,220]. Additionally, Section 2.2 describes how the parameters of the two-port model depend on the geometric parameters of the rhomboidal mechanism, and provide an experimental verification based on a similar piezo driven device. This provides insight on how to design a rhomboidal mechanism to achieve a given force and displacement specification. However, it is not immediately obvious how to account for multiple actuators in series that drive the input of the second layer. The question is whether five rhomboidal strain amplifiers connected in series can also be represented by a single two-port network model. If so, the input–output relationship of the tweezer mechanism as a whole could be assumed to be a two-port network.

Consider Figs. 4.34 and 4.35 where five PZT units are connected in series. The figures assume that each strain amplifier sees an identical input since the PZT actu-

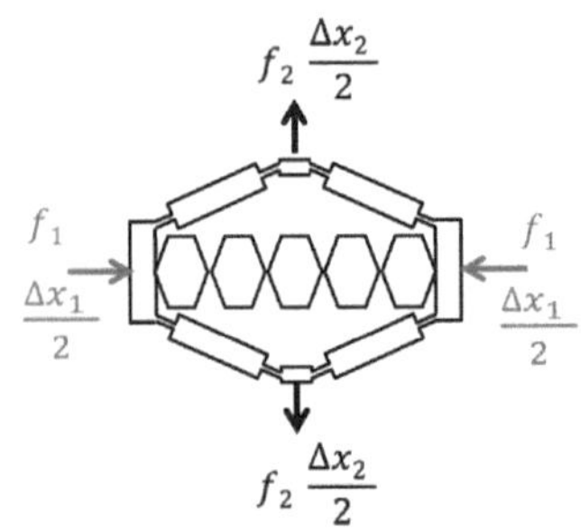

Figure 4.32 Schematic representation of second amplification layer. Five actuators in series that drive the input of the second layer. © 2014 IEEE, reprinted with permission

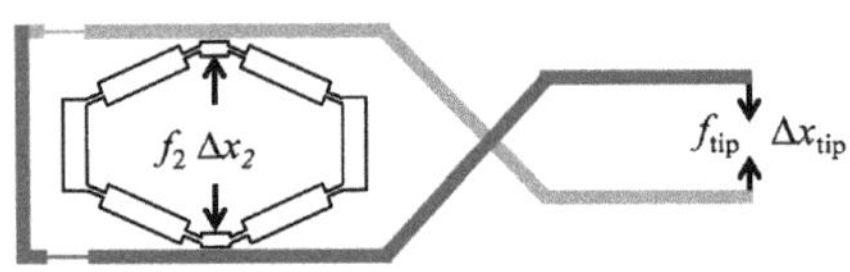

Figure 4.33 Schematic representation of the tweezer arms and second amplification layer. The output of the second layer drives and input and the lever action of the tweezer arms provides the final layer of strain amplification. © 2014 IEEE, reprinted with permission

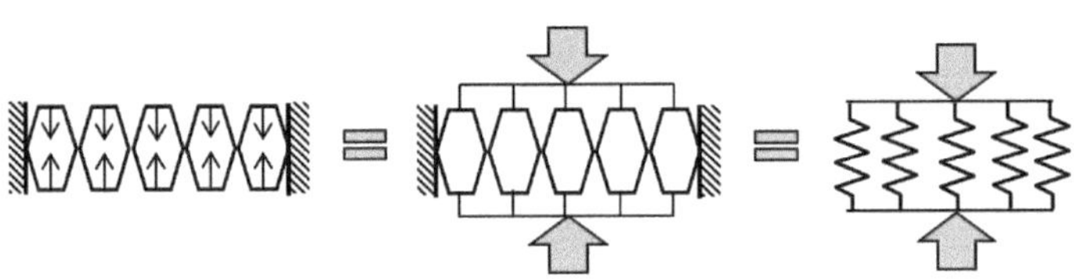

Figure 4.34 Mechanical analysis of five rhomboids in series, output blocked. When proportionally actuated, it is equivalent to five identical springs in parallel. © 2014 IEEE, reprinted with permission

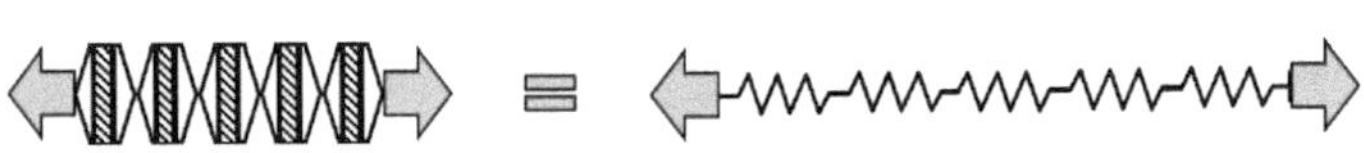

Figure 4.35 Mechanical analysis of five rhomboids in series, input blocked. When proportionally actuated, it is equivalent to five identical springs in series. © 2014 IEEE, reprinted with permission

ators are connected electrically in parallel. When the output is blocked, there are the equivalent of five springs in parallel at the input, meaning the effective s_1 would be five times that of a single rhomboid. When the input is blocked, the rhomboids act as springs in series in the output direction, so the effective s_2 is one-fifth that of a single rhomboid. Finally, when the output is blocked, a given input displacement creates the same output force for a singe rhombus as for five in series. The reaction forces at connection points cancel, leaving the output force unchanged. It is now seen that the series connection of five actuators can be modeled with a single two-port network. The input force to the combined model is $5f_{pzt}$, and the input displacement is Δx_{pzt}, since the five PZT stack actuators act in parallel on the input.

Eq. (4.53) can be rearranged to show outputs in terms of inputs, yielding

$$\begin{bmatrix} \Delta x_1 \\ f_1 \end{bmatrix} = \begin{bmatrix} \dfrac{-s_1}{s_3} & \dfrac{1}{s_3} \\ s_3 - \dfrac{s_1 s_2}{s_3} & \dfrac{s_2}{s_3} \end{bmatrix} \begin{bmatrix} \Delta x_{pzt} \\ f_{tip} \end{bmatrix}. \tag{4.59}$$

Similarly, (4.54) can be written as

$$\begin{bmatrix} \Delta x_2 \\ f_2 \end{bmatrix} = \begin{bmatrix} \dfrac{-s_4}{s_6} & \dfrac{1}{s_6} \\ s_6 - \dfrac{s_4 s_5}{s_6} & \dfrac{s_5}{s_6} \end{bmatrix} \begin{bmatrix} \Delta x_1 \\ f_1 \end{bmatrix}, \tag{4.60}$$

and (4.55) as

$$\begin{bmatrix} \Delta x_{tip} \\ f_{tip} \end{bmatrix} = \begin{bmatrix} \dfrac{-Q_1}{Q_3} & \dfrac{1}{Q_3} \\ Q_3 - \dfrac{Q_1 Q_2}{Q_3} & \dfrac{Q_2}{Q_3} \end{bmatrix} \begin{bmatrix} \Delta x_2 \\ f_2 \end{bmatrix}. \tag{4.61}$$

Eqs. (4.59), (4.60), and (4.61) can now be combined to represent f_{tip} and Δx_{tip} in terms of f_{pzt} and Δx_{pzt} as

$$\begin{aligned} \begin{bmatrix} \Delta x_{tip} \\ f_{tip} \end{bmatrix} &= \begin{bmatrix} \dfrac{-Q_1}{Q_3} & \dfrac{1}{Q_3} \\ Q_3 - \dfrac{Q_1 Q_2}{Q_3} & \dfrac{Q_2}{Q_3} \end{bmatrix} \begin{bmatrix} \dfrac{-s_4}{s_6} & \dfrac{1}{s_6} \\ s_6 - \dfrac{s_4 s_5}{s_6} & \dfrac{s_5}{s_6} \end{bmatrix} \\ &\times \begin{bmatrix} \dfrac{-N_1 s_1}{s_3} & \dfrac{1}{s_3} \\ s_3 - \dfrac{s_1 s_2}{s_3} & \dfrac{s_2}{N_1 s_3} \end{bmatrix} \begin{bmatrix} \Delta x_{pzt} \\ N_1 f_{pzt} \end{bmatrix}. \end{aligned} \tag{4.62}$$

Recall that if each layer of the device can be written as a two-port network, which is just shown above, the entire device can be modeled by a single two-port network model,

$$\begin{bmatrix} N_1 f_{pzt} \\ f_{tip} \end{bmatrix} = \begin{bmatrix} S_1 & S_3 \\ S_3 & S_2 \end{bmatrix} \begin{bmatrix} \Delta x_{pzt} \\ \Delta x_{tip} \end{bmatrix}, \tag{4.63}$$

where S_1,S_2, and S_3 are the parameters describing the combined effects of all three amplification layers. N_1 is the number of actuators connected in series in the first layer, five in this case. This equation can then be rearranged to give outputs in terms of inputs. Note that the first layer matrix and input vector have been written in terms of a single rhomboid and PZT actuator, but the s-parameters and input force have been appropriately modified as discussed above. Equivalently, the relationship can be represented by

$$\begin{bmatrix} \Delta x_{tip} \\ f_{tip} \end{bmatrix} = \begin{bmatrix} \dfrac{-S_1}{S_3} & \dfrac{1}{S_3} \\ S_3 - \dfrac{S_1 S_2}{S_3} & \dfrac{S_2}{S_3} \end{bmatrix} \begin{bmatrix} \Delta x_{pzt} \\ N_1 f_{pzt} \end{bmatrix}. \tag{4.64}$$

4.3.4 Combined electromechanical model of the tweezer device

Eqs. (1.1) and (1.3) can be written in a matrix form, shown in (4.65), as

$$\begin{bmatrix} \Delta x_{pzt} \\ f_{pzt} \end{bmatrix} = \begin{bmatrix} \dfrac{1}{\alpha} & \dfrac{-C_{pzt}}{\alpha} \\ \dfrac{-k_{pzt}}{\alpha} & \alpha + \dfrac{C_{pzt} k_{pzt}}{\alpha} \end{bmatrix} \begin{bmatrix} q_{pzt} \\ V_{pzt} \end{bmatrix}. \tag{4.65}$$

Note we have negated F since we wish to represent the force supplied by the PZT actuator rather than the external force on the actuator, since the latter is the input to the two-port network model representing the tweezers. Combining (4.65) with (4.64), the force and displacement at the tip can be written in terms of q_{pzt} and V_{pzt}.

$$\begin{aligned} \begin{bmatrix} \Delta x_{tip} \\ f_{tip} \end{bmatrix} &= \begin{bmatrix} \dfrac{-S_1}{S_3} & \dfrac{1}{S_3} \\ S_3 - \dfrac{S_1 S_2}{S_3} & \dfrac{S_2}{S_3} \end{bmatrix} \begin{bmatrix} \dfrac{1}{N_1\alpha} & \dfrac{-C_{pzt}}{\alpha} \\ \dfrac{-k_{pzt}}{\alpha} & N_1\alpha + \dfrac{N_1 C_{pzt} k_{pzt}}{\alpha} \end{bmatrix} \begin{bmatrix} N_1 q_{pzt} \\ V_{pzt} \end{bmatrix} \\ &= \begin{bmatrix} \dfrac{-N_1 k_{pzt} - S_1}{N_1 S_3 \alpha} & \dfrac{N_1 C_{pzt} k_{pzt} + C_{pzt} S_1 + N_1 \alpha^2}{S_3 \alpha} \\ \dfrac{S_2(-N_1 k_{pzt} - S_1) + S_3^2}{N_1 S_3 \alpha} & \dfrac{S_2(N_1 C_{pzt} k_{pzt} + C_{pzt} S_1 + N_1 \alpha^2) - C_{pzt} S_3^2}{S_3 \alpha} \end{bmatrix} \\ &\quad \times \begin{bmatrix} N_1 q_{pzt} \\ V_{pzt} \end{bmatrix}. \end{aligned} \tag{4.66}$$

Note that the matrix of (4.65) has been modified to reflect that connection of actuators electrically in parallel, but constants α, C_{pzt}, and k_{pzt} are with respect to a single actuator. For N_1 actuators connected in parallel, α, C_{pzt}, and the stiffness will all increase by a factor of N_1, which is reflected in (4.66). Additionally, the charge will be N_1 times that of a single actuator. As described earlier, here N_1 is five. By noticing the similarity between the first and second rows of the matrix, (4.66) can be expressed as

$$\begin{bmatrix} \Delta x_{tip} \\ f_{tip} \end{bmatrix} = \begin{bmatrix} P_1 & P_2 \\ S_2 P_1 + S_3/N_1\alpha & S_2 P_2 - C_{pzt} S_3/\alpha \end{bmatrix} \begin{bmatrix} N_1 q_{pzt} \\ V_{pzt} \end{bmatrix}, \tag{4.67}$$

where P_1 and P_2 are defined by

$$P_1 = \frac{-N_1 k_{pzt} - S_1}{N_1 S_3 \alpha}, \tag{4.68}$$

$$P_2 = \frac{N_1 C_{pzt} k_{pzt} + C_{pzt} S_1 + N_1 \alpha^2}{S_3 \alpha}. \tag{4.69}$$

Recalling that $V_{pzt} = V_{total} - H(q_{pzt})$, it is now seen that the force and displacement at the tweezer tip can be sensed simultaneously if the driving voltage, charge, and hysteresis operator $H(q_{pzt})$ are known.

4.3.5 On-site calibration procedure

In prior work on piezoelectric modeling and self-sensing, identification of model parameters has been relatively straightforward [87,15] where unamplified piezo actuators were used. In contrast, the addition of strain amplification mechanisms considered in this book complicates the problem. For a singular, unamplified PZT actuator, the parameters α, k_{pzt}, and C_{pzt}, and the hysteresis operator $H(q_{pzt})$ can be identified by taking three measurements, namely the maximum force generated by the blocked actuator, the maximum displacement of the free actuator, and the charge vs. voltage profile of the free actuator as the input voltage is varied from its minimum to maximum value [15]. When PZT actuators are nested inside several layers of strain amplification, it is practically impossible to recreate the necessary loading conditions. Therefore a different approach must be taken.

The parameters S_1, S_2, and S_3 can be estimated by the following equations (also see Section 1.3.2):

$$S_1 = \frac{f_{pzt}^{block}}{\Delta x_{pzt}^{block}}, \tag{4.70}$$

$$S_2 = \frac{\Delta x_{pzt}^{free}}{\Delta x_{tip}^{free}} \frac{f_{tip}^{block}}{\Delta x_{pzt}^{block}}, \tag{4.71}$$

$$S_3 = -\frac{f_{tip}^{block}}{\Delta x_{pzt}^{block}}. \tag{4.72}$$

Recall f_{pzt} and Δx_{pzt} are the input force and displacement supplied by the PZT actuator, while f_{tip} and Δx_{tip} are the force and displacement at the tweezer tip.

As discussed above, f_{tip} and Δx_{tip} can be measured easily, while f_{pzt} and Δx_{pzt} must be estimated from catalog data. Catalog data can also be used to estimate α, k_{pzt}, and C_{pzt}. However, it is desirable to reduce the number of model parameters to calibrate, especially for those which depend on difficult to measure quantities. To this end, consider the case of the free tweezer tip, i.e., $f_{tip} \equiv 0$, which yields

$$0 = f_{tip} = \left(S_2 P_1 N_1 q_{pzt} + S_2 P_2 V_{pzt}\right) + \left(\frac{S_3}{\alpha} q_{pzt} - \frac{C_{pzt} S_3}{\alpha} V_{pzt}\right). \tag{4.73}$$

Note that

$$\left(S_2 P_1 N_1 q_{pzt} + S_2 P_2 V_{pzt}\right) = S_2 \Delta x_{tip} \tag{4.74}$$

and that

$$S_2 = -\frac{\Delta x_{pzt}^{free}}{\Delta x_{tip}^{free}} S_3, \tag{4.75}$$

yielding

$$0 = \frac{S_3}{\alpha}\left(q_{pzt} - C_{pzt} V_{pzt}\right) - \beta S_3 \Delta x_{tip} \tag{4.76}$$

where

$$\beta = \frac{\Delta x_{pzt}^{free}}{\Delta x_{tip}^{free}}. \tag{4.77}$$

The term S_3 can then be canceled and the equation rearranged to yield an expression for V_{pzt},

$$V_{pzt} = \frac{q_{pzt}}{C_{pzt}} - \frac{\alpha\beta}{C_{pzt}} \Delta x_{tip}. \tag{4.78}$$

Combining (4.78) with (1.2) and (1.4) gives

$$H(q_{pzt}) = V_{total} - \frac{q_{pzt}}{C_{pzt}} + \frac{\alpha\beta}{C_{pzt}} \Delta x_{tip}. \tag{4.79}$$

Therefore knowledge of C_{pzt}, Δx_{tip}, α, Δx_{pzt}^{free}, and Δx_{tip}^{free} is required to calibrate $H(q_{pzt})$. C_{pzt} and Δx_{tip} can be measured directly, while α and Δx_{pzt} can be determined from published catalog data. Once $H(q_{pzt})$ is known, P_1, P_2, $S_2 P_1 + S_3/N_1\alpha$, and $S_2 P_2 - C_{pzt} S_3/\alpha$ can be easily determined by regression.

One further simplifying step can be taken. When the tweezer tip is free, Δx_{tip} is directly proportional to q_{pzt} [87]. Therefore, once the appropriate scale factor A_q is determined, one can write

$$H(q_{pzt}) = V_{total} - \frac{1}{C_{pzt}}\left(1 - \alpha\beta A_q\right) q_{pzt}. \tag{4.80}$$

This is beneficial because after initial determination of A_q the displacement measurement is not needed to calibrate $H(q_{pzt})$. The model parameters in (4.80) are shown in Table 4.3 (from [2]). $H(q_{pzt})$ can be experimentally modeled by a phenomenological model introduced in Appendix A.1.

4.3.6 Electrical circuit

The inputs to the sensing model are charge and voltage. The actuators are driven by a Cedrat CA45 high voltage amplifier. The amplifier shows good linearity up to 500 Hz,

Table 4.3 Model Parameters for Calibration of $H(q_{pzt})$. © 2014 IEEE, reprinted with permission

α	2.35 N/V
β	1.59×10^{-3}
C_{pzt}	0.3 F
A_q	84 m/C
Δx_{pzt}^{free}	10 μm
Δx_{tip}^{free}	6.3 mm

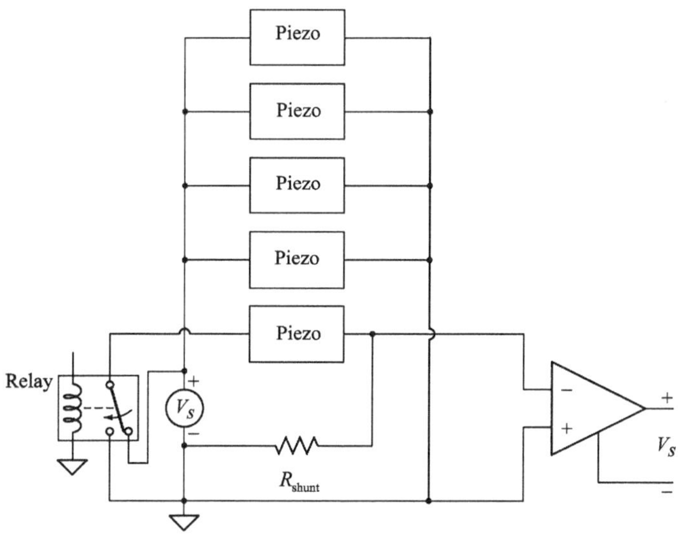

Figure 4.36 Charge Measurement Circuit. V_c is the voltage from the amplifier and V_s is the sensed voltage across the resistor. For simplicity, the gain of the instrumentation amplifier is assumed to be 1. © 2014 IEEE, reprinted with permission

well above the quasistatic operation considered here. This means that in practice the supply voltage does not need to be measured directly, but can be assumed as the amplifier gain times the input voltage to the amp. This eliminates the need for high voltage measurement equipment. The voltage across the actuator is given by

$$V_{total} = V_c - V_s \tag{4.81}$$

where V_c and V_s are the supply voltage from the amplifier and measured voltage across the resistor, seen in Fig. 4.36.

Charge can be accurately measured by placing a shunt resistance in series with a PZT actuator and measuring the voltage drop across it [15]. The charge is then calculated by integrating the current over time, namely,

$$q_{pzt}(t) = \frac{1}{R_{shunt}} \int_0^t V_s d\tau, \tag{4.82}$$

where $q_{pzt}(t)$ is the charge on the PZT actuator, R_{shunt} is the shunt resistance, and C_s is the measured voltage across the resistance. An instrumentation amp measures a differential voltage with a high impedance input on both the positive and negative terminals and low impedance at the output. This ensures minimal interaction between data acquisition hardware and the PZT actuators. It also ensures that the data acquisition hardware measures a low impedance source, which will give the best accuracy [178]. The relay allows the leads of the PZT actuators to be shorted to ground. Hysteresis is dependent on initial conditions [31]. Shorting the leads allows a consistent initial condition to be established so that hysteresis modeling will be more accurate.

4.3.7 Results

To assess the force and displacement self-sensing two experimental setups are used. Though the self-sensing technique can estimate force and displacement simultaneously, they are evaluated separately here because it simplifies taking the reference measurement. For displacement, a Micro-Epsilon OptoNCDT 1300 Laser Displacement sensor with a range of 20 mm and resolution of 4 μm is used as a reference measurement. Force and displacement self-sensing method is tested under a variety of loading conditions from fully blocked tweezer tips to fully free, as well as with various sizes of rigid objects used as a disturbances. First, the hysteresis operator $H(q_{pzt})$ was calibrated using experimental data from the free case and the Prandtl–Ishlinskii model introduced in Appendix A.1.

Fig. 4.37 shows the calibrated operator hysteresis $H(q_{pzt})$. Fifteen elementary play, superposition, and creep operators were used. $H(q_{pzt})$ shows a good match to the experimental data, with a maximum error of 3.34 V and an average error of 0.81 V, or 2.4% and 0.58%, respectively, of the output range. If the actuator is discharged before reaching its maximum charge, capacity minor loops will form. Also, note that charge is now the independent variable, as dictated by the model in Fig. 1.1.

Table 4.4 summarizes the performance of the self-sensing technique. Fig. 4.38 shows the self-sensed displacement and force versus the reference measurement for a variety of loading conditions. Note that it shows the estimation of $\Delta x_{pzt}/2$ since the laser sensor measures only one side of the tweezers. The displacement of the other side is assumed to be identical. Two simple techniques were implemented to slightly increase the accuracy. The force and displacement measurements were limited to their respective maximum and minimum values. Additionally, for the displacement measurement slight drift was removed with a sliding DC offset. The offset reset as the current displacement measurement whenever the input voltage was zero. Fig. 4.39 shows the effect of these techniques. Two input signals were used, one purely sinusoidal and one being a mixture of trapezoidal and sinusoidal inputs. The pure

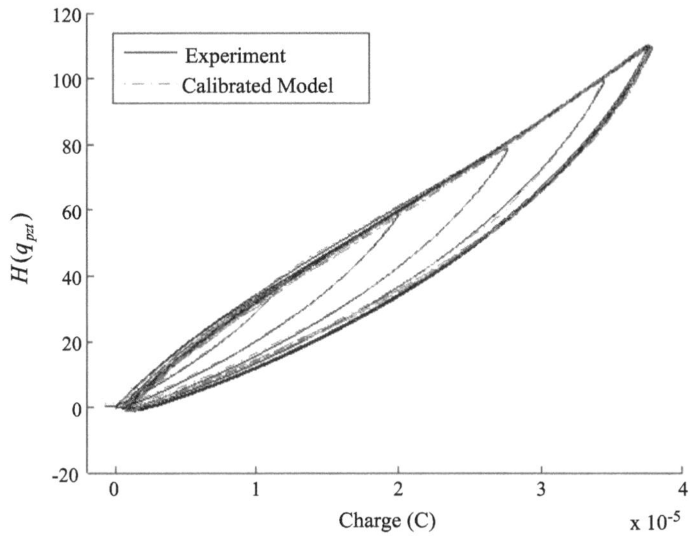

Figure 4.37 Hysteresis operator $H(q_{pzt})$ calibration. The loop is traversed clockwise with increasing charge. Minor loops form if less charge is collected on the actuator when it discharges. © 2014 IEEE, reprinted with permission

Table 4.4 Performance of self-sensing technique. © 2014 IEEE, reprinted with permission

Performance of Self-Sensing Technique			
Measurement	**Mean Error**	**Mean Error Reported by Kurita et al. [142]**	**Mean Error Reported by Badel et al. [15]**
Displacement	0.4 mm (12%)	0.39 mm (11%)	0.69 μm (2%)
Force	0.012 N (12%)	0.086 N (11%)	20 N (2%)

sinusoidal input was also used in the calibration of the hysteresis operator. The average accuracy of the self-sensed measurement is 12% of the dynamic range.

4.3.8 Discussion

Although error is present, the accuracy of the self-sensing method is comparable to that reported Kurita et al. in an earlier study of the same device [142]. The sensing method developed in that work relied on prior knowledge of the loading condition as either fixed or free. Due to the constant boundary conditions, i.e., fully blocked tip, the force (or displacement) at the tip depends only on the measured charge. Since hysteresis is generally not observed between charge and force (or displacement), the sensing accuracy is limited only by the charge measurement. The method presented in this section has lower accuracy but extends the sensing capability to situations with

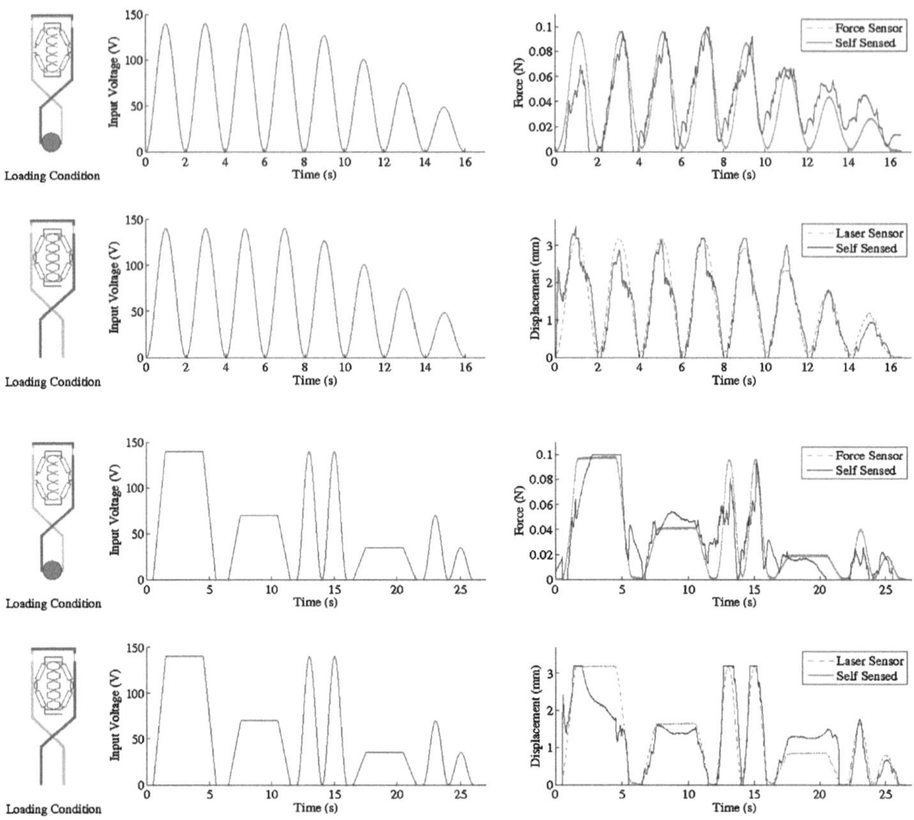

Figure 4.38 Blocked Force and Free Displacement. © 2014 IEEE, reprinted with permission

unknown tip conditions. This is beneficial since in many applications the condition at the tip cannot be known a priori.

To meaningfully compare the accuracy between different studies, the accuracy as a percent of the full scale output should be compared. Note that the absolute error reported here is lower than that in [142]. This is due to the fact that the version of the tweezers used in this study produced less output force due to different manufacturing process. If greater output force was produced, the measurement would be scaled up, but the signal-to-noise ratio of the charge and voltage measurements would remain the same. Since the noise would be scaled up as well, the error as a percentage of full scale output should remain the same.

The accuracy reported by Badel et al. [15] was achieved by using a solid PZT stack actuator. On the other hand, the tweezer mechanism is very compliant. This means that a large change in displacement at the tip corresponds to a small change at the PZT actuator that is much more rigid than the structure. Consider a simple mechanical system in Fig. 4.40, which represents a PZT actuator in series with a stiffness and serves to illustrate the effect in question. If the endpoint is free, the displacement of

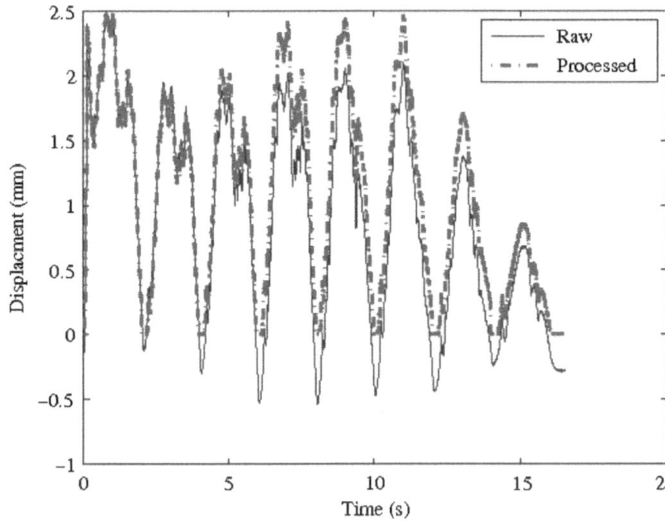

Figure 4.39 Effect of limiting and sliding DC offset. © 2014 IEEE, reprinted with permission

the midpoint is

$$\Delta x_{pzt} = \frac{f_{pzt}}{k_{pzt}}. \tag{4.83}$$

If the endpoint is blocked, the displacement becomes

$$\Delta x_{pzt} = \frac{f_{pzt}}{k_{pzt} + k_1}. \tag{4.84}$$

If k_1 is small relative to k_{pzt}, the displacements in the blocked and free cases will be very close to each other. The tweezer mechanism is more complicated than the system in Fig. 4.40, but the basic effect is the same, namely the displacement of the displacement of the PZT actuator will not vary much between the blocked and free cases. Charge is directly related to the displacement on the actuator and the applied voltage, based on (1.1). This means that for the same driving voltage a small change in the displacement, and subsequently charge, will cover the entire range of loading conditions at the tip. In effect, the signal-to-noise ratio is worsened, meaning that not only the charge and voltage measurements but also the hysteresis model must be accurate.

In fact, despite the less than 1% average error of the hysteresis model, this mismatch is likely the main cause of error. The force predicted by the self-sensing technique in the free case shows the effect of mismatch in the hysteresis model since this is the calibration case for the hysteresis operator. Therefore the effect of the mismatch on force prediction can be shown by subtracting this case from any other, given the same input voltage. Fig. 4.41 shows a self-sensed force measurement, and the same measurement when the error due to model mismatch has been subtracted. This significantly im-

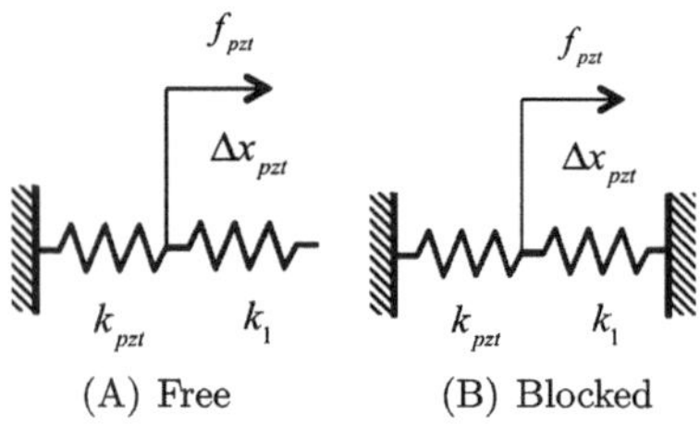

Figure 4.40 A simple model of a PZT actuator and a stiffness in series. © 2014 IEEE, reprinted with permission

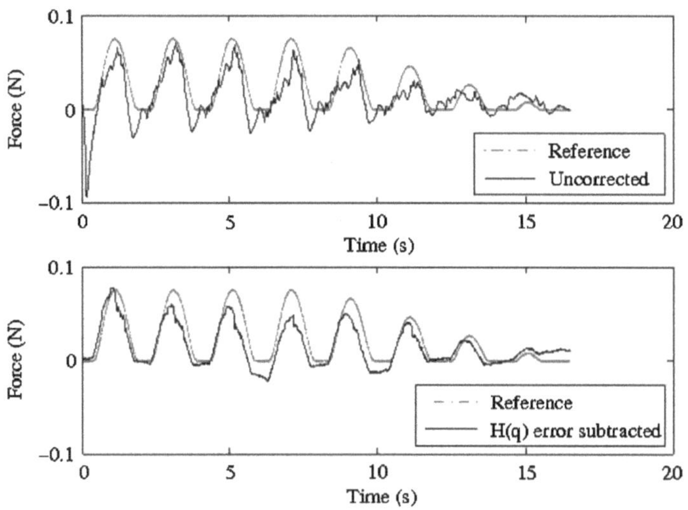

Figure 4.41 The effect of hysteresis model mismatch on accuracy. © 2014 IEEE, reprinted with permission

proves the measurement, indicating that despite its relatively low error the hysteresis operator is the main limiting factor of accuracy, rather than charge or voltage measurement. This indicates that for a robot with a set of preprogrammed motions, the model error could be calculated in advance and used to improve the sensing accuracy. For example, if the tweezer end effector was controlled by the push of a button to toggle between open or closed, the self-sensing method could be used to obtain useful information such as the size of a grasped object. If the command is not known in advance, increasing the number elementary operators in the hysteresis model could provide a slight increase in accuracy, but this becomes computationally intensive and will reach a point of diminishing returns. Additionally, the creep effect is another source of error. Prior work on self-sensing, even that assuming quasistatic operation, has been limited to a time scale of milliseconds, over which the effect of creep is negligible. For a robotic end effector, the time scale of interest is seconds or minutes, at which point creep can no longer be ignored.

To improve the on-site simultaneous calibration of hysteric and structural parameters for self-sensing, Han and Ueda [94] proposed a modified version of the self-sensing estimation to eliminate the need of identifying system constants in advance and to obtain a minimum set of coefficients purely from an on-site calibration procedure.

4.4 Biologically inspired robotic camera orientation system

4.4.1 Robotic realization of saccades and smooth-pursuit

4.4.1.1 Biologically inspired robotic vision

One research approach in biologically inspired robotics is to apply principles known in physiology to reproduce the same functionality in an artificial system. The human eye represents a good platform for demonstrating engineering challenges inherent in the neuromotor system. It has two important modes of operation: fast moves (saccades) and slow moves (smooth pursuit) [59,209,69,285]. Perception and eye movements are known to be linked in the visual system. However, little engineering study has been made on the oculomotor control for fast dynamic vision. A robotic camera positioning mechanism operated by the piezoelectric cellular actuators can be used as a research platform to demonstrate the effectiveness of this ocular physiology-inspired approach. The platform enables dynamics-based image processing methods in coordination with inherently discrete and rapid ocular movements that can reproduce saccades and smooth-pursuit. A panoramic image of a scene can be generated from multiple images acquired during saccade-type movements by removing motion blur and stitching images within a frame rate. Muscle-inspired impulsive actuation realizes tracking of a moving object with various velocity.

4.4.1.2 Saccades and smooth-pursuit

Saccades are known to be the fast and rapid eye movements enabling quick scanning the field of view and high-resolution fovea at the region of interest. Despite of saccadic eye motion, we are almost unaware of potentially blurry visual input. The neural system blocks images with substantial motion blur in the course of the eye motion, termed saccadic masking [168,113]. The equivalent frame rate of the human eye is reported to be at most 24 frames per second (fps) or 42 ms/frame, which varies depending on situation. Research shows that a typical saccadic (point-to-point) eye movement completes within 20–80 ms, comparable to the rate of visual perception. Saccadic masking plays a key role when the motion of the eye is not negligible in visual perception. Inversely, motion control of the image acquisition device must be planned to ease this masking process. Saccades occur much more rapidly than proprioceptive, vestibular, or visual feedback that can be returned to the brain [207]. This observation indicates that saccades are performed in an open-loop manner. This may be implemented as feedforward control based on the oculomotor dynamics.

Smooth-pursuit is another known eye movement. Unlike saccades, the eye continuously follows an object with relatively slow velocity. One might think that such velocity control of the eye is fairly easy. This is true when an electromagnetic servo motor is used. However, that is not the case in the neuromotor system primarily due to recruitment, or discrete activation of muscle bundles [243,70]. Muscles are essentially quantized actuators, and hence cannot produce continuous force. Despite of quantized force production, we are not aware of this discrete actuation in the eye and able to perceive the features of a moving object. Although possible neuromotor mechanisms in smooth-pursuit are debatable, from an engineering point of view, this problem can be seen as velocity matching control of an impulsive dynamical system.

4.4.1.3 Camera positioners inspired by the human ocular system

There are many devices developed for automatic camera positioning. Several were designed to be driven by traditional servo motors with rigid links [145,149]. These mechanisms successfully generated a fast motion that is comparable to saccade, but have little in common with biological muscle actuation. The human eye is oriented by means of antagonistic pairs of muscles. Also, heavy servo motors have a marked discrepancy from extraocular muscles. There are other positioning devices that are positioned by means of antagonistic pairs of alternative actuators such as piezoelectric or pneumatic actuators [282,147]. However, these devices still have a discrepancy from muscle actuators because of their rigidity, an artifact of the use of traditional servo motors, in contrast with the contractility and compliance of extraocular muscles. Further, most of the mechanisms [145,282,147] are controlled in a closed-loop manner, whereas physiological evidence indicates that a saccade is completed in an open-loop manner [123,298,207]. Schultz and Ueda have developed a single DOF camera positioner oriented by an antagonistic pair of compliant cellular actuators that are controlled by an open-loop switching method [219] that is presented in Section 1.3.6.

4.4.1.4 Image de-blurring technique

Various approaches have been studied in the computer vision field to restore blurry degraded images. Motion blur is caused by relative motion between an image acquisition system and an object during the exposure window. To restore an image that was degraded by motion blur, the motion path must be estimated. The estimated path can be represented as a point spread function (PSF). The PSF is a blur kernel that describes the camera motion during an exposure window. If the motion blur is shift invariant, an unblurred or latent image can be estimated with a deconvolution algorithm, given the estimated PSF and the obtained blurry image. Some studies proposed the estimation of the blur kernel from a single image [73,117,228], others from multiple images [230, 17]. Although these approaches successfully estimate the latent image, much computation time is required. In particular, the computation time greatly increases as the image size increases or as the blur kernel size increases.

Other approaches do not directly analyze inherent properties of blurry images but use external devices to estimate a PSF. One study uses an extra-high-speed camera with a low resolution to track the camera motion [24]. This approach requires post-image processing to process data from two camera systems. Another study uses a gyro sensor and an accelerometer to track camera shake or motion [121]. The method involves computations that estimate the path of a camera shake from noisy gyro and accelerometer sensors. Raskar et al. proposed a fluttered shutter approach that modified the exposure time of image acquisition to minimize motion blur [205].

4.4.2 Dynamics-based oculomotor-visual coordination in rapid camera movements

From a control point of view, saccadic eye movements cannot be performed by continuous visual feedback [298,123]. Although detailed neural mechanisms behind saccadic masking are not known, it is a natural consequence that motion control and image processing must be tightly synchronized. Eye movements are primarily self-induced, i.e., generated by the extraocular muscles receiving commands from the oculomotor system. It may be hypothesized that there is a neural mechanism that makes a prediction of such self-induced eye movements based on the dynamic model of the eye, enabling effective image de-blurring and panoramic image stitching, or in short, *dynamics-based visual processing*. Self-induced motion can be compensated in a sensorless fashion based on the dynamics; therefore, no image processing is necessary to obtain key information for image de-blurring.

This dynamics-based visual processing approach inspired by the ocular physiology is the key enabler of real-time panoramic image stitching with simultaneous motion de-blurring in a dynamic vision system, allowing for generic image sensors with a standard frame rate and significantly less computational load, and requiring no motion sensors. The vision system is rich in bio-inspired principles itself, and a rich synergy emerges by using the cellular actuator technology in a vision system. The camera positioner with muscle-inspired discrete-switching piezoelectric cellular actuators introduced in Section 1.3.6 is used as an experimental platform for investigating and demonstrating the method inspired by the oculomotor physiology.

4.4.3 Switching control of camera positioner

The developed camera orientation system shown in Fig. 1.49A achieves discrete angular displacements by specifying a number of ON actuators associated with a desired tilting angle. For example, when six PZT actuators need to be turned on, PZT actuators 1 to 6 are activated while the rest of the actuators are deactivated. While a number of combinations can create the same angular displacement due to the redundancy of cellular PZT actuation, the deviations in displacement created by different combinations were small, therefore negligible as reported in [219] and [267].

Because of the flexible nature of the amplification mechanisms, the camera positioner naturally has an oscillatory response when inputs are activated. The camera orientation mechanism exhibits a single dominant frequency at 14.4 Hz with a damping ratio of 0.19. However, if the natural frequencies of the resonant modes are known, nearly vibration-free moves can be made by MSDSVS [219] introduced in Section 3.1. The mechanical parameters are experimentally identified by the observation of step responses. The dynamics of the system can be represented as a transfer function of a linear second-order system

$$G(s) = \frac{R}{s^2 + 2\zeta\omega_n s + \omega_n^2} \tag{4.85}$$

where R is the residue, $\omega_n = 90.5$ rad/s, or 14.4 Hz, is the natural frequency, and $\zeta = 0.19$ is the damping coefficient. The value of R is 6650.3. This dynamic model can also be used to estimate the motion of the system when switching commands are given. When a total of n discrete (ON or OFF) force inputs are applied from the PZT actuators to this system, the time response of the angle $\psi(t)$ is represented as a superposition of step responses of a lightly damped system (4.85) and given by [132]

$$\psi(t) = \sum_{i=1}^{n} \frac{A_i R}{\omega_n^2} \left[1 - e^{-\zeta\omega_n(t-t_i)} \left(\cos\omega_d(t-t_i) + \frac{\zeta}{\sqrt{1-\zeta^2}} \sin\omega_d(t-t_i) \right) \right] \times 1(t-t_i) \tag{4.86}$$

where $\omega_d = \omega_n\sqrt{1-\zeta^2}$ is the damped natural frequency, A_i is the amplitude of each integer MSDSVS command, t_i is the time when A_i is applied, and $1(t)$ is the unit step function.

Fig. 4.42 shows the experimental results of camera positioning. Fig. 4.42A shows response to step commands with no vibration suppression. When the MSDSVS technique is applied, the response becomes almost vibration-free as shown in Fig. 4.42B. The overshoot caused the third layer to collide with the first layer units [219]. Although a slight nonlinearity was observed, approximately a variation of 12% of the natural frequency, when the desired step was beyond 77% of the full stroke, the repeatability of the response is overall very high. Also as shown in Fig. 4.42, the estimated response shows good agreement with the actual response with small errors after settling possibly due to a flexibility in the cellular actuators or mechanical friction. In addition, the settling time with MSDSVS is approximately within 55 ms that is comparable to the speed of response of human saccadic eye motion.

The discrete switching control is performed by a real-time controller based on a National Instruments (NI) cRIO-9024 system (see Appendix A.3). The main function of the real-time controller is to synchronize the camera and the motion controller, which is achieved by sharing trigger signals. Poor synchronization would lead to misestimating the motion during the exposure window, which results in poor estimation of the PSF and the latent image. For the vibration suppression control, a 1 kHz sampling rate is required due to a possible switching delay in commands. This FPGA-based motion

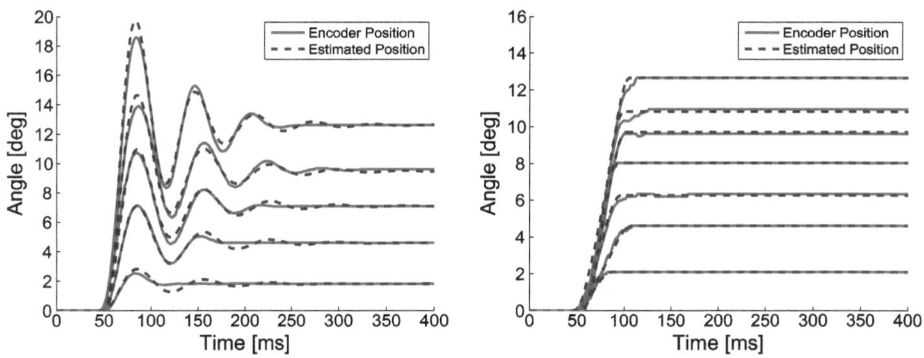

(A) Uncompensated step response (B) MSDSVS response

Figure 4.42 Time response of the camera orientation system given discrete switching commands. © 2015 SAGE, reprinted with permission

control system to control multiple discrete PZT actuators and a GPU-based image processing system to achieve image de-blurring at a realistic rate have been integrated into one system [131].

4.4.4 Dynamics-based blur kernel estimation for motion de-blurring

While high frame rate cameras, as fast as 1000 fps or more, are available, generic image sensors acquire images at a rate of 30 fps (33.3 ms) or 60 fps (16.6 ms). The exposure time of such generic image sensors is comparable to the settling time of the camera positioning system. The robotic camera positioning mechanism is suitable for acquiring multiple images in a relatively short period for quick scanning the field of view to generate a panoramic image. Rather than holding off image acquisition until the camera becomes completely stationary, an image may be acquired during a settling motion. As shown in Fig. 4.42, the residual vibration has almost diminished, but not completely, resulting in some blurriness in the image. This blurriness can be removed by using dynamic-based motion de-blurring.

The PSF, or the blur kernel, $\boldsymbol{K}_{br}$, can be computed by the solution of self-induced camera motion. The trajectory of the camera estimated by (4.86) is sampled and projected onto a grid of image pixels. The element of $\boldsymbol{K}_{br}$ at the pixel (x_k, y_k) is essentially proportional the total time that $\psi(t)$ has stayed within the pixel. When a single exposure window is closed, the sum of all the elements of the blur kernel is normalized to one for intensity conservation. In the system introduced above, the number of sampling periods within a single exposure window that corresponds to the element value is counted by FPGA for each of the pixels. The algorithm for the kernel estimation runs at 100 kHz on the FPGA, the blur kernel is generated in 30 μs after an image has been acquired [131]. This kernel estimation is significantly faster than existing image-based or sensor-based methods.

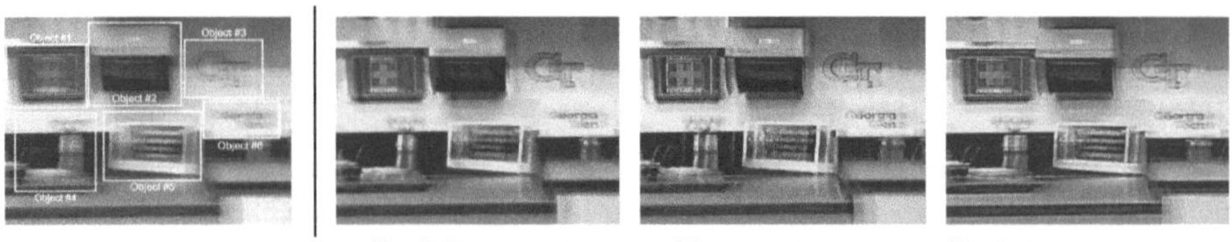

Figure 4.43 Representative image de-blurring results. © 2016 IEEE, reprinted with permission

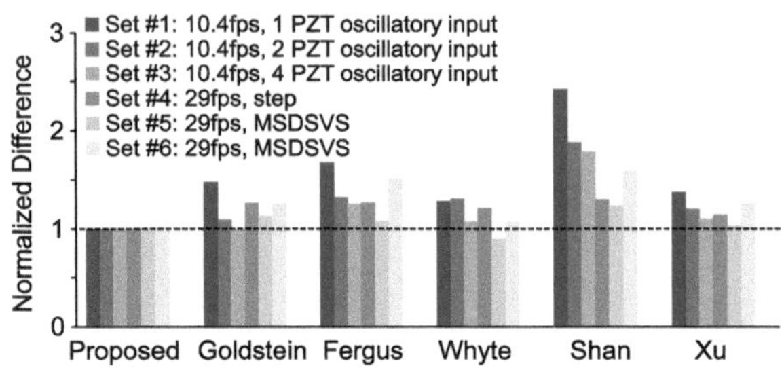

Figure 4.44 Dynamics-based image de-blurring and a comparison of image quality.

Considering a pure rotational mechanism such as the human eye as well as the camera positioning mechanism shown in Fig. 1.49, a spatially-invariant blurred image $\boldsymbol{B}$ can be represented by convoluting the shift-invariant blur kernel $\boldsymbol{K}_{br}$ and a latent sharp image $\boldsymbol{I}$ plus noise $\boldsymbol{N}$, $\boldsymbol{B} = \boldsymbol{K}_{br} * \boldsymbol{I} + \boldsymbol{N}$, where $*$ is the convolution operator. Once the blur kernel is obtained, the latent image can be estimated by means of deconvolution such as Richardson–Lucy algorithm.

Fig. 4.43 shows representative de-blurring results using the proposed method and conventional methods [89,73,286,228,291]. For a blurry raw image shown in Fig. 4.43A, the proposed method in Fig. 4.43B-1 successfully restored the image overall although minor blurriness remained for objects #3 and #4. The result of Xu's method [291] in Fig. 4.43B-2 shows a good contrast on object #5 with minor blurriness, but the image is still blurry overall. Goldstein's method [89] in Fig. 4.43B-3 has more blurriness than the other two. Fig. 4.44 shows a quantitative comparison where lower values are greater in performance, and the proposed method achieved mostly the best image quality. In the benchmark shown in the figure, Shan's [228] and Xu's methods run complied executables, and other methods including the proposed method run on MATLAB. Nevertheless, the proposed dynamics-based approach is one to two orders of magnitude faster than other methods as shown in Fig. 4.45 since no image processing for kernel estimation is necessary. In the proceeding research, the proposed algorithm has been implemented into the FPGA-based system for faster computation [131].

The results shown in Fig. 4.46 justify why vibration suppression should be used to improve the speed of scanning. There is a trade-off between the scanning time and image quality. Scans with vibration suppression complete faster than the ones with uncompensated step moves. Also standard deviations have improved.

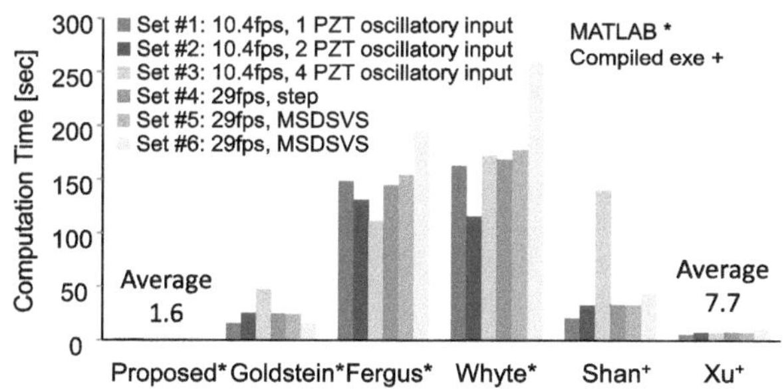

Figure 4.45 Comparison of computational time.

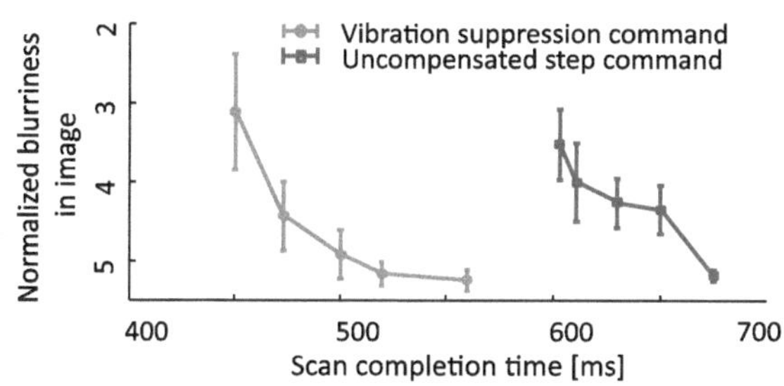

Figure 4.46 Improvement of scanning speed by the coordination of motion and vision. The camera scans five different equally spaced positions for a total travel distance.

4.4.5 Dynamics-based fast panoramic image stitching

The dynamics-based approach can be extended to image stitching. To stitch images for panoramic image generation, reasonable amount of overlap among images must be found. This is another computationally expensive process in conventional feature-point matching methods. In contrast, the dynamic-based method utilizes predicted camera positions to make the best guess of image overlapping. The camera positioner of interest is purely rotational and thus the homography matrix is a function of a static camera angle that can be obtained as the center of gravity of the blur kernel [131].

Fig. 4.47 shows the result of the dynamic-based image stitching and comparison with existing methods. The proposed method showed no geometric distortion or misalignment. Microsoft Image Composite Editor (MS ICE) [172] and PTGui [203] show some misalignment. Other comparisons including SIFT and RANSAC were conducted [131] where some methods failed to stitch images due to significant motion blur. Results indicate that the de-blurring technique is beneficial for image stitching, and the dynamics-based image stitching is the most robust as it does not rely on image features.

4.4.6 Discussion

This section has introduced a dynamics-based method for coordinating image processing and motion control for a robotic vision system whose speed of movement

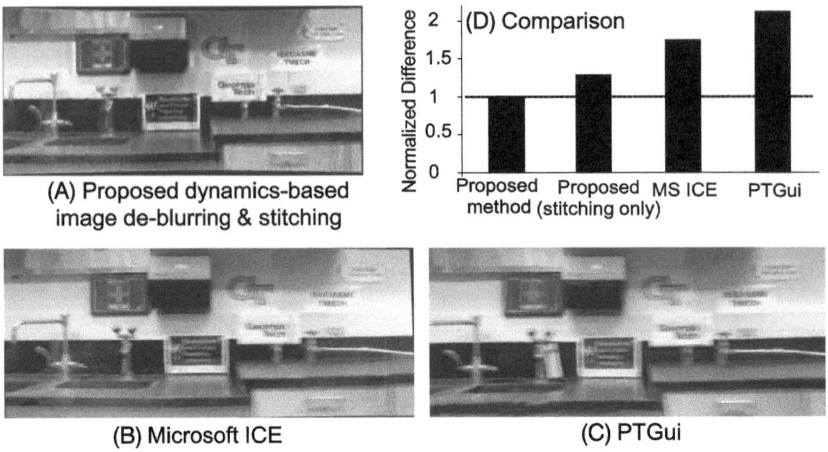

Figure 4.47 Panoramic image generation. The robotic camera system scanned the environment to the right and left and acquired a total of 21 images for each panoramic image. © 2016 IEEE, reprinted with permission

is as fast as the image acquisition rate. The presented work is an early step toward multidisciplinary approach to design a robotic vision system inspired by the ocular physiology. Existing image de-blurring algorithms typically investigate image features or use external sensors to recover blurry images, thus computationally expensive. The dynamics-based approach enables much faster computation without adding sensors. A panoramic image can be generated by acquiring multiple images in the course of motion. The dynamics-based image stitching method is robust to degraded blurry images.

The current research develops a discrete switching method inspired by smooth-pursuit [133]. The method employs a non-minimum discrete switching controller for object tracking and vibration suppression. The dynamics-based approach assumes that the camera movements are self-induced. Therefore, the method itself would not be robust to external disturbance such as mechanical shock or vibrations, for instance, when the camera system is mounted on a mobile platform. Recall that the camera module is supported by compliant actuators in the camera positioner. The viscoelasticity in the actuators is expected to work as a shock absorber. Unless external disturbance precisely excites the resonance of the camera–actuator system, mechanical disturbance is absorbed and its influence on image quality should be limited. Furthermore, unlike other general camera positioning mechanisms, the inertia of the moving component is very small, which makes the resonant frequency much higher than that of typical camera platforms. This impedance *mismatch* between the camera (eye) and housing (body) may isolate the oculomotor system from the body. Indeed, the piezoelectric cellular actuators are designed to be much faster than general electromagnetic motors, inspired by the fact that the extraocular muscles contract much faster than other voluntary muscles. Future research includes the investigation of vibration transmissibility of the camera platform and evaluation of the sensitivity to external disturbance.

Conclusion

5

5.1 Summary and future directions

5.1.1 Brief summary

This book has presented the cellular actuator architecture, a paradigm for generating robotic movement inspired by the way human musculature is organized. This allows robotic systems to capture two key advantages of biological systems, namely, high resiliency to damage (or graceful degradation) and natural smooth human-like movement. Motion is produced by the collective action of individual units contracting independently in an ON–OFF (or boolean) manner. The reason that this works, both in biology and in the cellular actuation system, is because there is elastic material joining the units together. At any given instant, this allows certain individual units to be contracted (active), and others inactive. In fact, the activation pattern can be completely arbitrary. When the cellular units are identical, the force produced depends only on how many units are active, not which are active.

Building cellular actuators from traditional servo motors (multi-series-elastic-actuators), while conceivable, does not package nicely and tends to be very large. A survey of active materials that contract in response to a voltage or current does not yield one perfect material. Each has a down side. Piezoelectric ceramics, specifically PZT's, downside is the small displacement produced. By using compliant mechanisms, the displacement can be scaled up by using multi-stage nested rhomboidal strain amplifiers. While strain amplification in general is a large body of research in the precision engineering community, this book has reviewed the most relevant and significant works pertaining to the cellular actuator concept. Chapter 1 has presented the amplification concept under some idealized assumptions, which provides reasonable accuracy if the deformation of the stack is small and there are few stages. The amplification concept was presented both for amplification of the axial movement of the stack, and a buckling concept, which provides very large displacement amplification while being a completely boolean paradigm. Chapter 2 has provided a deeper look at the underlying mechanics to more accurately predict the force–displacement behavior when the deformation of the individual strain amplification mechanisms and the energy stored therein are non-negligible. Quantitative methods to predict the point of diminishing returns, resiliency to damage of individual units, and the stiffness of an actuator from the point of view of the environment are presented.

Controlling the cellular actuator presents an entirely different paradigm from traditional servo motors due to its highly redundant architecture. Chapter 3 has covered the various control methods that have been developed for cellular actuation systems that proceed from the high redundancy and boolean operation. Piezoelectric ceramics *can* be operated in a proportional manner, but require hysteresis compensation, which requires a large characterization effort. The segmented binary control method presented in this book exploits the redundancy to avoid this problem entirely. This chapter also presented methods to characterize the system dynamics of cellular ac-

Cellular Actuators, DOI: 10.1016/B978-0-12-803687-7.00005-3
Copyright © 2017 Elsevier Inc. All rights reserved.

tuators, as they tend to be oscillatory due to the thin metallic structures of the strain amplifiers. Minimum Switching Discrete Switching Vibration Suppression is a control method presented that uses strategically phased switching in the open loop to prevent oscillation following a point-to-point move. As cellular actuation systems grow large, cabling, wiring, and addressing is a non-trivial problem. The stochastic broadcast feedback control method presented in this chapter allows each cell (equipped with a microprocessor) to act as an independent agent in response to a single error value, which can be transmitted wirelessly. The cellular architecture's intrinsic properties are also capable of naturally being used to generate signal-dependent noise, which makes it a suitable platform to implement physiological ideas about natural movement in the human body.

This book has presented a few applications as guidance for those who wish to build robotic or engineering devices using cellular actuators, detailed in Chapter 4. A camera-positioning mechanism driven by an antagonist pair of cellular actuators has been presented, and an associated image processing algorithm plausibly similar to what may go on in the human brain is shown to de-blur and stitch together shaky images into a panoramic image in response to a saccadic motion. It uses knowledge of the system dynamics of the positioner to achieve its goals. A pair of actuated tweezers with integrated self-sensing has also been presented. A highly damage-tolerant robotic arm has been described as well.

After having read this book, the reader doubtlessly has many of his own ideas for applications as well. The authors hope that the guidelines and modeling techniques presented are useful in realizing the ideas that are hatched in the imaginations of those who read this book. At the very least, the exposure to the concepts and examples presented may help those who use this technology to become comfortable with the new paradigm and use it effectively. The major thing to keep in mind is that just like muscles, cellular actuators have a length-dependent force generation property, and so just like in human beings, this means that particular poses must be generated for particular tasks in order to take advantage of the optimal muscle length for the task. In general, the force drops as the muscle length increases. This represents a sharp departure from servo motor driven robots whose force properties depend only on the Jacobian and the manipulator dynamics. It may also produce some key advantages with clever implementations, moment arms, and so on.

5.1.2 Future work

The work in this book represents a beginning of exploration of this dynamic and challenging area that in a very crude way moves the nature of operation of robotic systems closer to humans in the physiological sense. There are plenty of exciting areas for future work that will bring us even closer.

The cellular actuation units presented in this book are all identical. This was by design, as some of the mathematical results presented only became tractable when this assumption was made. However, it would be interesting to see some work built on these results that include the variability present in a human muscle, such as strong and weak motor units, and fast and slow-twitch (red and white) muscle fibers. As the human body strategically uses the diversity within muscle systems for enhanced performance, cellular actuator systems may be able to do so as well. Multi-material systems, such as those combining PZT and SMA, may produce better performance than either one alone.

Other control methods may be developed that exploit the cellular nature of the hardware. Minimum Switching Discrete Switching Vibration suppression is entirely an open-loop phenomenon. A particularly fruitful area would be to develop feedback control algorithms that use switching methods to eliminate oscillations generated by external disturbances. This will require some modification of existing algorithms as the published methods in quantized control are not directly applicable. It is likely that this will emerge as some sort of a predictive–corrective method. The role of embedded sensing or self-sensing within the array of cellular units is likely to be a major factor. With this will come a need for analysis of sensor fusion and observability for systems of this type.

As more researchers become interested in this topic, it is likely that new applications will emerge that can perform missions of long duration in unstructured environments. One particularly interesting area to explore would be legged locomotion. Topics of interest include activation patterns for efficient gait, detection of failure, and the ability to autonomously make abort/return vs. continue the mission decisions. For precision applications, quantifying the loss of accuracy due to degradation is important. A device may have acceptable resolution with boolean inputs when it ships, but as individual units degrade, what are the consequences for accuracy, and decisions about the merits of down time vs. continued operation? Finally the cellular actuation paradigm provides an avenue for new thinking about repair and adaptation to new environments. Human muscle both self-heals and hypertrophies in response to damage and strenuous environments. The cellular architecture opens up possibilities to quarantine malfunctioning units and add additional units *in situ* to increase capacity when needed (or move it from one part of a device to another). Self-healing is not beyond the pale of this technology, and a substantial amount of work has been completed in recent years in self-healing materials. Integration of self-healing components with the cellular actuation technology may provide yet another quantum leap in producing more human-like robotic systems.

Nomenclature

α	Electromechanical transduction factor
α_A, α_B	Quotient of expectation to squared error
$\bar{\alpha}$	Parameter taking a value between 0 and 1
$\bar{\eta}$	Mean of η
$\bar{k}_B$	Effective stiffness in the constrained space
β	Ratio between PZT and tip displacements
$\boldsymbol{a}_K$	Vector of creep eigenvalues
$\boldsymbol{B}$	Blurred image
$\boldsymbol{C}$	$\boldsymbol{K}/k$
$\boldsymbol{F}$	Matrix containing piezoelectrically generated forces
$\boldsymbol{G}$	Incoming connection matrix
$\boldsymbol{g}$	Additional element to $\boldsymbol{G}$
$\boldsymbol{H}$	Outgoing connection matrix
$\boldsymbol{h}$	Additional element to $\boldsymbol{H}$
$\boldsymbol{I}$	Sharp image
$\boldsymbol{K}$	Stiffness matrix
$\boldsymbol{K}_{bf}$	Blur kernel
$\boldsymbol{L}$	Actuator length matrix
$\boldsymbol{M}$	Mass matrix
$\boldsymbol{N}$	Noise matrix
$\boldsymbol{Q}$	Hysteresis and creep parameter matrix
$\boldsymbol{r}_K$	Vector of thresholds
$\boldsymbol{S}$	Stiffness matrix
$\boldsymbol{U}$	Matrix for inequality constraints
$\boldsymbol{w}_K$	Vector of weights
$\boldsymbol{x}$	Vector of displacements of cellular units
$\boldsymbol{X}_t$	Aggregate system state
$\boldsymbol{Z}_t$	Output observations
$\boldsymbol{Z}_{K0}$	Matrix of initial conditions
χ	$N_{film} \cdot d_{33}$
Δe_t	Change of e from time $t-1$ to t
ΔL	Elongation of an actuator
$\Delta x, \Delta y$	Displacement in the x-direction, displacement in the y-direction
Δx_k^i	Displacement of the ith unit of the kth layer
Δx_1	Displacement of a load
Δx_k	$d_k - d_k'$
$\Delta x_{pzt,max}$	Maximum displacement of a PZT stack actuator
Δx_{pzt}	Displacement of a PZT stack actuator
Δy_t	Change of y from time $t-1$ to t
δ_A	Displacement of a middle point A of the tweezer-shape structure
δ_B	Displacement of the tip B of the tweezer-shape structure
Δ_S	Determinant of the stiffness matrix
Δ_{FL}	Step size of floating point quantization
δ_{in}	Input of two-port network
δ_j	Displacement of a structure at a point j
δ_{out}	Output of two-port network

Δ_{UN}	Smallest step size of quantization
ϵ, ϵ_{pzt}	Strain, strain of a PZT actuator
ϵ_k	Strain of the kth layer unit
η	Displacement of a binary controlled actuator
$\Gamma[x](t)$	Modified discrete Prandtl–Ishlinskii operator
γ_k	Dimensionless stiffness
γ_m	Dimensionless mass
$\hat{\delta}_A$	Displacement of a middle point A of the tweezer-shape structure when $W_B = 0$
$\hat{a}, \hat{a}_k$	Estimate of a and that of the kth layer rhomboidal mechanism
$\hat{k}_J$	Estimate of k_J
$\hat{k}_{BI}, \hat{k}_{BO}, \hat{k}_{BIk}, \hat{k}_{BOk}$	Estimates of k_{BI} and k_{BO}, and those of the kth layer rhomboidal mechanism
$\hat{W}_A$	Force at a middle point A of the tweezer-shape structure when $W_B = 0$
κ	Effective amplification gain
κ_{total}	Total effective amplification gain
$\lambda_{1,2}$	Eigenvalues of linearized state equations
$\lambda_{max}, \lambda_{min}$	Eigenvalues of stiffness matrix
$\lambda_{nodamping1,2}$	Eigenvalues of undamped linearized state equations
$\mathbb{Z}$	Set of integers
$\mathbf{D}$	Set of input amplitudes
μ	Bound of a stochastic Lyapunov function
ν	Poisson's ratio
ν_{FL}	Quantized error of a floating-point quantizer
ν_{SDN}	Signal dependent noise
ν_{UN}	Quantized error of the uniform quantizer
ω_d	Damped natural frequency
ω_n	Natural frequency
ω_n	Vibration frequency
Φ	Function of actuator displacements and stiffnesses
ϕ_j	Phase angle corresponding to an impulse given at t_j
$\psi(t)$	Orientation of camera positioner
ρ	Bit of mantissa
σ_{FL}	Standard deviation of floating-point quantization
σ_{SDN}	Standard deviation of proportional signal-dependent noise
τ	$\max(n_F, n_B)$
θ	Angle of the oblique beam of the rhomboidal mechanism
θ_i	Orientation of segment i in the undeformed configuration
$\tilde{f}, \tilde{f}_k$	Effective force viewed from the output port of an amplification mechanism and that of the kth layer mechanism
$\tilde{k}, \tilde{k}_k$	Resultant stiffness viewed from the output port of an amplification mechanism and that of the kth layer mechanism
$\tilde{b}$	Damping coefficient of stroke amplification mechanism
$\tilde{c}$	Buckling force term
$\tilde{c}_0, \tilde{c}_1$	Elastic forcing terms
$\tilde{c}_S$	Series compliance
$\tilde{g}$	Stiffness between lumped mass and ON cell
$\tilde{g}_p(e), \tilde{g}_q(e)$	State transition probability functions of e
$\tilde{k}_S$	Stiffness of an N-unit strand
$\tilde{k}_{limiter}$	Stiffness of stroke limiting structure
$\tilde{m}$	lumped mass
$\tilde{m}_{limiter}$	Mass of stroke limiting structure
$\tilde{u}$	Compressed signal
$\tilde{u}'$	Uniformly quantized compressed signal
ξ	Muscle pennation angle
ζ, ζ_i	Damping coefficient, damping coefficient of the ith mode

$^{i}\boldsymbol{P}_{i-1}$ Projection matrix from segment $i-1$ to segment i
a, a_k Amplification gain, amplification gain of the kth layer unit
A_i Amplitude of the jth impulsive input
a_i, b_i Constants of continued fractions
A_k, B_k Parameters of recursive expression of stiffness
$B_1, \ldots, B_{n_B}$ Intermediate states in backward loading
b_J Width of a compliant joint
B_R Ready states in backward loading
c_k $a_k - \frac{b_{k+1}}{c_{k+1}}$
$C_{11}, C_{12}, C_{21}, C_{22}$ Quartic factors of geometric parameters
$C_{A1}, C_{A2}, C_{A3}, C_{A4}, C_{B1}, C_{B2}, C_{B3}, C_{B4}, C_{B5}$ Structural coefficients of tweezer-shape structure
C_{pzt} Clamped capacitance of piezoelectric actuator
d_k Initial gap between the surface of the internal $(k-1)$th layer unit and the apex of the rhombus mechanism of the kth layer unit
d'_k d_k after activation of the actuator unit
D_M Denominator
$d_\theta(z)$ Function for a deflection angle
E, E_{pzt} Young's modulus, Young's modulus of PZT material
$e_{x,y}$ Error between model and experiment
F Force acting to a cell
f_k^i Force of the ith unit of the kth layer
$F_1, \ldots, F_{n_F}$ Intermediate states in forward loading
f_I Net force applied to a rhomboidal mechanism from the internal unit
F_j Force applied to a point j
f_O Reaction force from an external load to a rhomboidal mechanism
F_R Ready state transitions in forward loading
F_x, F_y Force in the x-direction, force in the y-direction
F_{block} Force of PZT when the rhomboidal mechanism is blocked
f_{CDF} Cumulative distribution function
F_{out} Output force of a rhomboidal mechanism
FOM Figure of merit
$G(s)$ Transfer function
g_p, g_q Transition controller gains
$H(q_{pzt})$ Hysteresis operator
h_J Thickness of a compliant joint
h_k Height of the kth layer unit
h_{pzt} Height of a PZT stack actuator
$H_{r_H}[x](t)$ Weighted sum of play operators
I_1, I_2 Current values of a two-port network model
i_C Current charged to a SMA cell
I_i Moment of inertia of segment i
i_{Cd} Desired current of a SMA cell
J_E Dissipated energy
K Number of layers
k_c Gain
k_J, k_{Jk} Joint compliance of a rhomboidal mechanism and that of the kth layer rhomboidal mechanism
k_p General proportional gain
k_s Stiffness of a redirecting spring
k_ϕ Rotational stiffness of a compliant joint
$k_{BI}, k_{BO}, k_{BIk}, k_{BOk}$ Beam compliances of a rhomboidal mechanism and those of the kth layer rhomboidal mechanism
k_{load} Spring constant of a load

$k_{passive}$	Stiffness of passive actuator array
k_{pzt}	Stiffness of PZT stack actuator
$K_{r_K a_K}[x](t)$	Creep operator
k_{SDN}	Constant for signal-dependent noise
$K_{xx}, K_{xy}, K_{yx}, K_{yy}$	Elements of stiffness matrix
L	Actuator length
l	Length of a cell
L_J	Length of the gap of a compliant joint
l_{muscle}	Length of muscle
l_{tendon}	Length of tendon
$l_{preload}$	Preload displacement
l_{pzt}	Length of a PZT stack actuator
m	$\lfloor t/\tau \rfloor$
M_i	Internal moment load for segment i
M_k	Number of actuator units connected in parallel, or bundles, in the kth layer
n	Integer counter
$n'_{l1}, \ldots, n'_{ln}$	Positive integer values
N^{dead}	Number of nonfunctional cells
N^d	Number of ON cells required to achieve reference r
$N^{OFF,base}$	Number of consecutive OFF units connected to the base
N^{OFF}	Number of OFF cells
N^{ON}	Number of ON cells
N_a	Command of the number of active cells
n_B	Number of intermediate state transitions in backward loading
n_F	Number of intermediate state transitions in forward loading
N_i	Internal axial load for segment i
N_k	Number of actuator units connected in series, or stacks, in the kth layer
N_m	Number of significant vibration modes
N_{film}	Number of PZT films along the actuation direction
n_{li}	Number of cellular units in the ith lumped mass
N_{seg}	Number of straight segments in rhomboidal layer
p	Transition probability from OFF to ON
p_0	Transition probability for the minimum of the parabola
P_1, P_2	Parameters of electromechanical model
p_k	Sequence of continued fraction
q	Transition probability from ON to OFF
Q_1, Q_2, Q_3	Elements of stiffness matrix of the tweezer-structure
$q_{11}, q_{12}, q_{21}, q_{13}, q_{31}, q_{22}, q_{23}, q_{32}, q_{33}$	Elements of Q_i
Q_{FL}	Floating-point quantizer
Q_i	Integration of projection matrix ${}^iP_{i-1}$
q_{pzt}	Electric charge
Q_{UN}	Uniform quantizer
r	Reference input
R, R_i	Residue, residue of the ith vibration mode
r_H, r_S	Threshold value of a dead zone
R_δ	Ratio of absolute displacement
R_F	Ratio of absolute force
R_{shunt}	Shunt resistance
s	Laplace operator
S_0, S_1, S_2, S_3	States in hysteresis loop
s_1, s_2, s_3	Elements of stiffness matrix
s_4, s_5, s_6	Elements of stiffness matrix for the second and third layers
$S_{r_S}[x](t)$	Zone function

$State$	State space
T	Temperature of a SMA cell
t	Time
t_a	Thickness of rhomboidal mechanism
T_F, T_N, T_{FR}, T_{NR}	Branch and transition temperatures of a SMA cell
t_i	Time when the jth impulsive input is applied
t_s	Sampling interval
t_{charge}	Charge/discharge time
U	Potential energy
u	Control input
u'	Quantized control input
u_C	Binary input to a single cellular unit
u_F, u_N, u_{FR}, u_{NR}	Inputs to a hysteretic system
u_{OFF}	OFF state of hysteresis loop
u_{ON}	ON state of hysteresis loop
U_{red}	Potential energy associated with redirecting stiffness
V^D	Deterministic Lyapunov function
V^i	Voltage of the ith unit
V^S	Stochastic Lyapunov function
V_1, V_2	Voltage values of a two-port network model
V_c	Supply voltage
V_H	Charge induced voltage with hysteresis
V_i	Internal shear force for segment i
v_q, v_q	White noise
V_s	Measured voltage across the resistor
V_{pzt}	Mechanically induced piezoelectric voltage
V_{total}	Input voltage
W	Width of a cell
W_A	Force at a middle point A of the tweezer-shape structure
W_B	Force at the tip B of the tweezer-shape structure
w_k	Width of the kth layer unit
w_{pzt}	Width of a PZT stack actuator
X_1, X_2, X_3, X_4	Parameters representing compliance of rhomboidal mechanism
x_i	Displacement of the ith lumped mass in a strand of N cellular units
x_j	Approximated stiffness
x_k, y_k	Location of a pixel
x_s, y_s	Location of the other end of directing stiffness
y	Aggregate output of a cellular actuator array
$y(t)$	Output of the Prandtl–Ishlinskii operator
y_c	Critical value of y
y_g	Integer-valued desired displacement
y_i	Displacement of the ith cellular unit
y_j	Convergent stiffness
z_i	Distance to an arbitrary point on segment i
z_k	Sequence of continued fraction expansions for lumped stiffness

Appendix

A.1 Modeling of hysteresis

A.1.1 Hysteresis in piezoelectric actuators

Piezoelectric ceramics have been widely used in applications such as sound and ultrasound transduction, high voltage generation, and acceleration sensing [279]. As actuators, piezoceramics possess desirable qualities, such as high efficiency, high bandwidth, low noise, and no backlash. They have become widely used in micropositioning applications, most notably in atomic force microscopy [141].

A.1.2 Hysteresis modeling

A.1.2.1 The modified Prandtl–Ishlinskii hysteresis model

Many phenomenological models are available for modeling hysteresis, such as the Prandtl–Ishlinskii operator [139], Preisach operator [31], Generalized Maxwell Slip Operator [87], and differential equations method [16]. Here, the modified Prandtl–Ishlinskii approach developed by Kuhnen will be used [140]. It has several useful advantages, such as the ability to model asymmetric loops and minor loops, an automatic identification procedure, and extendability to creep modeling.

A schematic representation of the play operator is shown in Fig. A.1. The input is the position of element A, and the output is the position of element B. r_H is the threshold value that characterizes the operator. Such a system will display a simple, rate independent hysteresis between input and output, shown in Fig. A.2. Let $H_{r_H}[x, y_0](t)$ be a play operator. Mathematically, the play operator can be represented recursively by

$$y(t_0) = H_{r_H}(x(t_0), y_0, r_H), \tag{A.1}$$

$$y(t) = H_{r_H}[x, y_0](t) \quad \text{for} \quad t_i < t \leq t_{i+1}, \quad 0 \leq i \leq N-1, \tag{A.2}$$

for given initial time t_0, initial output position y_0, and monotonicity partition $t_0 < t_1 < \cdots < t_N$ such that the input x is monotonic on every $\left[t_i, t_{i+1}\right]$ [31]. H is defined as

$$H_{r_H}(x, y, r_H) = \max(x - r_H, \min(x + r_H, y)) \tag{A.3}$$

which describes a sliding symmetric dead zone. r_H is the threshold of the dead zone, analogous to width of element B in Fig. A.1.

Linear superposition of an infinite number of play operators with thresholds varying from 0 to ∞ can model more complex hysteresis loops than that shown in Fig. A.2.

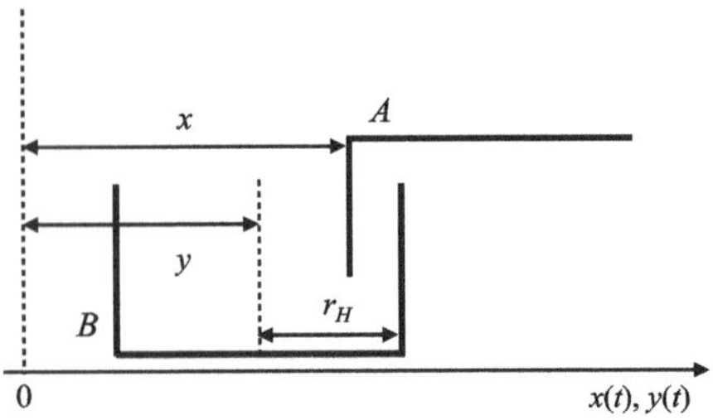

Figure A.1 Schematic representation of the play operator.

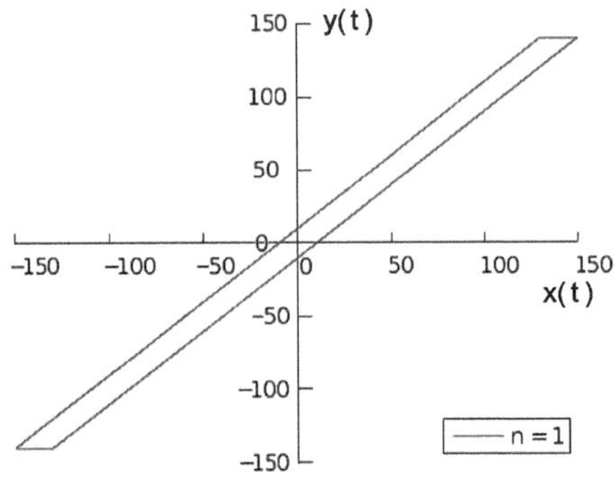

Figure A.2 The simple hysteretic behavior of the play operator.

In practice, this can be approximated by the weighted sum of a finite number of play operators, which is expressed as the product

$$H_{r_H}[x](t) := \boldsymbol{w}_H^T \cdot \boldsymbol{H}_{\boldsymbol{r}_H}[x, z_{H0}](t) \tag{A.4}$$

where $\boldsymbol{w}_H^T = [w_{H0}\ w_{H1} \ldots w_{Hn}]$ is a vector of weights, $\boldsymbol{r}_H^T = [r_{H0}\ r_{H1} \ldots r_{Hn}]$ is a vector of threshold values, $\boldsymbol{z}_{H0}^T = [z_{H00}\ z_{H01} \ldots z_{H0n}]$ is a vector of initial conditions, and $\boldsymbol{H}_{\boldsymbol{r}_H}[x, z_{H0}](t)^T = [H_{r_{H0}}[x, z_{H00}](t)\ \ H_{r_{H1}}[x, z_{H01}](t) \ldots H_{r_{Hn}}[x, z_{H0n}](t)]$ is a vector of play operators. The threshold values are subject to $0 = r_{H0} < r_{H1} < \cdots < r_{Hn} < \infty$. Fig. A.3 shows how play operators combine to form a more complex hysteresis.

This formulation can only model symmetric hysteresis loops, but the introduction of a memory free superposition operator allows asymmetric loops to be modeled [139]. The superposition operator is defined by a one-sided dead zone function

$$y(t) = S_{r_S}[x](t) = \begin{cases} \max(x(t) - r_S, 0), & \text{if } r_S > 0, \\ x(t), & \text{if } r_S = 0, \\ \min(x(t) - r_S), & \text{if } r_S < 0, \end{cases} \tag{A.5}$$

where r_S is the threshold value of the dead zone. In this application the input is the charge on the PZT actuator, which is assumed to be nonnegative since the applied

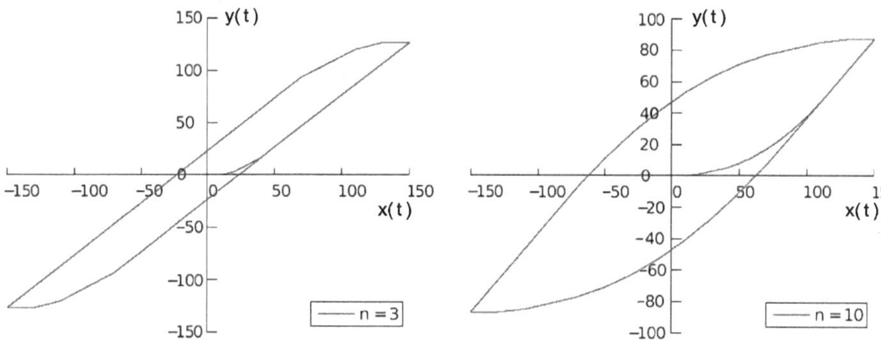

(A) Superposition of three play operators

(B) Superposition of ten play operators

Figure A.3 Summation of play operators.

voltage is always nonnegative. Therefore for this case we can simplify (A.5) to

$$y(t) = S_{r_S}[x](t) = \max(x(t) - r_S, 0). \tag{A.6}$$

In a similar fashion to (A.4), a finite number of superposition operators can be summed to model more complex behavior,

$$S_{r_S}[x](t) := \boldsymbol{w}_S^T \cdot \boldsymbol{S}_{\boldsymbol{r}_S}[x](t), \tag{A.7}$$

where $\boldsymbol{w}_S^T = [w_{S0}\ w_{S1} \dots w_{Sl}]$ is the vector of weights and $\boldsymbol{r}_S^T = [r_{S0}\ r_{S1} \dots r_{Sl}]$ is the vector of thresholds, subject to $r_{S0} = 0 < r_{S1} < \cdots < r_{Sl} < \infty$. The modified discrete Prandtl–Ishlinskii operator combines the play and superposition operators and is defined as

$$\Gamma[x](t) := \boldsymbol{w}_S^T \cdot \boldsymbol{S}_{\boldsymbol{r}_S}[\boldsymbol{w}_H^T \cdot \boldsymbol{H}_{\boldsymbol{r}_H}[x, \boldsymbol{z}_{H0}]](t). \tag{A.8}$$

As shown in [139], the operator can model asymmetric loops.

Fig. A.4 shows that in addition to hysteresis there is also creep between voltage and charge. The modified Prandtl–Ishlinskii operator can be extended to model this phenomenon as well with the addition of a creep operator that models the creep effect as $\log(t)$-type creep with a hysteretic equilibrium state [140]. The elementary creep operator is written as

$$y(t) = K_{r_K a_K}[x, y_{K0}](t) \tag{A.9}$$

and it is defined as the solution to the differential equation

$$\frac{d}{dt} y(t) = a_K \max(x(t) - y(t) - r_K, \min(x(t) - y(t) + r_K, 0)). \tag{A.10}$$

Though this definition is somewhat abstruse, (A.10) can be easily solved numerically. Using a zeroth-order hold to approximate the integral, its numerical solution takes the

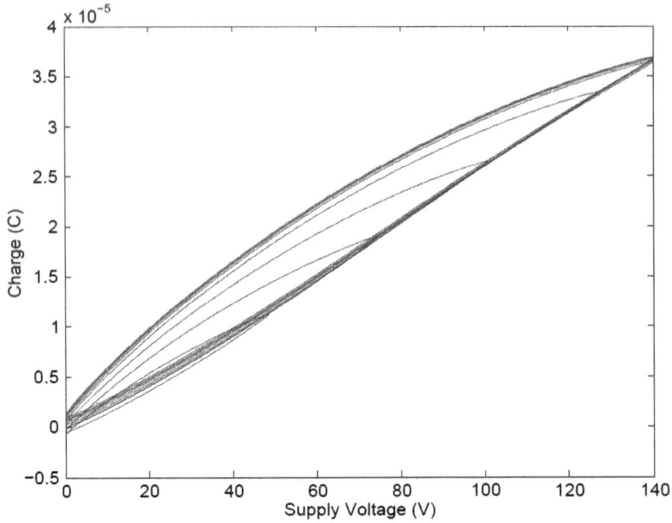

Figure A.4 Hysteresis between input voltage and charge.

form

$$y_k = \begin{cases} e^{-a_K t_s} y_{k-1} + (1 - e^{-a_K t_s})(x_{k-1} - r_K), & \text{if } x_{k-1} - y_{k-1} \geq r_K, \\ e^{-a_K t_s} y_{k-1} + (1 - e^{-a_K t_s})(x_{k-1} + r_K), & \text{if } x_{k-1} - y_{k-1} \leq -r_K, \\ y_{k-1}, & \text{if } -r_K < x_{k-1} - y_{k-1} < r_K, \end{cases} \tag{A.11}$$

where t_s is the sampling time, r_K is the threshold, and a_K is called the creep eigenvalue. log(t)-type creep refers to creep behavior that has a linear step response when viewed with respect to a logarithmic time scale. The summation of elementary creep operators produces this behavior when the creep eigenvalues are

$$a_{K_j} = \frac{1}{10^{j-1} t_s} \quad \text{for } j = 1, 2, \ldots, m, \tag{A.12}$$

and is called the log(t)-type creep operator [140]. As before, a finite number of log(t)-type creep operators will be summed, giving

$$K_{r_K a_K}[x](t) := \boldsymbol{w}_K^T \cdot \boldsymbol{K}_{\boldsymbol{r}_K \boldsymbol{a}_K}[x, \boldsymbol{Z}_{K0}](t) \cdot \boldsymbol{i} \tag{A.13}$$

where i is a $n \times 1$ vector of ones, $\boldsymbol{w}_K^T = [w_{K0}\ w_{K1} \ldots w_{Kn}]$ is a vector of weights, $\boldsymbol{r}_K^T = [r_{K0}\ r_{K1} \ldots r_{Kn}]$ is a vector of thresholds subject to $0 = r_{K0} < r_{K1} < \cdots < r_{Kn} < \infty$, and $\boldsymbol{a}_K = [a_{K1} \ldots a_{Km}]$ is the vector of creep eigenvalues according to (A.12). $\boldsymbol{K}_{\boldsymbol{r}_K \boldsymbol{a}_K}[x, \boldsymbol{Z}_{K0}](t)$ is a matrix of elementary creep operators, one for each combination of creep eigenvalue and threshold,

$$\boldsymbol{K}_{\boldsymbol{r}_K\boldsymbol{a}_K}[x, \boldsymbol{Z}_{K0}] = \begin{bmatrix} K_{r_{K0}a_{K1}}[x, z_{K001}] & \dots & K_{r_{K0}a_{Km}}[x, z_{K00m}] \\ K_{r_{K1}a_{K1}}[x, z_{K011}] & \dots & K_{r_{K1}a_{Km}}[x, z_{K01m}] \\ \vdots & \ddots & \vdots \\ K_{r_{Kn}a_{K1}}[x, z_{K0n1}] & \dots & K_{r_{Kn}a_{Km}}[x, z_{K0nm}] \end{bmatrix}. \tag{A.14}$$

Similarly, $\boldsymbol{Z}_{K0}$ is a matrix of initial conditions corresponding to each elementary creep operator,

$$\boldsymbol{Z}_{K0} = \begin{bmatrix} z_{K001} & \dots & z_{K00m} \\ z_{K011} & \dots & z_{K01m} \\ \vdots & \ddots & \vdots \\ z_{K0n1} & \dots & z_{K0nm} \end{bmatrix}. \tag{A.15}$$

Now, a combined hysteresis model is formed by adding the creep operator to the hysteresis operator in (A.8)

$$\Gamma[x](t) := \boldsymbol{w}_S^T \cdot \boldsymbol{S}_{\boldsymbol{r}_S}[\boldsymbol{w}_H^T \cdot \boldsymbol{H}_{\boldsymbol{r}_H}[x, \boldsymbol{z}_{H0}] + \boldsymbol{w}_K^T \cdot \boldsymbol{K}_{\boldsymbol{r}_K\boldsymbol{a}_K}[x, \boldsymbol{Z}_{K0}] \cdot \boldsymbol{i}](t). \tag{A.16}$$

This operator will be used as $H(q_{pzt})$ in (1.2) as follows:

$$V_{pzt} = H(q_{pzt}) = \Gamma[q_{pzt}](t). \tag{A.17}$$

A.1.2.2 Model identification

One drawback of the modified Prandtl–Ishlinskii approach is the relatively large number of parameters that characterize the model. However, by making certain assumptions about the threshold values, the identification of the weights can be formulated as a quadratic programming problem [140]. The weights are then optimized by a numerical search such that they minimize the error squared between model and experiment.

$x(t)$ is a known input and $y(t)$ is measured output. The error between model and experiment can be expressed as

$$\begin{aligned} e_{x,y}(t) &= H_{r_H}[x](t) + K_{r_K a_K}[x](t) - S_{r_S}^{-1}[y](t) \\ &= \boldsymbol{w}_H^T \cdot \boldsymbol{H}_{\boldsymbol{r}_H}[x, \boldsymbol{z}_{H0}](t) + \boldsymbol{w}_K^T \cdot \boldsymbol{K}_{\boldsymbol{r}_K\boldsymbol{a}_K}[x, \boldsymbol{Z}_{K0}](t) \cdot \boldsymbol{i} - \boldsymbol{w}_S'^T \cdot \boldsymbol{S}_{\boldsymbol{r}'_S}[y](t). \end{aligned} \tag{A.18}$$

Note that $S_{r_S}^{-1}$ has the same form as S_{r_S} but with different weights. The threshold values are determined by the range of measured input and output values according to Eqs. (A.19) and (A.20):

$$r_{Hi} = r_{Ki} = \frac{i}{n+1}\|x(t)\|_\infty \quad \text{for } i = 0, 1, \dots, n, \tag{A.19}$$

$$r'_{Si} = \begin{cases} 0, & \text{for } i = 0, \\ \dfrac{i - \frac{l}{2}}{l}\|y(t)\|_\infty, & \text{for } i = 1, 2, \dots, l. \end{cases} \tag{A.20}$$

Recall the creep eigenvalues are defined in Eq. (A.12). Let t_0 be the initial time and t_f be the final time of the measurement data. Appropriate weights can be identified by finding those that minimize the square of the error integrated from t_0 to t_f. This gives

$$\boldsymbol{w}^* = \arg\min(J(\boldsymbol{w})), \tag{A.21}$$

$$J(\boldsymbol{w}) = \frac{1}{2}\int_{t_0}^{t_f} e_{x,y}(t)dt \tag{A.22}$$

$$= \frac{1}{2}\boldsymbol{w}^T \cdot \int_{t_0}^{t_f} \boldsymbol{Q}\boldsymbol{Q}^T dt \cdot \boldsymbol{w} \tag{A.23}$$

with

$$\boldsymbol{w} = \begin{bmatrix} \boldsymbol{w}_H \\ \boldsymbol{w}'_S \\ \boldsymbol{w}_K \end{bmatrix} \in \mathbb{R}^{2n+l+3}, \tag{A.24}$$

$$\boldsymbol{Q}(t) = \begin{bmatrix} \boldsymbol{H}_{\boldsymbol{r}_H}[x, z_{H0}](t) \\ \boldsymbol{S}_{\boldsymbol{r}'_S}[y](t) \\ \boldsymbol{K}_{\boldsymbol{r}_K \boldsymbol{a}_K}[x, \boldsymbol{Z}_{K0}](t) \cdot \boldsymbol{i} \end{bmatrix} \in \mathbb{R}^{2n+l+3}, \tag{A.25}$$

where $\boldsymbol{w}^*$ is the optimal set of weights. The optimization is subject to the inequality constraints

$$\boldsymbol{U} \cdot \boldsymbol{w} - \boldsymbol{u} \leq \boldsymbol{o}, \tag{A.26}$$

$$\boldsymbol{U} = \begin{bmatrix} \boldsymbol{U}_H & \boldsymbol{0} & \boldsymbol{0} \\ \boldsymbol{0} & \boldsymbol{U}_S & \boldsymbol{0} \\ \boldsymbol{0} & \boldsymbol{0} & \boldsymbol{U}_K \end{bmatrix} \in \mathbb{R}^{2n+l+3 \times 2n+l+3}, \tag{A.27}$$

$$\boldsymbol{u} = \begin{bmatrix} \boldsymbol{u}_H \\ \boldsymbol{u}_S \\ \boldsymbol{u}_K \end{bmatrix} \in \mathbb{R}^{2n+l+3}, \tag{A.28}$$

$$\boldsymbol{U}_H = \boldsymbol{U}_K = \begin{bmatrix} -1 & 0 & \dots & 0 \\ 0 & -1 & \dots & 0 \\ \vdots & \vdots & \ddots & \vdots \\ 0 & 0 & \dots & -1 \end{bmatrix} \in \mathbb{R}^{n+1 \times n+1}, \tag{A.29}$$

$$\boldsymbol{U}_S = \begin{bmatrix} -1 & -1 & \dots & -1 \\ 0 & -1 & \dots & -1 \\ \vdots & \vdots & \ddots & \vdots \\ 0 & 0 & \dots & -1 \end{bmatrix} \in \mathbb{R}^{l+1 \times l+1}, \tag{A.30}$$

$$\boldsymbol{u}_H^T = [-\varepsilon \ 0 \ \dots \ 0] \in \mathbb{R}^{n+1}, \tag{A.31}$$

$$\boldsymbol{u}_S^T = [-\varepsilon \ -\varepsilon \ \dots \ -\varepsilon] \in \mathbb{R}^{l+1}, \tag{A.32}$$

$$\boldsymbol{u}_K^T = [0 \ 0 \ \dots \ 0] \in \mathbb{R}^{n+1}. \tag{A.33}$$

$\varepsilon > 0$ defines a lower bound for the weights, and can be any small number. $\boldsymbol{o}$ and $\mathbf{0}$ are vectors and matrices of zeros of appropriate size. The optimization is also subject to the equality constraint

$$\boldsymbol{g}^T \cdot \boldsymbol{w} - g = 0 \tag{A.34}$$

where

$$\boldsymbol{g} = \begin{bmatrix} \|x\|_\infty \boldsymbol{i} - \boldsymbol{r}_H \\ \boldsymbol{o} \\ m\|x\|_\infty \boldsymbol{i} - m\boldsymbol{r}_K \end{bmatrix} \in \mathbb{R}^{2n+l+3} \tag{A.35}$$

and

$$g = \|x\|_\infty. \tag{A.36}$$

The optimization returns the weights of the hysteresis and creep operators, and the weights of the inverse superposition operator.

$$\boldsymbol{w}_S = \begin{cases} w_{S0} = \dfrac{1}{w'_{S0}}, \\ w_{Si} = -\dfrac{w'_{Si}}{\left(w'_{S0} + \sum\limits_{j=1}^{i} w'_{sj}\right)\left(w'_{S0} + \sum\limits_{j=1}^{i-1} w'_{sj}\right)}, \quad i = 1, 2, \ldots, l. \end{cases} \tag{A.37}$$

Additionally, the thresholds are determined by

$$r_{Si} = \sum_{j=0}^{i} w'_{Sj}(r'_{Si} - r'_{Sj}), \quad i = 0, 1, \ldots, l. \tag{A.38}$$

If the calibration data is taken starting from a fully relaxed starting point, all initial conditions can be set to zero. In this case, however, the preload on the PZT actuator prevents this from occurring. The initial conditions can be set experimentally be starting with zeros, and then incrementing z_{H0} until the initial slope matches the experimental data. All terms in z_{H0} were assumed to be the same for simplicity. The weights were optimized numerically using the MATLAB command *quadprog*.

A.2 Structural parameters of tweezer-style end-effector

Fig. A.5 shows a schematic diagram of the tweezer-style end-effector. It is assumed that the end-effector is fixed at point C and the force generated from the actuator module is applied at the point A. From Castigliano's theorem using a Bernoulli–Euler beam model, $C_{A1} \sim C_{A4}$ and $C_{B1} \sim C_{B5}$ in Eqs. (1.49) and (1.50) can be written as follows:

$$
\begin{aligned}
C_{A1} = {} & -\frac{2}{3}\left\{l_{CD}^3 + (l_{AC} - l_{CD} - l_{DE})^3\right\} \\
& + (l_{CF} + l_{FG}\cos\theta_F + l_{BG}\cos\theta_G)l_{CD}(l_{CD} - 2l_{AC}) + l_{AC}l_{CD}^2 \\
& - (l_{CF} + l_{FG}\cos\theta_F + l_{BG}\cos\theta_G - l_{CD} - l_{DE})(l_{AC} - l_{CD} - l_{DE})^2 \\
& + (l_{AC} - l_{CD} - l_{DE})^3, \qquad \text{(A.39)}
\end{aligned}
$$

$$
C_{A2} = \frac{2}{3}\left\{l_{CD}^3 + (l_{AC} - l_{CD} - l_{DE})^3\right\} + 2l_{AC}l_{CD}(l_{AC} - l_{CD}), \qquad \text{(A.40)}
$$

$$
\begin{aligned}
C_{A3} = {} & -\frac{2}{3}l_{DE}^3 + (l_{AC} - l_{CD})l_{DE}^2 + (l_{CF} + l_{FG}\cos\theta_F + l_{BG}\cos\theta_G - l_{CD}) \\
& \times l_{DE}(l_{DE} - 2l_{AC} + 2l_{CD}), \qquad \text{(A.41)}
\end{aligned}
$$

$$
C_{A4} = \frac{2}{3}l_{DE}^3 + 2(l_{AC} - l_{CD})l_{DE}(\theta_G - l_{CD} - l_{DE}), \qquad \text{(A.42)}
$$

$$
\begin{aligned}
C_{B1} = {} & \frac{2}{3}\left\{l_{CD}^3 + (l_{AC} - l_{CD} - l_{DE})^3\right\} \\
& + \frac{2}{3}(l_{CF} - l_{AC})^3 + \frac{2}{3}l_{BG}^3\cos^3\theta_G \\
& + 2l_{CD}(l_{CF} + l_{FG}\cos\theta_F + l_{BG}\cos\theta_G) \\
& \times (l_{CF} + l_{FG}\cos\theta_F + l_{BG}\cos\theta_G - l_{CD}) \\
& + 2(l_{CF} + l_{FG}\cos\theta_F + l_{BG}\cos\theta_G - l_{CD} - l_{DE})(l_{AC} - l_{CD} - l_{DE}) \\
& \times (l_{CF} + l_{FG}\cos\theta_F + l_{BG}\cos\theta_G - l_{AC}) \\
& + 2(l_{CF} - l_{AC})(l_{CF} + l_{FG}\cos\theta_F + l_{BG}\cos\theta_G - l_{AC}) \\
& \times (l_{FG}\cos\theta_F + l_{BG}\cos\theta_G), \qquad \text{(A.43)}
\end{aligned}
$$

$$
\begin{aligned}
C_{B2} = {} & -\frac{2}{3}\left\{l_{CD}^3 + (l_{AC} - l_{CD} - l_{DE})^3\right\} \\
& + l_{AC}l_{CD}^2 + (l_{AC} - l_{CD} - l_{DE})^3 \\
& + l_{CD}(l_{CF} + l_{FG}\cos\theta_F + l_{BG}\cos\theta_G)(l_{CD} - 2l_{AC}) \\
& - (l_{CF} + l_{FG}\cos\theta_F + l_{BG}\cos\theta_G - l_{CD} - l_{DE}) \\
& \times (l_{AC} - l_{CD} - l_{DE})^2, \qquad \text{(A.44)}
\end{aligned}
$$

$$
\begin{aligned}
C_{B3} = {} & \frac{2}{3}l_{FG}^3\cos^3\theta_F \\
& + 2l_{FG}\cos\theta_F l_{BG}\cos\theta_G(l_{FG}\cos\theta_F + l_{BG}\cos\theta_G), \qquad \text{(A.45)}
\end{aligned}
$$

$$
\begin{aligned}
C_{B4} = {} & \frac{2}{3}l_{DE}^3 + 2l_{DE}(l_{CF} + l_{FG}\cos\theta_F + l_{BG}\cos\theta_G - l_{CD}) \\
& \times (l_{CF} + l_{FG}\cos\theta_F + l_{BG}\cos\theta_G - l_{CD} - l_{DE}), \qquad \text{(A.46)}
\end{aligned}
$$

$$
\begin{aligned}
C_{B5} = {} & -\frac{2}{3}l_{DE}^3 + (l_{AC} - l_{CD})l_{DE}^2 + l_{DE}(l_{CF} + l_{FG}\cos\theta_F + l_{BG}\cos\theta_G - l_{CD}) \\
& \times (l_{DE} - 2l_{AC} + 2l_{CD}), \qquad \text{(A.47)}
\end{aligned}
$$

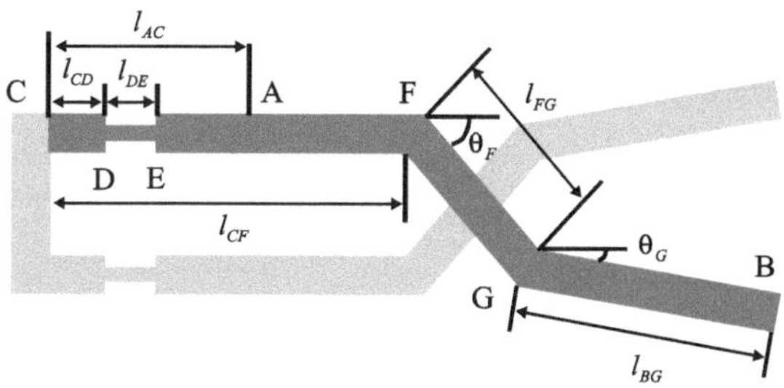

Figure A.5 Schematic of the tweezer-style end-effector.

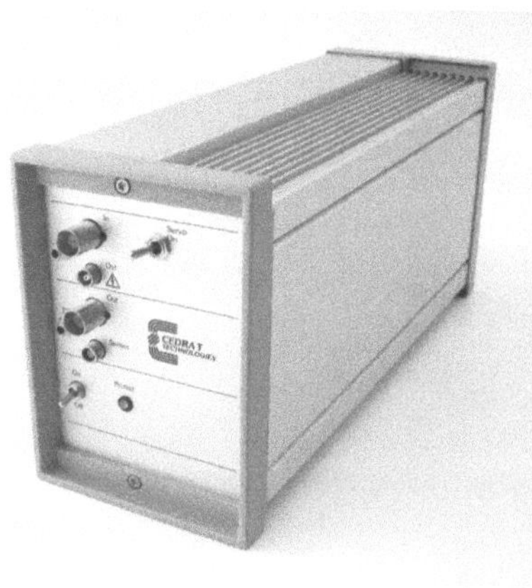

Figure A.6 Cédrat CA45 Standalone Linear Amplifier (Picture courtesy of Cédrat).

where l_{CD}, l_{DE}, and l_{AC} are the length between C and D, D and E, C and A, respectively, l_{CF}, l_{FG}, l_{BG} are the length between C and F, F and G, G and B, respectively, and θ_F, θ_G are the angles shown in Fig. A.5.

A.3 Piezoelectric driving circuit and control system

A.3.1 Cédrat charge amplifiers

A number of linear piezoelectric amplifiers developed for piezoelectric transducers are commercially available. The operating principle of a linear voltage amplifier is that it charges a transducer and regulates the voltage across it to a value found by multiplying a command voltage by a certain gain (i.e., linearly). For example, Cédrat Technologies' CA45 stand-alone linear amplifier shown in Fig. A.6 for piezoelectric actuators has a gain of 20. The principle of operation of a linear amplifier is simple; as shown in Fig. A.7, a command voltage is multiplied by a gain factor (i.e., 20) and is induced across the piezoelectric transducer. Fig. A.8 shows another smaller footprint amplifier provided by Cédrat Technologies.

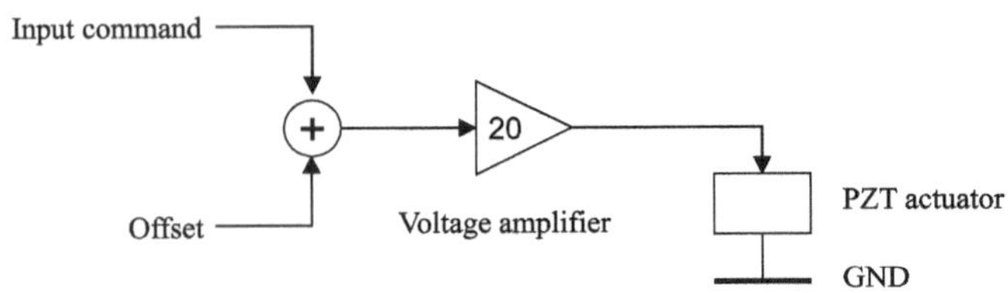

Figure A.7 Operation principle of a linear amplifier.

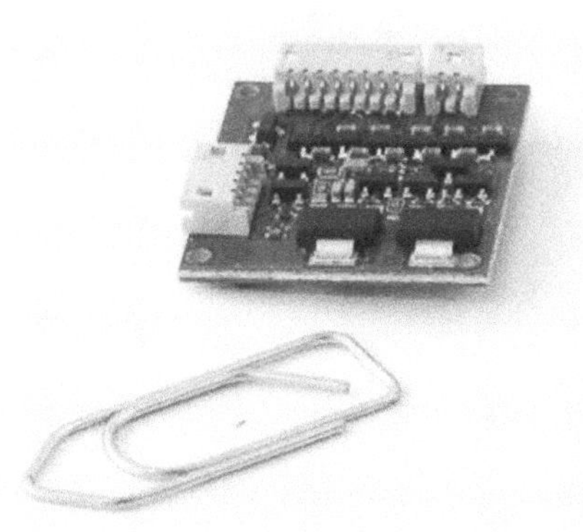

Figure A.8 Cédrat CAu10 miniature linear amplifier (Picture courtesy of Cédrat).

A.3.2 Discrete switching piezoelectric drive circuit

The piezoelectric stacks at the heart of the Cédrat APA50XS actuators require a drive voltage of 150 V to produce the displacements described in their datasheet. As mentioned in the main text, great advantages come from using the amplified piezoelectric stacks in switching rather than proportional mode. As most prior applications use individual PZT stacks in proportional mode, there are not many switching drivers commercially available in a small form factor. For this reason the switching drive circuit described in this appendix was developed. The circuit as presented in the schematic will drive 4 piezoelectric stacks (meaning 8 boards were used in total to drive the camera positioner). This circuit can be adapted to the application as needed.

Since options for commercially available 150 V power supplies and voltages this high can produce safety certification or arcing concerns if transmitted over cables, this circuit generates the necessary 150 V on the board. The A5-150 Module from American Power Design Inc. is a convenient pin-through-hole package that can be soldered directly to the board to produce the driving voltage from a more conventional DC supply, such as 24 V. Any supply voltage between 9 and 36 V is acceptable; appropriate voltage references are on-board the module so that the output voltage does not depend on the input voltage. The 33 mA output was determined to be sufficient to drive four APA50XS actuators based on the capacitance of each piezoelectric stack. The 150 V rail was stabilized with 47 μF and 220 pF capacitors in parallel. If it is unlikely that multiple stacks will be activated from the same board in rapid succession, or the capacitances are changed, it may be possible to drive more than 4 stacks from a single A5-150 Module. The module itself is large by electronic component standards; it requires a 50.8 × 22.9 mm footprint and is almost 10.5 mm thick. For large numbers of stacks, a dedicated high voltage power supply with associated safety measures may be more appropriate. For convenience, an Oki 78SR-5 DC–DC converter is used to generate logic-level voltages on-board rather than having a separate connector.

Piezoelectric devices operate on a somewhat different paradigm than copper-coil based actuators (servo motors and solenoids). To produce motion, rather than supply coil current, we must provide a charge. Disconnecting the stack will not cause it to go back to its original state (over a period of tens of seconds, parasitic leakage through the stack *will* cause the stack to return, but this is much too slow for our application). To return to its original state, the stack must be discharged and allowed to run into the ground plane. For this reason, a circuit is needed to switch between charging and discharging modes. This is accomplished better by a relay than drive transistors typical of motor or solenoid drive applications. A mechanical relay would work fine, but most mechanical relays have mechanical reed components whose system dynamics is too slow to be used in this application. High voltage switching is accomplished by the Avago ASSR302C solid state relay, which has a much faster response time than mechanical relays. One end of the stack is always connected to the ground plane. A standard TTL logic level signal can be used to cause the relay to switch states. In charging mode, the solid state relay connects the opposite end directly to the 150 V rail, allowing it to charge as quickly as the stack capacitance will allow. In discharging mode, the relay shorts the stack, allowing it to discharge quickly. The series resistance of the solid state relay is minimal; significantly less than mechanical relays.

The circuit includes "break before make" delay protection using digital logic with an included RC delay so that the 150 V rail will never be connected directly to ground through the ASSR302C bridge, as this condition will destroy the IC. The RC values that produced the appropriate amount of delay were empirically determined for the best performance–safety tradeoff for this particular solid state relay. Any adaptation of this circuit must include this delay protection. The 270 Ω resistance at the input of the drive circuit is not a critical value. It was chosen mainly to prevent spurious activation due to leakage currents from the controller hardware. If the controller hardware produces low current at the output pins this value can be increased without any problem.

A simplified version of a single channel (drives one PZT stack) showing only the signal path is shown in Fig. A.9 to understand basic functionality. Components between the input and the ASSR302-C block are used to implement the switching delay. The full schematic for the discrete PZT switching drive board as used in the camera positioner is shown in Figs. A.10–A.14.

The schematic uses the Diptrace hierarchical block function for clarity of representing repeated circuits. The version of the drive circuit shown in the schematic includes status LEDs for signal visualization and debugging purposes. These can be removed if desired. As this was a prototype board, it was made with only two layers and the layout was not optimal to make the board easier to repair if necessary. It is likely that the form factor can be greatly condensed.

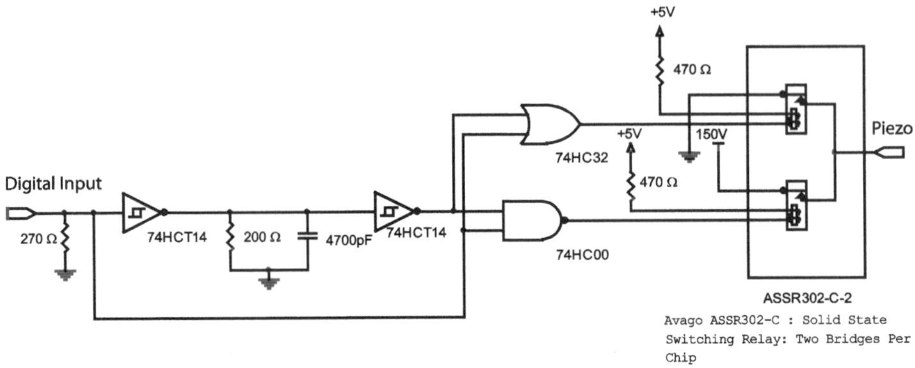

Figure A.9 Piezoelectric dual-bridge switching drive circuit.

Fig. A.15 shows fabricated circuit boards vertically stacked. Each board can drive 4 piezoelectric stacks, so 8 boards were used in total to drive the piezoelectric actuator array with 32 individual actuator units.

A.3.3 Hardware configuration of real-time controller

The camera orientation system was operated by a National Instruments (NI) CompactRio (cRIO) platform. The platform consists of cRIO-9024 and cRIO-9118, which consist of a Real-Time (RT) controller and reconfigurable embedded chassis as illustrated in Fig. A.16. The RT controller runs RT Operation System (RTOS), VxWorks, at 1 kHz and the embedded chassis runs FPGAs (Field-Programmable Gate Array), Xilinx Virtex-5 LX 85, at 40 MHz [131].

In the FPGAs, a number of programs can be operated concurrently in a parallel manner because programs were compiled into physical hardware, and thus the FPGA allowed for parallel operation at a deterministic and fast rate [177]. The discrete switching commands required sampling frequency of 10 kHz at most to achieve a rapid point-to-point motion while suppressing residual vibrations. Therefore, the control algorithms were programmed in the FPGAs to meet the sampling frequency. By taking advantage of parallelism, multiple sub-functions were operated in a parallel manner at 100 kHz to determine activations of the PZT actuators [131].

The activation signals were generated from NI 9401 modules, which are digital I/O (DIO) modules with 5 V/TTL and 100 ns sampling. Because the PZT actuator achieved a full stroke when an input voltage was 150 V, the voltage signal from NI 9401 was amplified through a custom ON–OFF circuit board.

The RTOS was used for data logging and high-level motion planning because the FPGAs only support logical operations. The data was transferred between the FPGAs and RTOS using a high-speed PCI bus by a DMA (Direct Memory Access) FIFO (First In, First Out) method. A Host PC was used for UI (User Interface), and user commands were transferred to the RTOS via Ethernet.

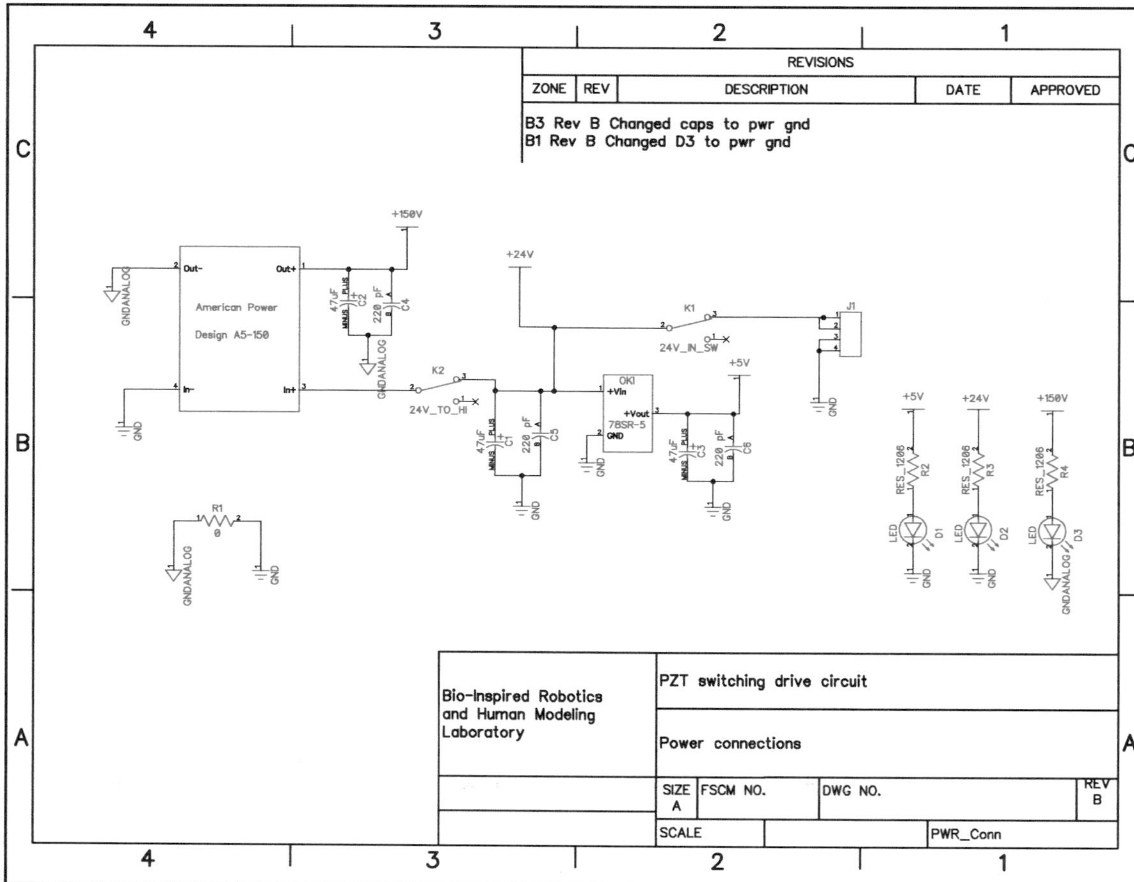

Figure A.10 Diagram of a discrete PZT driver circuit (1/5).

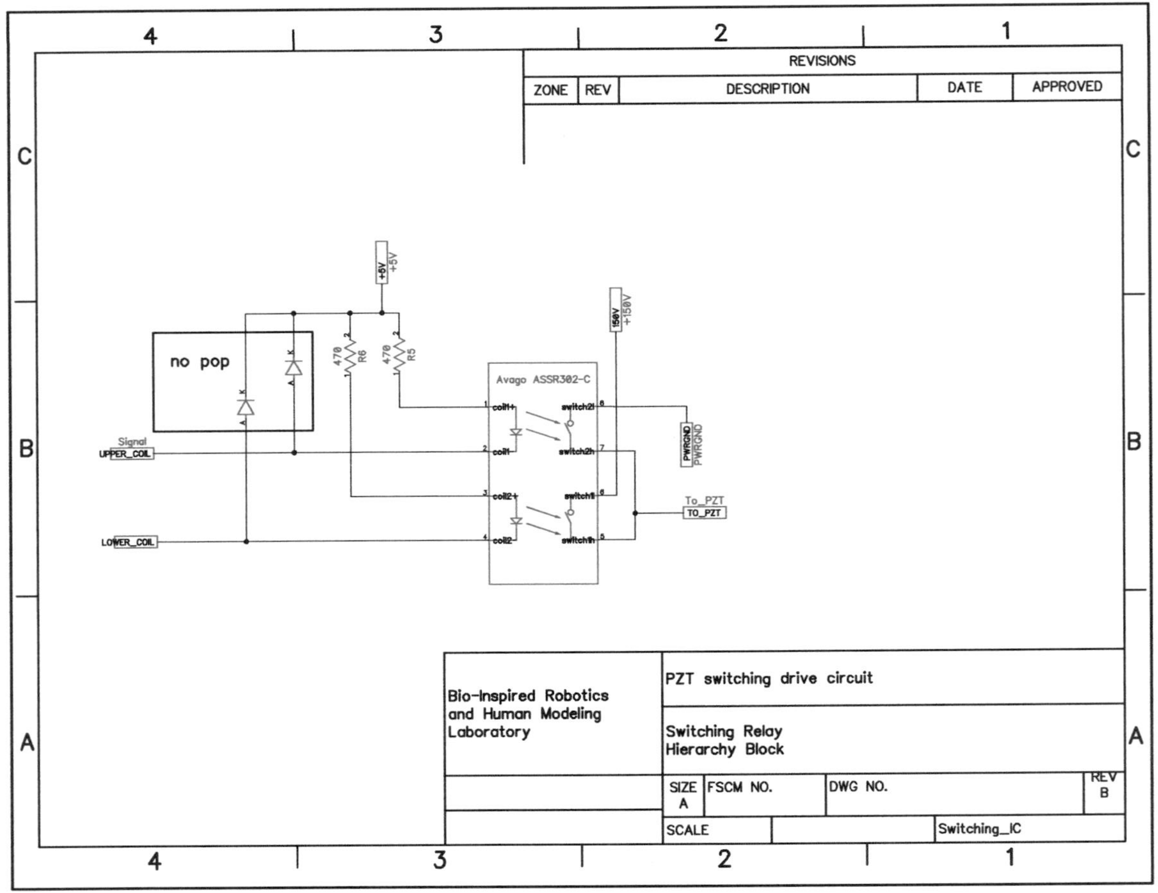

Figure A.11 Diagram of a discrete PZT driver circuit (2/5).

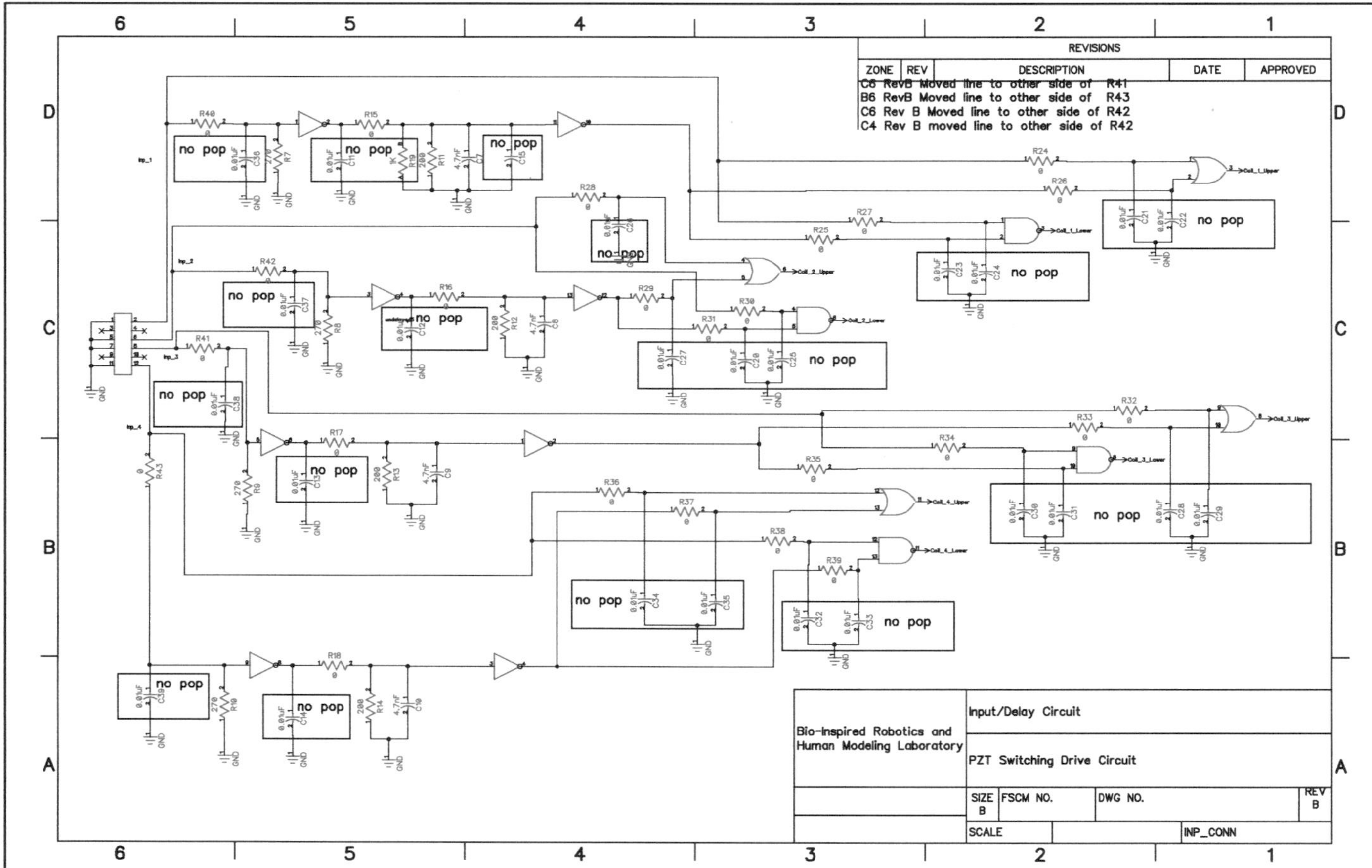

Figure A.12 Diagram of a discrete PZT driver circuit (3/5).

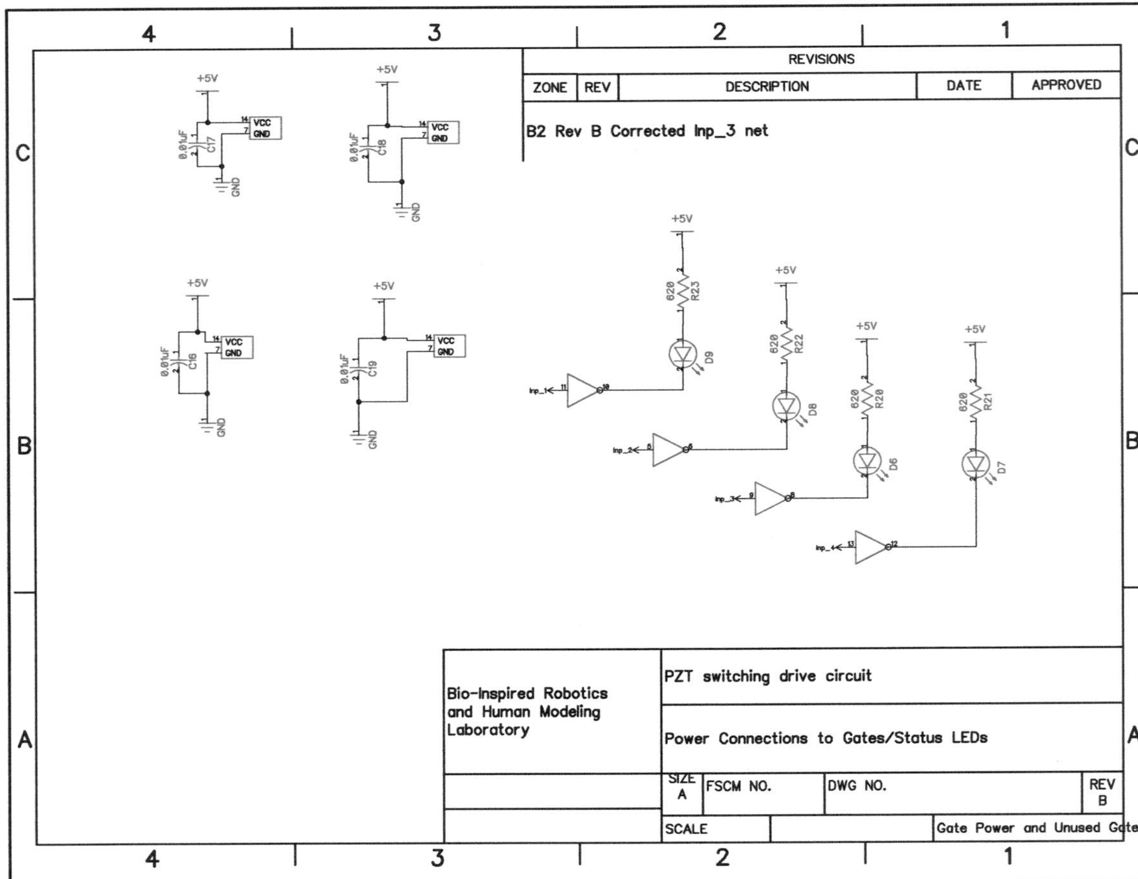

Figure A.13 Diagram of a discrete PZT driver circuit (4/5).

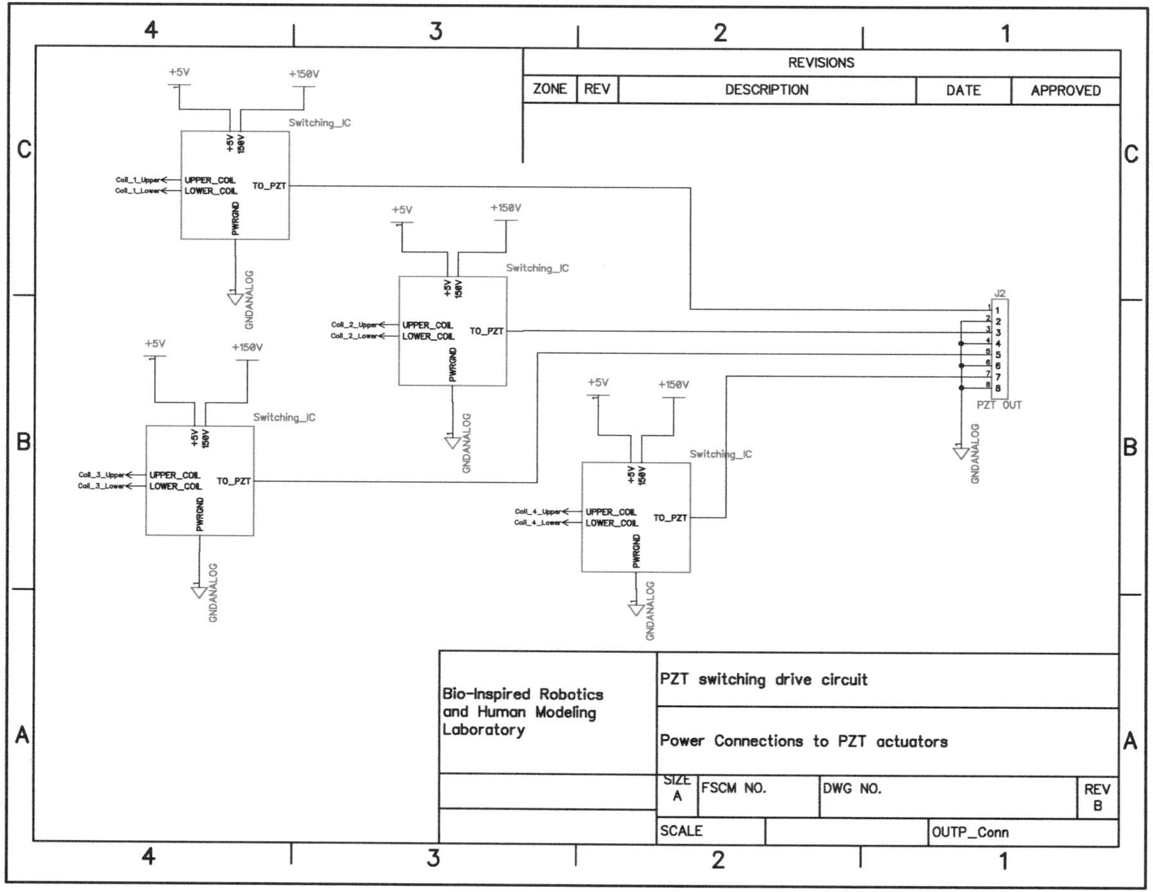

Figure A.14 Diagram of a discrete PZT driver circuit (5/5).

Figure A.15 Discrete switching control circuit boards.

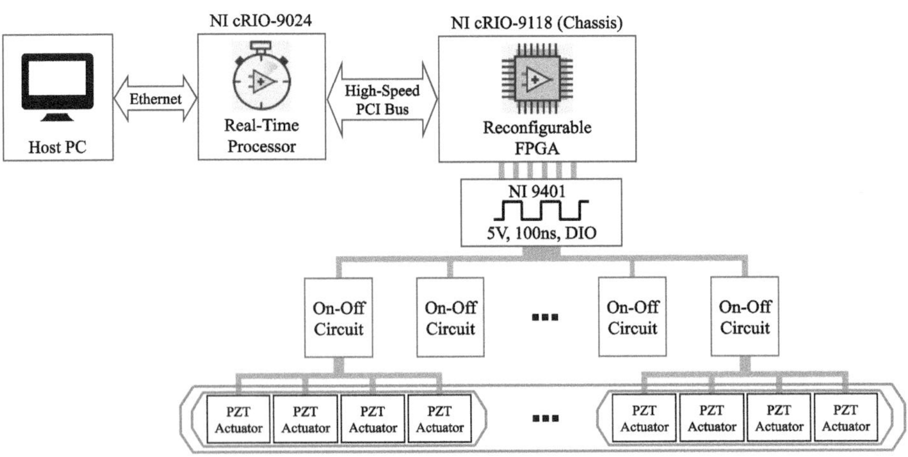

Figure A.16 Hardware configuration.

A.4 Compliance matrix elements in Section 2.2

$$\begin{aligned}
C_{11} =& -\csc^2\left(\frac{\theta}{2}\right)\sec^4\left(\frac{\theta}{2}\right)\Big(96d^4\cos(\theta) - 64d^4\cos(2\theta) + 32d^4\cos(3\theta) \\
&- 64d^4 - 64d^3h\sin(\theta) - 64d^3h\sin(3\theta) - 64d^3w\sin(\theta) - 64d^3w\sin(3\theta) \\
&+ 64d^2t^2\cos(2\theta) - 54.144d^2t^2\cos(3\theta) + 11.072d^2t^2\cos(4\theta) \\
&+ 16d^2\cos(\theta)\left(3.384t^2 + 3w^2\right) - 75.072d^2t^2 - 48d^2w^2\cos(3\theta) \\
&- 86dht^2\sin(\theta) + 11.072dht^2\sin(2\theta) + 34dht^2\sin(3\theta) \\
&- 5.536dht^2\sin(4\theta) - 6dht^2\sin(5\theta) + 2dht^2\sin(7\theta) \\
&- 129.216dt^2w\sin(\theta) + 11.072dt^2w\sin(2\theta) + 43.072dt^2w\sin(3\theta) \\
&- 5.536dt^2w\sin(4\theta) - 48dw^3\sin(\theta) + 16dw^3\sin(3\theta) - 5h^2t^2\cos(\theta) \\
&+ 14h^2t^2\cos(2\theta) + 5h^2t^2\cos(3\theta) - 4h^2t^2\cos(4\theta) - h^2t^2\cos(5\theta) \\
&+ 2h^2t^2\cos(6\theta) + h^2t^2\cos(7\theta) - 12h^2t^2 - hw\cos(\theta)\left(18.536t^2 + 8w^2\right) \\
&+ 2hw\cos(2\theta)\left(20.536t^2 + 8w^2\right) + 18.536ht^2w\cos(3\theta) - 4ht^2w\cos(4\theta) \\
&- ht^2w\cos(5\theta) + 2ht^2w\cos(6\theta) + ht^2w\cos(7\theta) - 39.072ht^2w \\
&+ 8hw^3\cos(3\theta) - 16hw^3 - 2w^2\cos(\theta)\left(6.768t^2 + w^2\right) \\
&+ 4w^2\cos(2\theta)\left(6.768t^2 + w^2\right) + 13.536t^2w^2\cos(3\theta) \\
&- 27.072t^2w^2 + 2w^4\cos(3\theta) - 4w^4\Big),
\end{aligned} \tag{A.48}$$

$$\begin{aligned}
C_{21} = C_{12} =& 2\Big(128d^4\csc^3(\theta) - 128d^4\csc(\theta) + 384d^4\cot^2(\theta)\csc(\theta) \\
&- 512d^4\cot(\theta)\csc^2(\theta) + 384d^3h\cot^2(\theta) + 1152d^3h\csc^2(\theta) \\
&- 512d^3(2h - w)\cot(\theta)\csc(\theta) - 1152d^3h - 12d^2h^2\cos^2(\theta) \\
&- 24d^2h^2\cos(\theta) + 9d^2h^2\cos(2\theta) + 120d^2h^2\cot^2(\theta) \\
&+ 258d^2h^2\csc^2(\theta) + 6d^2h^2\cos^2(\theta)\cot^2(\theta) - 24d^2h^2\cos(\theta)\cot^2(\theta) \\
&- 360d^2h^2\cot(\theta)\csc(\theta) - 249d^2h^2 - 541.44d^2t^2\sin(\theta) \\
&- 866.304d^2t^2\cot(\theta) + 324.864d^2t^2\cos(\theta)\cot(\theta) \\
&+ 64d^2\csc(\theta)\left(8.46t^2 - 6w^2\right) + 64dht^2\cos^4(\theta) - 128dht^2\cos^3(\theta) \\
&+ 128dht^2\cos^2(\theta) + 32dht^2\cos(2\theta) + 56dht^2\cos(4\theta) \\
&+ 384dht^2\sin^2(\theta)\cos(\theta) - 128dt^2(4.384h + 3.384w)\cos(\theta) \\
&- 24dht^2 - 384dhw^2 - 96h^2w^2 - 128ht^2(h + w)\sin(\theta)\cos^3(\theta) \\
&+ 128ht^2(h + w)\sin^3(\theta)\cos(\theta)\Big),
\end{aligned} \tag{A.49}$$

$$
\begin{aligned}
C_{22} = 2\Big(&512d^4\csc^2(\theta) - 2048d^4\cot(\theta)\csc(\theta) - 3072d^3h\cot(\theta) \\
&+ 1024d^3(h+w)\csc(\theta) + 192d^2h^2\sin(\theta) - 1440d^2h^2\cot(\theta) \\
&+ 576d^2h^2\csc(\theta) + 192d^2h^2\cos(\theta)\cot(\theta) + 1536d^2hw \\
&- 386.304d^2t^2\cos(\theta) + 256d^2t^2\cot^2(\theta) + 768d^2t^2\csc^2(\theta) \\
&- 32d^2t^2\cos(\theta)\cot^2(\theta) - 992d^2t^2\cot(\theta)\csc(\theta) \\
&+ 98.304d^2t^2 + 85dh^3\sin(\theta) + dh^3\sin(3\theta) \\
&- 2d\cot(\theta)\left(115h^3 + 322.58ht^2 + 256t^2w\right) \\
&+ 4d\csc(\theta)\left(43h^3 + 120.756ht^2 + 112t^2w\right) \\
&+ 16d\cos(\theta)\cot(\theta)\left(5h^3 + 12.46ht^2 + 4t^2w\right) \\
&- dh\left(h^2 + 1.692t^2\right)\cos^4(\theta)\cot(\theta) \\
&+ 2dh\left(h^2 - 126.308t^2\right)\sin(\theta)\cos^3(\theta) \\
&+ 4dh\left(h^2 + 1.692t^2\right)\cos^3(\theta)\cot(\theta) \\
&- dh\cos^2(\theta)\cot(\theta)\left(96dh + 25\left(h^2 + 1.692t^2\right)\right) \\
&+ dh\left(3h^2 + 261.076t^2\right)\sin^3(\theta)\cos(\theta) \\
&- dh\sin(\theta)\cos(\theta)\left(96dh + 29h^2 + 305.068t^2\right) \\
&+ 768dh^2w + 416.972dht^2\sin(\theta) + 33.692dht^2\sin(3\theta) \\
&+ 384dht^2\sin(\theta)\cos^2(\theta) + 241.152dt^2w\sin(\theta) \\
&+ 32h^4 + 128h^3w + 16h^2t^2\cos(2\theta) - 28h^2t^2\cos(4\theta) \\
&+ 260.576h^2t^2 - 32ht^2(h+w)\cos^4(\theta) + 16ht^2w\cos(2\theta) \\
&- 28ht^2w\cos(4\theta) + 388.576ht^2w + 128t^2w^2\Big).
\end{aligned}
\tag{A.50}
$$

A.5 SMA cellular actuators

A.5.1 *SMA cellular actuator design*

The primary choice of actuator material for cellular actuators in this book is PZT due to its fast speed of response and reliability. On the other hand, shape-memory alloy (SMA) is also a promising material to investigate methods for effective compensation of hysteresis discussed in Sections 3.3 and 3.4. Three designs of cellular actuators made from SMA wire are introduced. First, a damped SMA actuator array is presented and its quasi-static response is discussed. Several guidelines are presented which could aid future designers building relatively slow actuator arrays. Then a dynamic SMA actuator array is presented that is used to validate the quasi dynamic aspects of the actuator. Finally, a two-degree-of-freedom robotic arm with SMA cellular actuators is developed and demonstrated.

A.5.2 Damped SMA array

The damped SMA array experiment shows a practical silicone-based SMA actuator array design that was inspired by the M-lines and Z-lines of sarcomeres in biological muscle and the interaction between the myosin and actin fibers which generate displacement and subsequent contractile force. The silicone rubber based SMA actuator array shown in Fig. A.17 capitalizes on the idea of recruitment to negate the effect of hysteresis, an inherent drawback with SMA actuators along with many other linear actuation technologies. The fingerprint of this actuator array is given by

$$\left[\begin{array}{c|c|c|c} -1 & \&1 & \&2 & \&3 \\ 2 & 1 & 1 & -1 \end{array}\right]. \tag{A.51}$$

When heated via electrical current, the SMA shifts from its martensite phase to its austenite phase which is roughly 3–4% less than its length in the martensite phase. The silicone connecting structure provides the translation between displacement and force based on external loading conditions and the length of other cells. As more force or displacement is desired, additional cells are activated causing additional preloading of the silicone "springs" and thus additional force or displacement to the external environment. Flexinol 100 μm HT SMA wire produced by Dynalloy Inc. is selected as the actuation material primarily for its balance between force and cooling time. Larger diameter SMA wires can generally produce higher forces, but take significantly longer to cool while smaller diameter wires produce significantly less force but also cool more rapidly. The spring constant of the Flexinol 100 μm HT SMA wire was determined experimentally to be 2.54 N/mm for the 75 mm length of wire used in the final cell array. This was determined using a Futek LBS200 5 lb force sensor and two 0.02 mm resolution micro-positioners with the SMA wire in its relaxed Martensite phase. While the spring constant of the SMA wire does vary greatly between the martensite (relaxed) and austenite (active) phase, the martensite's spring constant is lower than that of the austenite phase, and both values are an order of magnitude above that of the silicone connecting structure. For this reason, the value of the martensite phase was taken as the actuator's stiffness. During the testing phase the wire was subjected to approximately 8.45 N and remained in its elastic phase. This is roughly 5 times the force an individual SMA wire experiences in the final actuator array which suggests that, coupled with the silicone connective structure, the SMA wires remain linear and resistant to breaking. In order to increase the force output of a single cell, 4 wires were used in parallel. Likewise to increase the displacement, the wires were wrapped around a Z-line bracket as shown in Fig. A.18. The silicone chosen needed to have a

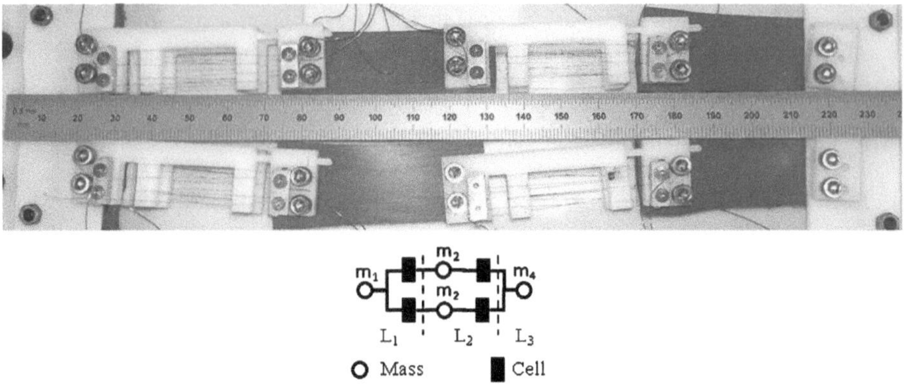

Figure A.17 Silicone rubber based damped SMA actuator array.

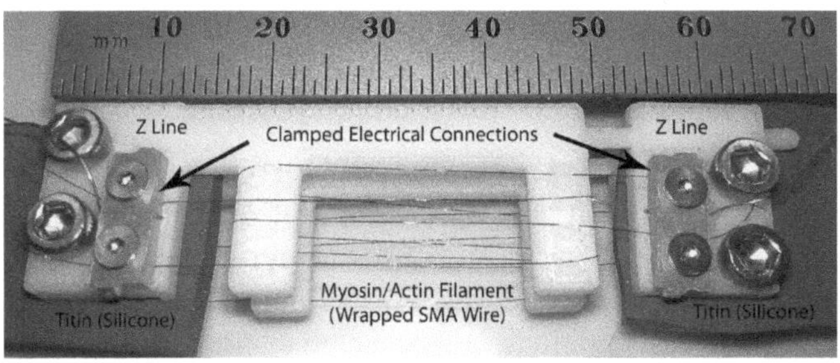

Figure A.18 Silicone rubber based actuator array cell.

stiffness less than that of the additive SMA wires force for a given displacement so that the SMA wires can always achieve their full displacement. The actual stiffness of the silicone connecting structure is dependent both on the stress–strain properties of the silicone and the geometry of the connecting structure. Several silicone rubber sheets were tested experimentally using the same sensor and micro-positioners as for the SMA testing, and it was determined that shore A 20 durometer silicone rubber provided favorable properties. A 30 mm × 62 mm × 6 mm section of silicone rubber, the size used for the final cells, was tested producing a final spring constant of 0.611 N/mm. A stiffer silicone sheet could have been used and would have increased the strength of the resultant muscle significantly, however, this was not done for experimental setup reliability reasons. The Z-line brackets were printed with a 3D rapid prototyper using ABS plastic for ease of construction and plastic's inherent electrical resistivity.

While the four parallel SMA wires in each cell for this experimental setup have a theoretical force capability of 18 N and a displacement capability of 5 mm, manufacturing errors caused the actual displacement to be 0.7 mm, and thus the effective control force of the cell was given by $(0.7\ \text{mm}) \cdot (2.54\ \text{N/mm}) = 1.7\ \text{N}$. The effective applied force of the cell was therefore $(0.7\ \text{mm}) \cdot (0.61\ \text{N/mm}) = 0.41\ \text{N}$. While this is significantly lower than what is needed for implementation in a full scale muscle, a different choice of silicone sheet and an improved design and manufacturing process could increase the properties. The current SMA wires also take roughly a second to fully heat and to fully cool; however, introducing additional cooling mechanisms such as forced air or submersion in an oil bath can significantly improve this as well. As the focus of the current research was to validate the presented theory and show viability of the experimental setup, these changes were left for future work.

The properties of a given actuator array are also highly dependent on the topology and choice of internal cell actuation properties (such as SMA wire size). The presented design is highly modular and, depending on the chosen topology, highly robust. Cells are connected together through the silicone structure thus reconfiguring the muscle is as easy as cutting/punching out a different silicone shape and connecting the Z-line elements to it. Likewise, if a cell fails it can simply be disconnected from the surrounding cells and replaced. Parallel structures add robustness, meaning if a cell fails either due to an electrical failure (the cell goes dead but still remains intact) or mechanical failure (cell physically breaks leaving zero stiffness) the remaining cells are still able to carry force. Thus an arm using the muscle may have a reduced force capacity but will still be able to function.

Experimental validation of the above mechanism was carried out in two stages. In the dynamic experiment with the highly damped silicone based SMA array, the 4 cells were activated from rest at four different levels: 1 cell on, 2 cells on, 3 cells on, and all cells. For each, the force activation profile was the Sigmoid function $\left(F \cdot \left(\frac{1}{1+e^{-5 \cdot t+6}}\right)\right)$ followed by a similar inverse Sigmoid function $\left(F \cdot \left(1 - \frac{1}{1+e^{-5 \cdot (t-t_d)+6}}\right)\right)$ where t_d is the deactivation time and F is the force of the cell. This function was determined experimentally by measuring the SMA wire contraction profile. Fig. A.19 shows sample results with the simulation results overlaid. Due to the system's high natural frequency and high damping, the results for the silicone based SMA actuator do not highlight dynamic effects, but they do show that predicted force levels of the theory and simulation match those of the experimental results.

A.5.3 Dynamic SMA array

In order to validate the dynamic aspects of SMA cellular actuation, a dynamic (under damped) actuator array was created. Each cell in the array consists of two Miga NanoMuscle 704 SMA actuators mounted to an ABS rapid prototyped shell. A compression spring (series elastic element) inside the shell connected to a rod which goes through the spring and out to form the outgoing connection point. The arms of the SMA actuators form the incoming connection points, and are activated together to cancel any moment that would otherwise be generated. The SMA actuators are significantly stiffer than the compression spring, but they do have some flexibility to them. They also take time to heat when activated and time to cool when deactivated. For these two reasons, they are approximated by a pure force acting across a stiff spring

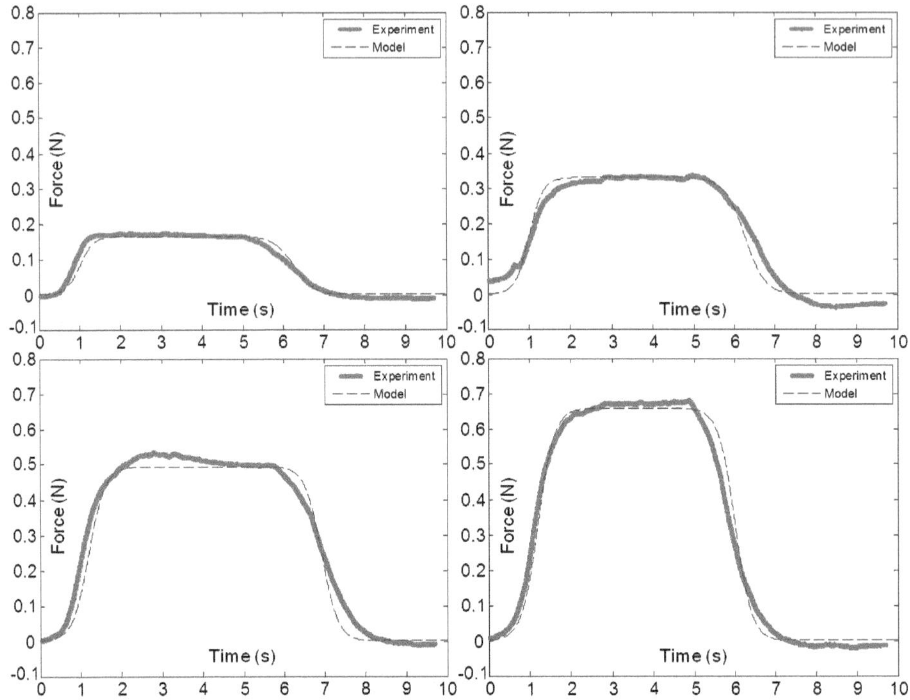

Figure A.19 Comparison of 4 cell actuator array physical system and simulated results.

(actuator elastic element) and a damper. Magnet wires were chosen to power the SMA actuators in order to reduce the effect of wire stiffness on the results; however, some stiffness remained and was accounted for as the parallel elastic element. The mass, stiffnesses, damping (measured from response time), and force outputs for each cell were measured using a calibrated Futek 9 N force transducer powered by an Omega signal conditioner and the results were viewed on a LeCroy 600 MHz scope. The spring constants for the parallel elastic, series elastic, and SMA actuator were found to be 20, 386.52, and 1000 N/m, respectively, by stretching each between two 0.02 mm resolution micro-positioners and measuring the output force. The damping coefficient was found to be 500 N/(m/s), and each mass was 15.83 grams. 6 cells were set up in an isometric contraction arrangement as shown in Fig. A.20. The fingerprint of this actuator array is given by

$$\left[\begin{array}{c|c|c|c|c} -1 & \&3 & \&1 & \&3 & \&1 \\ 2 & 1 & 2 & 1 & -1 \end{array}\right]. \tag{A.52}$$

A Hill-type model can be used to represent dynamically damped characteristics of a SMA actuator. The model is shown in Fig. A.21. While other Hill-type cell models exist, the explanations and examples throughout this chapter refer to this model. The

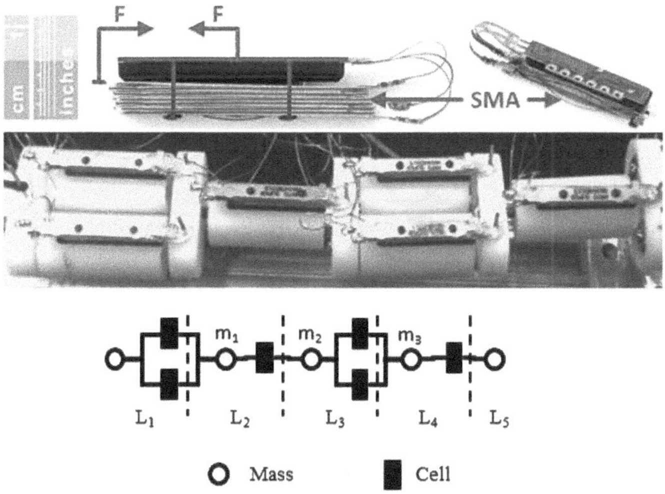

Figure A.20 Physical 6 cell dynamic SMA array actuator.

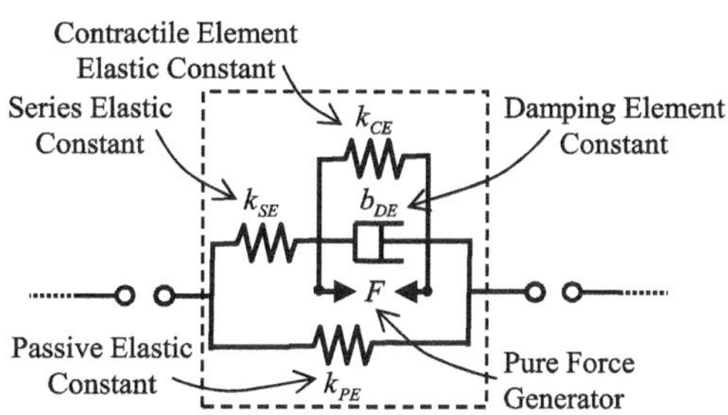

Figure A.21 SMA actuator cell model.

variables are defined based on Fig. 1.52(b). The damper connected in parallel with the series elastic element in the theory was negligible with this particular actuator, allowing the dynamics to play a greater role in validating the theory. Additionally, the simple Miga NanoMuscle SMA actuators can be represented accurately for the all-on all-off case as a spring and damper in parallel with a constant pure-force acting on it as an input. This allows for the contractile element to be less of a black box, again allowing for further validation of the theory.

The dynamic SMA array shows the dynamic effects much more clearly than the damped array. Fig. A.22 shows force results for the dynamic SMA array. In the figure, the upper trace shows the dynamic response when all cells were activated while the lower trace shows the response when half the cells were activated. For each experiment the cells were activated for 3 seconds and then deactivated. As can be seen in the graph, the results track the simulated values very closely including rises and falls in the graph due to individual cell inertia. The tracking is not perfect, partly because all mass elements were assumed to be uniform in the simulated results whereas in the physical system they differed between connection points with two outgoing cells and

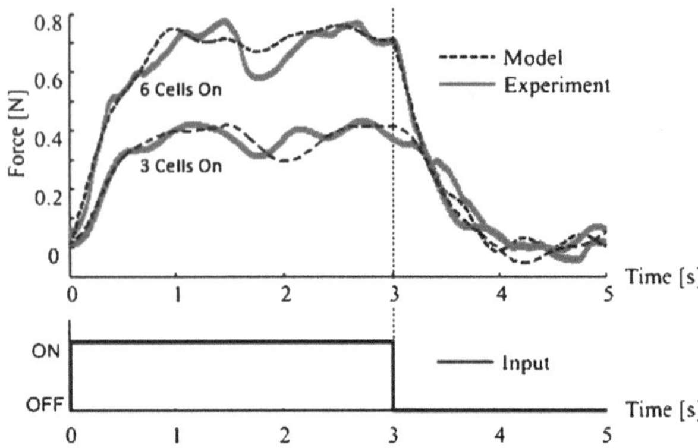

Figure A.22 Comparison of 6 cell actuator array physical system and simulated results. All cells were activated for 3 seconds and then deactivated.

those with only one outgoing cell. While the experimental setup could be upgraded to normalize the masses, usable actuator arrays will likely have non-uniform masses, and the results show that this difference can often be small enough to neglect. It should be noted that no scaling of the physical system or the simulated results was done, and all cell properties were calculated prior to conducting any trials to remove any unintended bias. The results validate the dynamic aspects of the presented theory as the system was not damped. Additionally, since the control input was simply a step acting on a damper, it can be seen that no input shaping was done to introduce a bias, which further validates the theory.

A.5.4 Implementation of floating-point quantization into dynamic SMA actuator array

Fig. A.23 shows isometric force responses when the array shown in Fig. A.24 (also introduced in Fig. 2.35) that approximately realizes the floating-point quantization discussed in Section 3.5 with mantissa = 1. The cells are non-uniformly grouped as Group 1 = {Cell 2}, Group 2 = {3}, Group 3 = {4, 8}, Group 4 = {9, 10, 11}, and Group 5 = {6, 12, 13, 14}. Cells 1, 5, 7 were not used. Each of the cells is represented by the dynamic SMA model discussed in the previous section. The progressive activation inspired by the size principle achieves a fine resolution for a small motor command and a more coarse resolution for a larger command. Note that there are several ways of connecting actuator units (mechanically, electrically, or a mixture of these) to realize such floating-point actuation. The responses in Fig. A.23 show fluctuations from convergence values, mainly due to underdamped modes in the system. Muscle forces during contractions also show fluctuations [119]. Although it has not been fully investigated yet, the identification of such fluctuations would provide an interesting insight into the neuromotor variability.

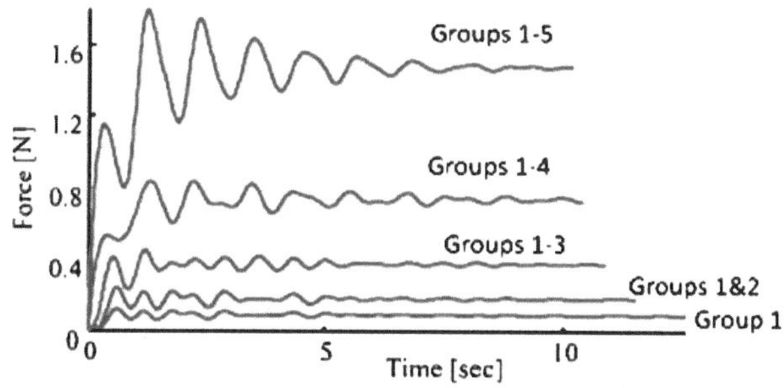

Figure A.23 Floating-point quantized actuation of an non-uniform actuator array.

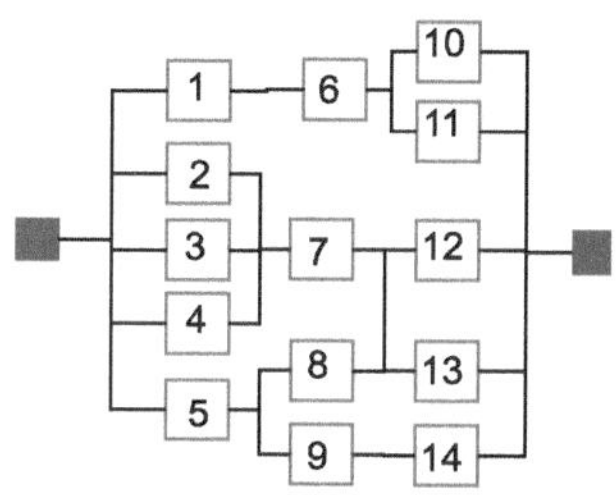

Figure A.24 Grouping of cells.

A.5.5 Robotic arm with SMA cellular actuators

A two degree of freedom robot arm driven by shape memory alloy (SMA) actuators has been designed and fabricated as shown in Fig. A.25. Compliant cellular actuator arrays have been designed and fabricated, utilizing linear shape memory alloy actuators (NanoMuscle NM706 by Miga Motors). The SMA actuators are "retract only" and thus require antagonistic pairing at each of the two joints for effective motion control. In addition, the effective strain of the NanoMuscle actuator is up to 4%, realized by a unique layered structure, which is still not large enough to generate a practical range of motion in the link mechanism. To further increase the strain, an actuator array shown in Fig. A.26 was designed. A single array or a cell, an array of a total of 8 NanoMuscle actuators, is configured to have 4 pairs of actuators in series. The pairs of actuators are essentially connected serially but folded. This design shortens the length of the array and increases the effective strain up to 16% (4% $\times$ 4). The fingerprint of this cell is given by

$$\left[\begin{array}{c|c|c|c} -1 & \&3 & \&2 & \&3 \\ 2 & 2 & 2 & -1 \end{array}\right]. \tag{A.53}$$

Fig. A.27 shows representative movements of the robot arm.

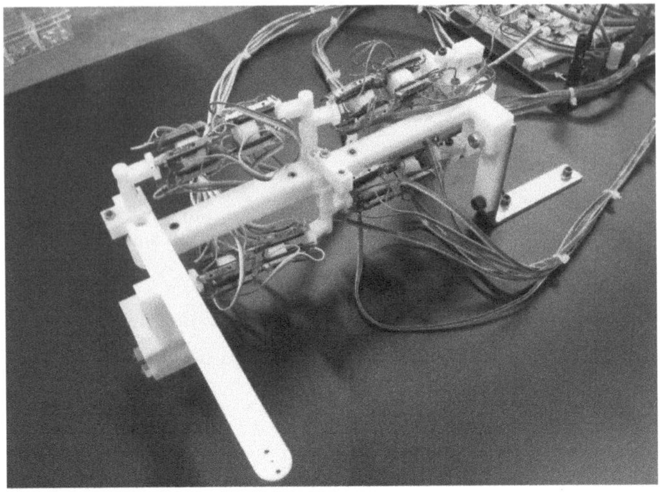

Figure A.25 Robot arm with SMA cellular actuators.

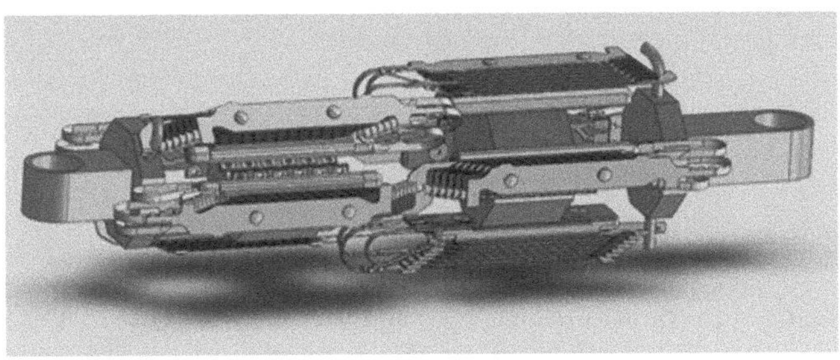

Figure A.26 SMA actuator array design.

A.6 Deterministic analysis and stability of expectation

The convergence of the expectation can be analyzed by assuming $e_t \to E[e_t]$ when $N \to \infty$. Let $V^D = e_t^2$ be a Lyapunov function, where the superscript "D" denotes the deterministic analysis. $\Delta V_t^D = V_{t+1}^D - V_t^D$ is given as follows:

$$\Delta V_t^D = e_{t+1}^2 - e_t^2$$
$$= \begin{cases} p(\eta N - \eta N_d - \eta N_{dead}^{OFF} + e_t) & \\ \times\{-2e_t + p(\eta N - \eta N_d - \eta N_{dead}^{OFF} + e_t)\}, & e_t > 0, \\ q(\eta N_d - \eta N_{dead}^{ON} - e_t) & \\ \times\{2e_t + q(\eta N_d - \eta N_{dead}^{ON} - e_t)\}, & e_t < 0. \end{cases} \tag{A.54}$$

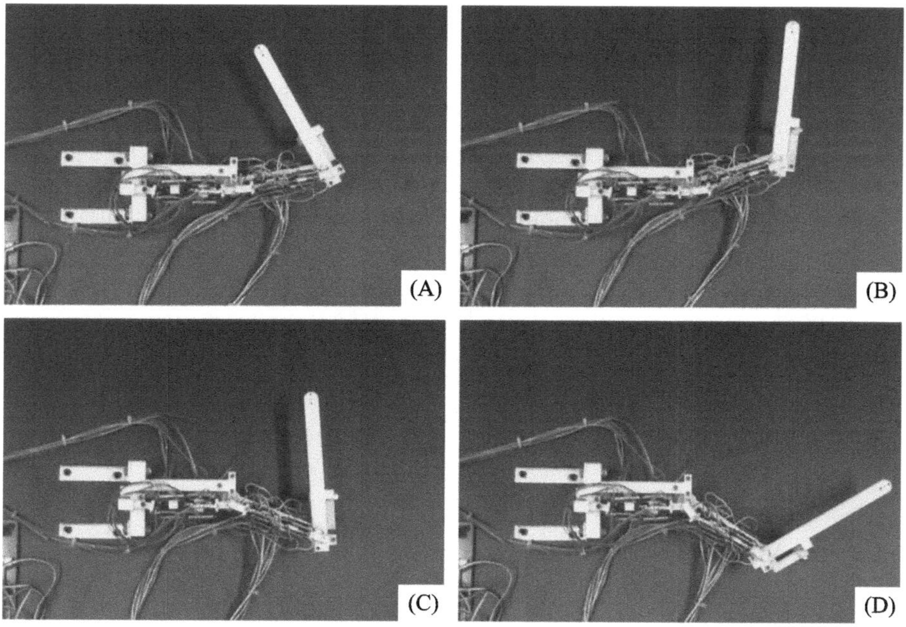

Figure A.27 Joint control of SMA robot arm.

$\Delta V_t^D < 0$ holds if

$$p(e) = \begin{cases} 0, & e \leq 0, \\ 0 < p(e) < \min(\frac{2e}{\eta(N-N_d-N_{dead}^{OFF})+e}, 1), & e > 0, \end{cases} \tag{A.55}$$

$$q(e) = \begin{cases} 0 < q(e) < \min(\frac{-2e}{\eta(N_d-N_{dead}^{ON})-e}, 1), & e < 0, \\ 0, & e \geq 0. \end{cases} \tag{A.56}$$

Recall that $L \geq \eta(N_d - N_{dead}^{ON}) - e$ and $L \geq \eta(N - N_d - N_{dead}^{OFF}) + e$ always hold. Then, we have the following transition probabilities for the cellular control system without using N^{ON}, N_{dead}^{ON}, N^{OFF}, or N_{dead}^{OFF} explicitly:

$$p(e) = \begin{cases} 0, & e \leq 0, \\ 0 < p(e) < \min(2e/L, 1), & e > 0, \end{cases} \tag{A.57}$$

$$q(e) = \begin{cases} 0 < q(e) < \min(-2e/L, 1), & e < 0, \\ 0, & e \geq 0. \end{cases} \tag{A.58}$$

A.7 Proof of Lemma 2 in Section 3.4

Without loss of generality, we can assume $e_t > 0$. First, we can show that there is at least one cell at the states on the forward branch of the hysteresis loop. Otherwise, all the cells are *on* and $y_t = N$. Then $e_t = r - y_t$ cannot be positive. Therefore, at least one cell must exist on the forward states, $F_1, \ldots, F_{N_F}$, and F_R. Although no ready cell is available at time t, $N_t^{FR} = 0$, other cells at states $F_1, \ldots, F_{N_F}$ can reach the ready state F_R in at most n_F time steps.

This implies that at least once in $n_F \le \tau$ time steps, ready cells appear, $N_t^{FR} \ge 1$, for which the following computations apply. If $1 \le N_t^{FR} < e_t/\eta$, then p_{min} from (3.78), and the quotient of $E[e_{t+1}^2|\boldsymbol{X}_t]$ to e_t^2 is given by

$$\alpha_A = \frac{E[e_{t+1}^2|\boldsymbol{X}_t]}{e_t^2} = \left(1 - \eta\left(\frac{N_t^{FR}}{e_t}\right)\right)^2 \le \left(1 - \frac{1}{N}\right)^2 < 1 \tag{A.59}$$

if $1/2 \le e_t/\eta \le N_t^{FR}$, $p_{min} = p_0$ and

$$\alpha_B = \frac{E[e_{t+1}^2|\boldsymbol{X}_t]}{e_t^2} = 1 - \left(\frac{N_t^{FR}}{N_t^{FR} - 1}\right)\left(1 - \frac{\eta}{2e_t}\right) \le 1 - \frac{1}{4}\left(\frac{N}{N-1}\right) < 1. \tag{A.60}$$

Both cases yield $E[e_{t+1}^2|\boldsymbol{X}_t] < e_t^2$. Let $\tilde{\alpha} = \max(\alpha_A, \alpha_B)$, then $E[e_{t+1}^2|\boldsymbol{X}_t] \le \tilde{\alpha} e_t^2$ at least once in $n_F < \tau$ time steps.

For $e_t < 0$, at least one cell can reach the backward ready state B_R in at most $n_B \le \tau$ time steps, and at that time $E[e_{t+1}^2|\boldsymbol{X}_t] < e_t^2$. Therefore, the expected squared error decreases at least once in $\tau = \max(n_F, n_B)$ time steps and the expected squared error is bounded with the same $\tilde{\alpha} < 1$ as in the case of $e_t > 0$, $E[e_{t+1}^2|\boldsymbol{X}_t] \le \tilde{\alpha} e_t^2$. □

A.8 Recursive computation of probability $\Pr(\mathbf{X}_t|\mathbf{X}_0)$

(1) Set $t = 1$ and compute

$$\Pr(N_t^{B_1}|\boldsymbol{X}_{t-1}) = \begin{cases} \binom{N_{t-1}^{FR}}{j} p^j (1-p)^{(N_{t-1}^{FR} - j)}, & e_{t-1} > 0, 0 \le j \le N_{t-1}^{FR}, \\ 0, & \text{otherwise}, \end{cases}$$

$$\Pr(N_t^{F_1}|\boldsymbol{X}_{t-1}) = \begin{cases} \binom{N_{t-1}^{BR}}{j} q^j (1-q)^{(N_{t-1}^{BR} - j)}, & e_{t-1} < 0, 0 \le j \le N_{t-1}^{BR}, \\ 0, & \text{otherwise}. \end{cases} \tag{A.61}$$

(2) Set $t + 1$ to t and compute

$$\Pr(\boldsymbol{X}_t|\boldsymbol{X}_0) = \sum_{State} \Pr(\boldsymbol{X}_{t-1}|\boldsymbol{X}_0)\Pr(\boldsymbol{X}_t|\boldsymbol{X}_{t-1}) \tag{A.62}$$

where $\Pr(\boldsymbol{X}_t|\boldsymbol{X}_{t-1})$ is evaluated using Eqs. (A.61), (3.70), and (3.71) with p and q determined from (3.78) evaluated at $\boldsymbol{X}_{t-1}$.
(3) If $t < t_{final}$, go back to (2), otherwise stop.

A.9 Proof of Lemma 2 in Section 4.1

Proof of Lemma 2.
Suppose that Lemma 2 is not true. That is, suppose that there exist positive integers $n'_{l1}, n'_{l2}, \ldots, n'_{lNON}$ such that $J(n'_{l1}, n'_{l2}, \ldots, n'_{lNON}) > J(1, 1, \ldots, 1, N - N^{ON} + 1)$. Then this implies

$$\begin{aligned} &n'_{l1}x_1^2 + n'_{l2}x_2^2 + \cdots + n'_{NON-1}x_{NON-1}^2 + n'_{NON}x_{NON}^2 \\ &> x_1^2 + x_2^2 + \cdots + x_{NON-1}^2 + (N - N^{ON} + 1)x_{NON}^2. \end{aligned} \tag{A.63}$$

Using the constraint $n'_{lNON} = N - N^{ON} - \sum_{i=1}^{NON-1} n'_{li}$ allows (A.63) to be rewritten as

$$\begin{aligned} &\left(\frac{n'_{l1} - 1}{1 - N^{ON} + n + \sum_{i=1}^{NON-1} n'_{li}}\right)\left(\frac{x_1}{x_{NON}}\right)^2 + \\ &\cdots + \left(\frac{n'_{lNON-1} - 1}{1 - N^{ON} + n + \sum_{i=1}^{NON-1} n'_{li}}\right)\left(\frac{x_{NON-1}}{x_{NON}}\right)^2 > 1. \end{aligned} \tag{A.64}$$

The inequality above can be more strictly stated as

$$\left(\frac{1 - N^{ON} + \sum_{i=1}^{NON-1} n'_{li}}{1 - N^{ON} + n + \sum_{i=1}^{NON-1} n'_{li}}\right)\left(\frac{x_{NON-1}}{x_{NON}}\right)^2 > 1, \tag{A.65}$$

which is a contradiction because of (4.14). □

Bibliography

[1] Lever actuator comprising a longitudinal-effect electroexpansive transducer. US Patent 4,435,666.

[2] Piezoelectric stack actuators. CEDRAT Inc., http://www.cedrat.com/.

[3] Shape memory alloy actuators and control methods. US Patent 6,574,958, 2003.

[4] Mostafa Abdalla, Mary Frecker, Zafer Gürdal, Terrence Johnson, Douglas K. Lindner, Design of a piezoelectric actuator and compliant mechanism combination for maximum energy efficiency, Institute of Physics Smart Materials and Structures 14 (nov 2005) 1421–1430.

[5] S. Abdelwahed, Rong Su, S. Neema, A feasible lookahead control for systems with finite control set, in: Proceedings of 2005 IEEE Conference on Control Applications. CCA 2005, IEEE, 2005, pp. 663–668.

[6] Anne M. Agur, Arthur F. Dalley, Grant's Atlas of Anatomy, 12th edition, Wolters Kluwer, Philadelphia, 2009.

[7] Alin Albu-Schaffer, Oliver Eiberger, Markus Grebenstein, Sami Haddadin, Christian Ott, Thomas Wimbock, Sebastian Wolf, Gerd Hirzinger, Soft robotics, IEEE Robotics & Automation Magazine 15 (3) (sep 2008) 20–30.

[8] Götz Alefeld, Jürgen Hertzberger, Introduction to Interval Computations, Academic Press, New York, 1983.

[9] P.W. Alexander, D. Brei, Piezoceramic telescopic actuator quasi-static experimental characterization, Journal of Intelligent Material Systems and Structures 14 (10) (2003) 643–655.

[10] Emil V. Ardelean, Daniel G. Cole, Robert L. Clark, High performance "V-stack" piezoelectric actuator, Journal of Intelligent Material Systems and Structures 15 (11) (2004) 879–889.

[11] H. Asada, J.J.E. Slotine, Robot Analysis and Control, Wiley, 1986, Chapter 7.

[12] H. Harry Asada, Jean-Jacques Slotine, Robot Analysis and Control, John Wiley and Sons, Inc., 1986.

[13] Samuel K. Au, Hugh M. Herr, Powered ankle-foot prosthesis, IEEE Robotics and Automation Magazine 15 (3) (2008) 52–59.

[14] Shun-ichi Azuma, Ryota Yoshimura, Toshiharu Sugie, Broadcast control of multi-agent systems, Automatica 49 (8) (2013) 2307–2316.

[15] Adrien Badel, Jinhao Qiu, Tetsuaki Nakano, Self-sensing force control of a piezoelectric actuator, IEEE Transactions on Ultrasonics, Ferroelectrics and Frequency Control 55 (2008) 2571–2581.

[16] Reinder Banning, Willem L. de Koning, Han J.M.T.A. Adriaens, Richard K. Koops, State-space analysis and identification for a class of hysteretic systems, Automatica 37 (2001) 1883–1892.

[17] L. Bar, B. Berkels, M. Rumpf, G. Sapiro, A variational framework for simultaneous motion estimation and restoration of motion-blurred video, in: Computer Vision, 2007. ICCV 2007. IEEE 11th International Conference on, 2007, pp. 1–8.

[18] Yoseph Bar-Cohen, Electroactive Polymer (EAP) Actuators as Artificial Muscles: Reality, Potential, and Challenges, vol. 136, SPIE Press, 2004.

[19] Patrick R. Barragán, Shinichiro Tsukahara, H. Harry Asada, Design of energy-saving PZT drive amplifiers for mobile and wearable physical assists, in: ASME 2011 Dynamic Systems and Control Conference and Bath/ASME Symposium on Fluid Power and Motion Control, American Society of Mechanical Engineers, 2011, pp. 783–790.
[20] Eric J. Barth, Michael Goldfarb, A control design method for switching systems with application to pneumatic servo systems, in: 2002 ASME International Mechanical Engineering Congress and Exposition (IMECE), ASME, New Orleans, LA, nov 2002, p. 33424.
[21] K.J. Bathe, Finite Element Procedures, Prentice Hall, 1996.
[22] Siegfried Bauer, Simona Bauer-Gogonea, Ingrid Graz, Martin Kaltenbrunner, Christoph Keplinger, Reinhard Schwödiauer, 25th anniversary article: a soft future: from robots and sensor skin to energy harvesters, in: Advanced Materials (Deerfield Beach, Fla.), nov 2013, pp. 149–162.
[23] A.D. Belegundu, S.D. Rajan, A shape optimization approach based on natural design variables and shape functions, Computer Methods in Applied Mechanics and Engineering 66 (1) (1988) 87–106.
[24] M. Ben-Ezra, S.K. Nayar, Motion-based motion deblurring, Pattern Analysis and Machine Intelligence, IEEE Transactions on 26 (6) (2004) 689–698.
[25] Smita Bharti, Mary Frecker, Compliant mechanical amplifier design using multiple optimally placed actuators, Journal of Intelligent Material Systems and Structures 18 (mar 2007) 209–217.
[26] Antonio Bicchi, Alessia Marigo, Benedetto Piccoli, On the reachability of quantized control systems, IEEE Transactions on Automatic Control 47 (4) (2002) 546–563.
[27] Antonio Bicchi, Giovanni Tonietti, Fast and "soft-arm" tactics, IEEE Robotics & Automation Magazine 11 (2) (2004) 22–23.
[28] Wayne Book, Controlled motion in an elastic world, Transaction of the ASME Journal of Dynamic Systems, Measurement, and Control 115 (1993) 252–261.
[29] D.J. Braun, Florian Petit, Felix Huber, Robots driven by compliant actuators: optimal control under actuation constraints, IEEE Transactions on Robotics 29 (5) (2013) 1085–1101.
[30] R.W. Brockett, D. Liberzon, Quantized feedback stabilization of linear systems, IEEE Transactions on Automatic Control 45 (7) (jul 2000) 1279–1289.
[31] Martin Brokate, Jürgen Sprekels, Hysteresis and Phase Transitions, Springer, 1996.
[32] Matthew Bryant, Michael A. Meller, Ephrahim Garcia, Variable recruitment fluidic artificial muscles: modeling and experiments, Smart Materials and Structures 23 (7) (2014) 74009.
[33] T. Buchanan, D. Shreeve, An evaluation of optimization techniques for the prediction of muscle activation patterns during isometric tasks, Journal of Biomechanical Engineering 118 (1996) 565.
[34] D.G. Caldwell, N. Tsagarakis, Biomimetic actuators in prosthetic and rehabilitation applications, Technology and Health Care 10 (2) (2002) 107–120.
[35] M. Calisti, M. Giorelli, G. Levy, B. Mazzolai, B. Hochner, C. Laschi, P. Dario, An octopus-bioinspired solution to movement and manipulation for soft robots, Bioinspiration & Biomimetics 6 (3) (oct 2011) 036002.
[36] S. Canfield, M. Frecker, Topology optimization of compliant mechanical amplifiers for piezoelectric actuators, Structural and Multidisciplinary Optimization 20 (4) (2000) 269–279.
[37] Vinod R. Challa, M.G. Prasad, Yong Shi, Frank T. Fisher, A vibration energy harvesting device with bidirectional resonance frequency tunability, Smart Materials and Structures 17 (1) (2008).

[38] M. Chandrashekar, P.H. Roe, G.J. Savage, A unified approach to modelling photovoltaic power systems, Modeling and Simulation 23 (4) (1993) 313.
[39] Junho Choi, Sunchul Park, Woosub Lee, Sung-Chul Kang, Design of a robot joint with variable stiffness, in: Proceedings – IEEE International Conference on Robotics and Automation, 2008, pp. 1760–1765.
[40] Kyung K. Choi, Kuang-Hua Chang, A geometry-based parameterization method for shape design of elastic solids, Mechanics of Structures and Machines 20 (2) (1992) 215–252.
[41] S.B. Choi, C.C. Cheong, H.C. Shin, Sliding mode control of vibration in a single-link flexible arm with parameter variations, Journal of Sound and Vibration 179 (5) (1995) 737–748.
[42] John Choma Jr., Electrical Networks: Theory and Analysis, John Wiley and Sons, Inc., New York, 1985.
[43] M.M. Churchland, A. Afshar, K.V. Shenoy, A central source of movement variability, Neuron 52 (6) (2006) 1085–1096.
[44] N.J. Conway, Sang-Gook Kim, Large-strain, piezoelectric, in-plane micro-actuator, in: 17th IEEE International Conference on Micro Electro Mechanical Systems (IEEE MEMS), 2004, pp. 454–457.
[45] N.J. Conway, Z.J. Traina, S.G. Kim, A strain amplifying piezoelectric MEMS actuator, Journal of Micromechanics and Microengineering 17 (4) (2007) 781–787.
[46] Face International Corporation, http://www.thunderandlightningpiezos.com/.
[47] Nikolaus Correll, Ç.D. Önal, H. Liang, Soft autonomous materials – using active elasticity and embedded distributed computation, Experimental Robotics 79 (2014) 227–240.
[48] John J. Craig, Introduction to Robotics, Mechanics and Control, third edition, Pearson Prentice-Hall, Upper Saddle River, 2005.
[49] Roy D. Crowninshield, Richard A. Brand, A physiologically based criterion of muscle force prediction in locomotion, Journal of Biomechanics 14 (11) (jan 1981) 793–801.
[50] P.M. Davies, Steps to Follow: The Comprehensive Treatment of Patients with Hemiplegia, Springer, 2000.
[51] S. Davis, D.G. Caldwell, Biologically inspired damage tolerance in braided pneumatic muscle actuators, Journal of Intelligent Material Systems and Structures 23 (3) (sep 2011) 313–325.
[52] D.F. Delchamps, Stabilizing a linear system with quantized state feedback, Automatic Control, IEEE Transactions on 35 (8) (1990) 916–924.
[53] Hua Deng, M. Krstic, Output-feedback stochastic nonlinear stabilization, Automatic Control, IEEE Transactions on 44 (2) (1999) 328–333.
[54] I.M. Díaz, E. Pereira, V. Feliu, J.J.L. Cela, Concurrent design of multimode input shapers and link dynamics for flexible manipulators, Mechatronics, IEEE/ASME Transactions on 15 (4) (aug 2010) 646–651.
[55] A. Diaz-Calderon, C.J.J. Paredis, P.K. Khosla, Automatic generation of system-level dynamic equations for mechatronic systems, Computer-Aided Design 32 (5–6) (may 2000) 339–354.
[56] Michael H. Dickinson, Farley T. Claire, Robert J. Full, M.A.R. Koehl, Rodger Kram, Steven Lehman, How animals move: an integrative view, Science 288 (5463) (apr 2000) 100–106.
[57] Carl F. DiSalvo, Francine Gemperle, Jodi Forlizzi, Sara Kiesler, All robots are not created equal: the design and perception of humanoid robot heads, in: Proceedings of the 4th Conference on Designing Interactive Systems, 2002, pp. 321–326.

[58] Kay Dittrich, Cellular actuator device and methods of making and using the same. US Patent 7,055,782, 2006.
[59] Raymond Dodge, Five types of eye movement in the horizontal meridian plane of the field of regard, American Journal of Physiology–Legacy Content 8 (4) (1903) 307–329.
[60] A. Dogan, K. Uchino, R.E. Newnham, Composite piezoelectric transducer with truncated conical endcaps "cymbal", Ultrasonics, Ferroelectrics and Frequency Control, IEEE Transactions on 44 (3) (1997) 597–605.
[61] A. Dogan, Q. Xu, K. Onitsuka, S. Yoshikawa, K. Uchino, R.E. Newnham, High displacement ceramic metal composite actuators (moonies), Ferroelectrics 156 (1) (1994) 1–6.
[62] J.L. Doob, Stochastic Processes, reprint edition, Wiley Classics Library, Wiley-Interscience, 1990.
[63] Jeffrey J. Dosch, Daniel J. Inman, Ephrahim Garcia, A self-sensing piezoelectric actuator for collocated control, Journal of Intelligent Material Systems and Structures 3 (1992) 166–185.
[64] John Doyle, Bruce Francis, Allen Tannenbaum, Feedback Control Theory, Macmillan Publishing, 1990.
[65] X. Du, R. Dixon, R.M. Goodall, A.C. Zolotas, Modelling and control of a high redundancy actuator, Mechatronics 20 (1) (feb 2010) 102–112.
[66] X. Du, R. Dixon, R.M. Goodall, A.C. Zolotas, Modelling and control of a highly redundant actuator, in: Control, vol. 44, 2006.
[67] Xinli Du, Roger Dixon, Roger M. Goodall, Argyrios C. Zolotas, LQG control of a high redundancy actuator, in: 2007 IEEE/ASME International Conference on Advanced Intelligent Mechatronics, IEEE, 2007, pp. 1–6.
[68] Bülent Ekmekci, Residual stresses and white layer in electric discharge machining (EDM), Applied Surface Science 253 (sep 2007) 9234–9240.
[69] John D. Enderle, James W. Wolfe, Time-optimal control of saccadic eye movements, Biomedical Engineering, IEEE Transactions on BME-34 (1) (1987) 43–55.
[70] R.M. Enoka, Neuromechanics of Human Movement, Human Kinetics Publishers, 2008.
[71] J.D. Ervin, D. Brei, Recurve piezoelectric-strain-amplifying actuator architecture, IEEE/ASME Transactions on Mechatronics 3 (4) (1998) 293–301.
[72] William Feller, An Introduction to Probability Theory and Its Applications, vol. 1, 3rd edition, Wiley, 1968.
[73] Rob Fergus, Barun Singh, Aaron Hertzmann, Sam T. Roweis, William T. Freeman, Removing camera shake from a single photograph, ACM Transactions on Graphics (TOG) 25 (3) (2006) 787–794.
[74] J. Fiene, G. Niemeyer, Toward switching motor control, Mechatronics, IEEE/ASME Transactions on 11 (1) (2006) 27–34.
[75] R. Filippini, S. Sen, A. Bicchi, Toward soft robots you can depend on, IEEE Robotics & Automation Magazine 10 (3) (2008) 31–41.
[76] T. Flash, N. Hogan, The coordination of arm movements: an experimentally confirmed mathematical model, The Journal of Neuroscience 5 (7) (1985) 1688–1703.
[77] Gene F. Franklin, David J. Powell, Abbas Emami-Naeini, Feedback Control of Dynamic Systems, 4th edition, Prentice Hall PTR, Upper Saddle River, NJ, USA, 2001.
[78] Mary Frecker, Recent advances in optimization of smart structures and actuators, Journal of Intelligent Material Systems and Structures 14 (2003) 207–216.
[79] Yongqing Fu, Hejun Du, Weimin Huang, Sam Zhang, Min Hu, TiNi-based thin films in MEMS applications: a review, Sensors and Actuators A: Physical 112 (2–3) (may 2004) 395–408.

[80] Eiichi Furukawa, Makoto Mizuno, Toshio Doi, Development of a flexure-hinged translation mechanism driven by two piezoelectric stacks, JSME International Journal. Ser. C, Dynamics, Control, Robotics, Design and Manufacturing 38 (4) (1995) 743–748.
[81] F. Gantmacher, M. Krein, Oscillation Matrices, revised edition, American Mathematical Society, 2002.
[82] Wenbin Gao, Hongguang Wang, An automatic dynamics generation method for reconfigurable modular robot, Advances in Reconfigurable Mechanisms and Robots I 5 (2012) 551–560.
[83] Garnett E. Simmers Jr., Jeffrey R. Hodgkins, David D. Mascarenas, Gyuhae Park, Hoon Sohn, Improved piezoelectric self-sensing actuation, Journal of Intelligent Material Systems and Structures 15 (2004) 941–953.
[84] Thomas Geijtenbeek, Antonie J. Van Den Bogert, Ben J.H. Van Basten, Arjan Egges, Evaluating the physical realism of character animations using musculoskeletal models, in: Lecture Notes in Computer Science, vol. 6459, 2010, pp. 11–22.
[85] Helen M.S. Georgiou, Ridha Ben Mrad, Electromechanical modeling of piezoceramic actuators for dynamic loading applications, ASME Journal of Dynamic Systems, Measurement, and Control 128 (2006) 558–567.
[86] G.M.L. Gladwell, Inverse Problems in Vibration, Martinus Nijhoff, Dordrecht, 1986.
[87] Michael Goldfarb, N. Celanovic, A lumped parameter electromechanical model for describing the nonlinear behavior of piezoelectric actuators, ASME Journal of Dynamic Systems, Measurement, and Control 119 (3) (1997) 478–485.
[88] Michael Goldfarb, Nikola Celanovic, Modeling piezoelectric stack actuators for control of micromanipulation, IEEE Control Systems (jun 1997) 69–79.
[89] Amit Goldstein, Raanan Fattal, Blur-kernel estimation from spectral irregularities, in: European Conference on Computer Vision (ECCV), Springer, 2012, pp. 622–635.
[90] M. Grossard, C. Rotinat-Libersa, N. Chaillet, M. Boukallel, Mechanical and control-oriented design of a monolithic piezoelectric microgripper using a topological optimization method, IEEE/ASME Transactions on Mechatronics 14 (1) (2009) 32–45.
[91] G.H. Haertling, Rainbow ceramics – a new type of ultra-high-displacement actuator, American Ceramic Society Bulletin 73 (1) (1994) 93–96.
[92] Gene H. Haertling, Ferroelectric ceramics: history and technology, Journal of the American Ceramic Society 82 (4) (apr 1999) 797–818.
[93] Antonia F. De C. Hamilton, Kelvin E. Jones, Daniel M. Wolpert, The scaling of motor noise with muscle strength and motor unit number in humans, Experimental Brain Research (Experimentelle Hirnforschung) (Expérimentation cérébrale) 157 (4) (aug 2004) 417–430.
[94] Chenlu Han, Jun Ueda, On-site simultaneous calibration of hysteric and structural parameters for self-sensing robotic tweezers with strain amplified piezoelectric actuators, in: American Control Conference (ACC), 2014, IEEE, 2014, pp. 3657–3662.
[95] H. Hanafusa, H. Asada, A robotic hand with elastic fingers and its application to assembly processes, in: Proceedings – FAC Symp. on Information Control Problems in Production Engineering, 1977, pp. 127–138.
[96] H. Hanafusa, H. Asada, Stable pretension by a robotic hand with elastic fingers, in: Proceedings – 7th Int. Symp. on Industrial Robots, 1977, pp. 361–367.
[97] Susumu Hara, Tetsuji Zama, Wataru Takashima, Keiichi Kaneto, Artificial muscles based on polypyrrole actuators with large strain and stress induced electrically, Polymer Journal 36 (2) (2004) 151–161.
[98] C.M. Harris, D.M. Wolpert, Signal-dependent noise determines motor planning, Nature 394 (6695) (1998) 780–784.

[99] W. Haustein, Considerations on Listing's Law and the primary position by means of a matrix description of eye position control, Biological Cybernetics 60 (6) (1989) 411–420.
[100] E. Henneman, Relation between size of neurons and their susceptibility to discharge, Science 126 (3287) (1957) 1345–1347.
[101] Francis B. Hildebrand, Advanced Calculus for Applications, Prentice-Hall, Inc., Englewood Cliffs, 1976.
[102] E. Hiris, Detection of biological and non-biological motion, Journal of Vision 7 (12) (2007) 1–16.
[103] N. Hogan, Adaptive control of mechanical impedance by coactivation of antagonist muscles, IEEE Transactions on Automatic Control 29 (8) (aug 1984) 681–690.
[104] Neville Hogan, Impedance control: an approach to manipulation: part I – theory, Journal of Dynamic Systems, Measurement and Control, Transactions of the ASME 107 (1) (1985) 1–7.
[105] Larry Howell, Compliant Mechanisms, 1st edition, John Wiley and Sons, New York, 2001.
[106] Shyh-Chour Huang, Wei-Liang Chen, Design of topologically optimal microgripper, in: IEEE International Conference on Systems, Man and Cybernetics, 2008.
[107] J. Hurst, A. Rizzi, Series compliance for an efficient running gait, IEEE Robotics and Automation Magazine 15 (3) (2008) 42–51.
[108] J.W. Hurst, J.E. Chestnutt, A.A. Rizzi, An actuator with physically variable stiffness for highly dynamic legged locomotion, in: IEEE 2004 International Conference on Robotics and Automation (ICRA 2014), 2004, pp. 4662–4667.
[109] D. Huston, B. Esser, G. Spencer, D. Burns, E. Kahn, Hierarchical actuator systems, Proceedings of SPIE 5762 (may 2005) 311–319.
[110] Dryver Huston, Brian Esser, Michael Werner, Hierarchical actuators, in: Proceedings of the First World Congress on Biomimetics and Artificial Muscles, Albuquerque, 2002.
[111] N. Iguchi, Y. Sakaguchi, F. Ishida, The minimum endpoint variance trajectory depends on the profile of the signal-dependent noise, Biological Cybernetics 92 (4) (2005) 219–228.
[112] Filip Ilievski, Aaron D. Mazzeo, Robert F. Shepherd, Xin Chen, George M. Whitesides, Soft robotics for chemists, Angewandte Chemie (International ed. in English) 50 (8) (feb 2011) 1890–1895.
[113] David E. Irwin, Joseph S. Brown, Jun-shi Sun, Visual masking and visual integration across saccadic eye movements, Journal of Experimental Psychology: General 117 (3) (1988) 276.
[114] K. Kurien Issac, Sunil K. Agrawal, An investigation into the use of springs and wing motions to minimize the power expended by a pigeon-sized mechanical bird for steady flight, Journal of Mechanical Design, Transactions of the ASME 129 (4) (2007) 381–389.
[115] Amir Jafari, Nikos G. Tsagarakis, Darwin G. Caldwell, A novel intrinsically energy efficient actuator with adjustable stiffness (AwAS), IEEE/ASME Transactions on Mechatronics 18 (1) (2013) 355–365.
[116] P. Janker, M. Christmann, F. Hermle, T. Lorkowski, S. Storm, Mechatronics using piezoelectric actuators, Journal of the European Ceramics Society 19 (6) (1999) 1127–1131.
[117] Jiaya Jia, Single image motion deblurring using transparency, in: Computer Vision and Pattern Recognition, 2007. CVPR'07. IEEE Conference on, IEEE, 2007, pp. 1–8.
[118] Gunnar Johansson, Visual perception of biological motion and a model for its analysis, Perception & Psychophysics 14 (2) (1973) 201–211.
[119] K.E. Jones, A.F.C. Hamilton, D.M. Wolpert, Sources of signal-dependent noise during isometric force production, Journal of Neurophysiology 88 (3) (2002) 1533.

[120] William B. Jones, W.J. Thron, Continued Fractions: Analytic Theory and Applications, Encyclopedia of Mathematics and Its Applications, vol. 11, Addison-Wesley, 1980.
[121] Neel Joshi, Sing Bing Kang, C. Lawrence Zitnick, Richard Szeliski, Image deblurring using inertial measurement sensors, ACM Transactions on Graphics (TOG) 29 (4) (2010) 30.
[122] A. Agung Julius, Ádám Halász, M. Selman Sakar, Harvey Rubin, Vijay Kumar, George J. Pappas, Stochastic modeling and control of biological systems: the lactose regulation system of escherichia coli, Automatic Control, IEEE Transactions on 53 (Special Issue) (2008) 51–65.
[123] R. Jürgens, W. Becker, H.H. Kornhuber, Natural and drug-induced variations of velocity and duration of human saccadic eye movements: evidence for a control of the neural pulse generator by local feedback, Biological Cybernetics 39 (2) (1981) 87–96.
[124] Robert C. Juvinall, Kurt M. Marshek, Fundamentals of Machine Component Design, 3rd edition, John Wiley and Sons, 2000.
[125] Mohammad Hasan H. Kani, Hamed Ali Yaghini Bonabi, Hamed Jalaly Bidgoyly, Mohammad Javad Yazdanpanah, Majid Nili Ahmadabadi, Design and implementation of a distributed variable impedance actuator using parallel linear springs, Journal of Mechanisms and Robotics 8 (734) (2016) 1–45.
[126] Andrei Kholkin, Bahram Jadidian, Ahmad Safari, Encyclopedia of Smart Materials, vol. 1, John Wiley and Sons, 2002.
[127] Soheil Kianzad, Milind Pandit, Johnathan D. Lewis, Alexander R. Berlingeri, Karl J. Haebler, John D.W. Madden, Variable stiffness structure using nylon actuators arranged in a pennate muscle configuration, in: SPIE Smart Structures and Materials+ Nondestructive Evaluation and Health Monitoring, International Society for Optics and Photonics, 2015, pp. 94301Z–94301Z.
[128] Charles J. Kim, Yong-Mo Moon, Sridhar Kota, A building block approach to the conceptual synthesis of compliant mechanisms utilizing compliance and stiffness ellipsoids, Journal of Mechanical Design 130 (2) (2008) 022308.
[129] Jun Hyung Kim, Soo Hyun Kim, Yoon Keun Kwak, Development of a piezoelectric actuator using a three-dimensional bridge-type hinge mechanism, American Institute of Physics Review of Scientific Instruments 74 (5) (may 2003) 2918–2924.
[130] Jun Hyung Kim, Soo Hyun Kim, Yoon Keun Kwak, Development and optimization of 3-D bridge-type hinge mechanisms, Sensors and Actuators A: Physical 116 (3) (2004) 530–538.
[131] M.D. Kim, J. Ueda, Real-time panoramic image generation and motion deblurring by using dynamics-based robotic vision, IEEE/ASME Transactions on Mechatronics 21 (3) (jun 2016) 1376–1387.
[132] Michael D. Kim, Jun Ueda, Dynamics-based motion de-blurring for a PZT-driven, compliant camera orientation mechanism, The International Journal of Robotics Research 34 (4–5) (2015) 653–673.
[133] Michael Duckjune Kim, Jun Ueda, Discrete switching commands for tracking and vibration suppression using a quantized, compliant camera orientation system, in: 2016 IEEE International Conference on Robotics and Automation (ICRA), IEEE, 2016, pp. 5177–5182.
[134] Kazuo Kitamura, Toshio Yanagida, Stochastic properties of actomyosin motor, Biosystems 71 (1–2) (sep 2003) 101–110.
[135] Roy Kornbluh, Ron Pelrine, Joseph Eckerle, Jose Joseph, Electrostrictive polymer artificial muscle actuators, in: Proceedings – IEEE International Conference on Robotics and Automation, vol. 3, 1998, pp. 2147–2154.

[136] A.I. Kostyukov, Muscle hysteresis and movement control: a theoretical study, Neuroscience 83 (1) (mar 1998) 303–320.

[137] S. Kota, K.-J. Liu, B. Kreiner, J. Trease, J. Arenas, J. Geiger, Design and application of compliant mechanisms for surgical tools, ASME Journal of Biomedical Engineering 127 (2005) 981–989.

[138] Vijaya Krishnamoorthy, Simon Goodman, Vladimir Zatsiorsky, Mark L. Latash, Muscle synergies during shifts of the center of pressure by standing persons: identification of muscle modes, Biological Cybernetics 89 (2) (aug 2003) 152–161.

[139] Klaus Kuhnen, Modeling, identification, and compensation of complex hysteretic nonlinearities a modified Prandtl–Ishlinskii approach, European Journal of Control 9 (2003) 407–418.

[140] Klaus Kuhnen, Modeling, identification, and compensation of complex hysteretic and Log(t)-type creep nonlinearities, Control and Intelligent Systems 33 (2005) 134–147.

[141] S. Kuiper, G. Schitter, Active damping of a piezoelectric tube scanner using self-sensing piezo actuation, Mechatronics 20 (2010) 656–665.

[142] Y. Kurita, F. Sugihara, J. Ueda, T. Ogasawara, Piezoelectric tweezer-type end effector with force- and displacement-sensing capability, Mechatronics, IEEE/ASME Transactions on 17 (6) (dec 2012) 1039–1048.

[143] Harold J. Kushner, On the stability of stochastic dynamical systems, Proceedings of the National Academy of Sciences 53 (1) (1965) 8–12.

[144] Harold J. Kushner, Stochastic Stability and Control, Mathematics in Science and Engineering, Academic Press, 1967.

[145] Yi-Chiao Lee, Chao-Chieh Lan, Cheng-Yu Chu, Chih-Ming Lai, Yi-Jie Chen, A pan-tilt orienting mechanism with parallel axes of flexural actuation, Mechatronics, IEEE/ASME Transactions on 18 (3) (2013) 1100–1112.

[146] Eli S. Leland, Paul K. Wright, Resonance tuning of piezoelectric vibration energy scavenging generators using compressive axial preload, Smart Materials and Structures 15 (5) (2006) 1413–1420.

[147] A. Lenz, S.R. Anderson, A.G. Pipe, C. Melhuish, P. Dean, J. Porrill, Cerebellar-inspired adaptive control of a robot eye actuated by pneumatic artificial muscles, Systems, Man, and Cybernetics, Part B: Cybernetics, IEEE Transactions on 39 (6) (2009) 1420–1433.

[148] Mirko Leomanni, Andrea Garulli, Antonio Giannitrapani, Fabrizio Scortecci, Minimum switching control for spacecraft precision pointing with on / off actuators, in: Conference on Decision and Control (CDC), 2015.

[149] M. Lesmana, D.K. Pai, A biologically inspired controller for fast eye movements, in: Robotics and Automation (ICRA), 2011 IEEE International Conference on, 2011, pp. 3670–3675.

[150] Sungung Lim, Homer Stevens, Jonathan P. How, Input shaping for multi-input flexible systems, ASME Journal of Dynamic Systems, Measurement, and Control 121 (sep 1999) 443–447.

[151] Nicolae Lobontiu, Compliance-based matrix method for modeling the quasi-static response of planar serial flexure-hinge mechanisms, Precision Engineering 38 (3) (2014) 639–650.

[152] Nicolae Lobontiu, Ephrahim Garcia, Analytical model of displacement amplification and stiffness optimization for a class of flexure-based compliant mechanisms, Computers & Structures 81 (32) (2003) 2797–2810.

[153] Nicolae Lobontiu, Michael Goldfarb, Ephrahim Garcia, Maximizing the resonant displacement of piezoelectric beams, Proceedings of SPIE – The International Society for Optical Engineering 3668 (I) (1999) 154–163.

[154] Nicolae Lobontiu, Jeffrey S.N. Paine, Design of circular cross-section corner-filleted flexure hinges for three-dimensional compliant mechanisms, ASME Journal of Mechanical Design 124 (sep 2002) 479–484.
[155] Nicolae Lobontiu, Jeffrey S.N. Paine, Ephrahim Garcia, Michael Goldfarb, Corner-filleted flexure hinges, Journal of Mechanical Design 123 (3) (2001) 346–352.
[156] A. De Luca, G. Di Giovanni, Rest-to rest motion of a two-link robot with a flexible forearm, in: Proceedings of ASME/IEEE International Conference on Advanced Intelligent Mechatronics, vol. 2, 2001, pp. 929–935.
[157] Ming Luo, Erik H. Skorina, Weijia Tao, Fuchen Chen, Cagdas D. Onal, Optimized design of a rigid kinematic module for antagonistic soft actuation, in: IEEE Conference on Technologies for Practical Robot Applications, TePRA, 2015-August, 2015.
[158] D. MacNair, J. Ueda, Dynamic cellular actuator arrays and the expanded fingerprint method for dynamic modeling, Robotics and Autonomous Systems 62 (7) (sep 2014) 1060–1072.
[159] D.L. MacNair, J. Ueda, A fingerprint method for variability and robustness analysis of stochastically controlled cellular actuator arrays, The International Journal of Robotics Research 30 (5) (mar 2011) 536–555.
[160] David Luke MacNair, Modeling cellular actuator arrays, Ph.D. Thesis, 2013.
[161] J.D.W. Madden, N.A. Vandesteeg, P.A. Anquetil, P.G.A. Madden, A. Takshi, R.Z. Pytel, S.R. Lafontaine, P.A. Wieringa, I.W. Hunter, Artificial muscle technology: physical principles and naval prospects, IEEE Journal of Oceanic Engineering 29 (3) (jul 2004) 706–728.
[162] D.R. Madill, D. Wang, Modeling and L 2-stability of a shape memory alloy position control system, IEEE Transactions on Control Systems Technology 6 (4) (1998) 473–481.
[163] Matteo Malosio, Zhen Gao, Multi-stage piezo stroke amplifier, in: IEEE/ASME International Conference on Advanced Intelligent Mechatronics, AIM, 2012, pp. 1086–1092.
[164] Elaine N. Marieb, Human Anatomy and Physiology, 5th edition, Benjamin Cummings, San Francisco, 2001.
[165] Michael J. Martell, Joshua A. Schultz, Multiport Modeling of Force and Displacement in Elastic Transmissions for Underactuated Hands, in: Proceedings of the 2014 IEEE/RSJ International Conference on Intelligent Robots and Systems, Chicago, IL, 2014, pp. 1074–1079.
[166] Frederic H. Martini, Edwin F. Bartholomew, Essentials of Anatomy & Physiology, 4th edition, Benjamin Cummings, 2006.
[167] Glenn Mathijssen, Joshua Schultz, Bram Vanderborght, Antonio Bicchi, A muscle-like recruitment actuator with modular redundant actuation units for soft robotics, Robotics and Autonomous Systems 74, Part A (jul 2015) 40–50.
[168] Ethel Matin, Saccadic suppression: a review and an analysis, Psychological Bulletin 81 (12) (1974) 899.
[169] J.J. McPhee, On the use of linear graph theory in multibody system dynamics, Nonlinear Dynamics 9 (1–2) (feb 1996) 73–90.
[170] T. McPherson, J. Ueda, A force and displacement self-sensing piezoelectric MRI-compatible tweezer end effector with an on-site calibration procedure, Mechatronics, IEEE/ASME Transactions on 19 (2) (apr 2014) 755–764.
[171] A.V. Mezheritsky, Invariants of electromechanical coupling coefficients in piezoceramics, Ultrasonics, Ferroelectrics and Frequency Control, IEEE Transactions on 50 (12) (2003) 1742–1751.
[172] Microsoft. Microsoft image conposite editor (ice).

[173] Chet T. Moritz, Benjamin K. Barry, Michael A. Pascoe, Roger M. Enoka, Discharge rate variability influences the variation in force fluctuations across the working range of a hand muscle, Journal of Neurophysiology 93 (5) (may 2005) 2449–2459.

[174] A.J. Moskalik, D. Brei, Quasi-static behavior of individual C-block piezoelectric actuators, Journal of Intelligent Material Systems and Structures 8 (7) (1997) 571–587.

[175] Patrick Mottard, Yves St-Amant, Analysis of flexural hinge orientation for amplified piezo-driven actuators, Smart Materials and Structures 18 (2009).

[176] Richard R. Murray, Zexiang Li, Shankara Shastry, A Mathematical Introduction to Robotic Manipulation, 1st edition, CRC Press, Boca Raton, 1994.

[177] National Instruments. Compactrio Developers Guide, http://www.ni.com/compactriodevguide/, 2009.

[178] National Instruments Corporation, 11500 North Mopac Expressway, Austin, TX 78759, USA. DAQ M Series Manual, jul 2008.

[179] Ali H. Nayfeh, Dean T. Mook, Nonlinear Oscillations, John Wiley & Sons, 2008.

[180] Devin Neal, H. Harry Asada, Design of cellular piezoelectric actuators with high blocking force and high strain, in: ASME 2008 Dynamic Systems and Control Conference, American Society of Mechanical Engineers, 2008, pp. 779–786.

[181] Dragomir N. Nenchev, Kazuya Yoshida, Prasert Vichitkulsawat, Masaru Uchiyama, Reaction null-space control of flexible structure mounted manipulator systems, IEEE Transactions on Robotics and Automation 15 (6) (dec 1999) 1011–1023.

[182] R.E. Newnham, A. Dogan, Q.C. Xu, K. Onitsuka, J. Tressler, S. Yoshikawa, Flextensional moonie actuators, in: 1993 IEEE Proceedings. Ultrasonics Symposium, vol. 1, 1993, pp. 509–513.

[183] Christopher Niezrecki, Diann Brei, Sivakumar Balakrishnan, Andrew Moskalik, Piezoelectric actuation: state of the art, The Shock and Vibration Digest 33 (4) (2001) 269–280.

[184] Daichi Nozaki, Kimitaka Nakazawa, Masami Akai, Muscle activity determined by cosine tuning with a nontrivial preferred direction during isometric force exertion by lower limb, Journal of Neurophysiology 93 (5) (may 2005) 2614–2624.

[185] Lael Odhner, Jun Ueda, H. Harry Asada, Feedback control of stochastic cellular actuators, in: Experimental Robotics, Springer, 2008, pp. 481–490.

[186] Katsuhiko Onitsuka, Aydin Dogan, James F. Tressler, Qichang Xu, Shoko Yoshikawa, Robert E. Newnham, Metal-ceramic composite transducer, the "moonie", Journal of Intelligent Material Systems and Structures 6 (4) (1995) 447–455.

[187] Rieko Osu, Naoki Kamimura, Hiroshi Iwasaki, Eri Nakano, Chris M. Harris, Yasuhiro Wada, Mitsuo Kawato, Optimal impedance control for task achievement in the presence of signal-dependent noise, Journal of Neurophysiology 92 (2) (aug 2004) 1199–1215.

[188] C.A. Ottenheijm, L.M. Heunks, R.P. Dekhuijzen, Diaphragm adaptations in patients with COPD, Respiratory Research 9 (1) (2008) 12.

[189] Erhan Oztop, David W. Franklin, Thierry Chaminade, Gordon Cheng, Human–humanoid interaction: is a humanoid robot perceived as a human?, International Journal of Humanoid Robotics 2 (4) (2005) 537–559.

[190] Lucy Y. Pao, Input shaping design for flexible systems with multiple actuators, in: Proceedings of the 13th World Congress of the International Federation of Automatic Control, San Francisco, jul 1996.

[191] J.M. Paros, L. Weisbord, How to design flexure hinges, Machine Design 37 (27) (nov 1965) 151–156.

[192] Florian Petit, Maxime Chalon, Werner Friedl, Markus Grebenstein, Alin Albu-sch, Bidirectional antagonistic variable stiffness actuation: analysis, design & implementation, in:

Proceedings of the 2010 IEEE International Conference on Robotics and Automation, Anchorage, 2010, pp. 4189–4196.

[193] Huy-Hoang Pham, I-Ming Chen, Stiffness modeling of flexure parallel mechanism, Precision Engineering 29 (2005) 467–478.

[194] Bruno Picasso, Antonio Bicchi, Hypercubes are minimal controlled invariants for discrete-time linear systems with quantized scalar input, Nonlinear Analysis: Hybrid Systems 2 (3) (aug 2008) 706–720.

[195] Bruno Picasso, Frédéric Gouaisbaut, Antonio Bicchi, Construction of invariant and attractive sets for quantized-input linear systems, in: Decision and Control, 2002, Proceedings of the 41st IEEE Conference on, vol. 1, IEEE, 2002, pp. 824–829.

[196] J.S. Plante, S. Dubowsky, On the nature of dielectric elastomer actuators and its implications for their design, in: Daniele Inaudi, Wolfgang Ecke, Brian Culshaw, Kara J. Peters, Eric Udd (Eds.), Smart Structures and Materials 2006: Smart Sensor Monitoring Systems and Applications, in: Proceedings of the SPIE, vol. 6168, 2006, pp. 424–434.

[197] N.S. Pollard, R.C. Gilbert, Tendon arrangement and muscle force requirements for human-like force capabilities in a robotic finger, in: Proceedings of the 2002 IEEE International Conference on Robotics and Automation, Washington, DC, 2002, pp. 3755–3762.

[198] José L. Pons, Emerging Actuator Technologies: A Micromechatronic Approach, John Wiley & Sons, 2005.

[199] M.J.D. Powell, Numerical Methods for Nonlinear Algebraic Equations, Gordon and Breach Science Publishers, 1970, pp. 115–161, Chapter 7.

[200] G.A. Pratt, M.M. Williamson, Series elastic actuators, in: Intelligent Robots and Systems 95. 'Human Robot Interaction and Cooperative Robots', Proceedings. 1995 IEEE/RSJ International Conference on, vol. 1, 1995, pp. 399–406.

[201] Eric F. Prechtl, Steven R. Hall, Design of a high efficiency, large stroke, electromechanical actuator, Smart Materials and Structures 8 (1) (1999) 13.

[202] B.I. Prilutsky, Coordination of two- and one-joint muscles: functional consequences and implications for motor control, Motor Control 4 (1) (2000) 1–44.

[203] PTGui, image stitching software.

[204] Marc Raibert, Legged Robots that Balance, 1st edition, MIT Press, Cambridge, MA, 1986.

[205] Ramesh Raskar, Amit Agrawal, Jack Tumblin, Coded exposure photography: motion deblurring using fluttered shutter, ACM Transactions on Graphics (TOG) 25 (3) (2006) 795–804.

[206] Darcy S. Reisman, John P. Scholz, Aspects of joint coordination are preserved during pointing in persons with post-stroke hemiparesis, Brain 126 (11) (nov 2003) 2510–2527.

[207] D.A. Robinson, The mechanics of human saccadic eye movement, The Journal of Physiology 174 (1964) 245–264.

[208] Andrew M. Rockett, Peter Szüz, Continued Fractions, World Scientific, 1992.

[209] C.G. Theod Ruete, Ocular physiology, Strabismus 7 (1) (1999) 43–60.

[210] Augustin Sanchez, Vincent Mahout, Bertrand Tondu, Nonlinear parametric identification of a McKibben artificial pneumatic muscle using flatness property of the system, in: Proceedings of the 1998 IEEE International Conference on Control Applications (Cat. No. 98CH36104), vol. 1, 1998, pp. 70–74.

[211] L. Sass, J.M.C. McPhee, C. Schmitke, P. Fisette, D. Grenier, A comparison of different methods for modelling electromechanical multibody systems, Multibody System Dynamics 12 (3) (2004) 209–250.

[212] Marcus Scherrer, John Mcphee, Dynamic modelling of electromechanical multibody systems, Multibody System Dynamics 9 (1) (2003) 87–115.

[213] R. Schiavi, G. Grioli, S. Sen, A. Bicchi, VSA-II: a novel prototype of variable stiffness actuator for safe and performing robots interacting with humans, in: Proceedings – IEEE International Conference on Robotics and Automation, 2008, pp. 2171–2176.

[214] Joshua Schultz, Glenn Mathijssen, Bram Vanderborght, Antonio Bicchi, Toward motor-unit-recruitment actuators for soft robotics, in: Proceedings of the IEEE International Conference on Biomedical Robotics and Biomechatronics, São Paulo, 2014, pp. 887–892.

[215] Joshua Schultz, Glenn Mathijssen, Bram Vanderborght, Antonio Bicchi, A selective recruitment strategy for exploiting muscle-like actuator impedance properties, in: Intelligent Robots and Systems (IROS), 2015 IEEE/RSJ International Conference on, IEEE, 2015, pp. 2231–2237.

[216] Joshua Schultz, Jun Ueda, Intersample discretization of control inputs for flexible systems with quantized cellular actuation, in: The 2010 Dynamic Systems and Control Conference (DSCC'10), American Society of Mechanical Engineers, 2010.

[217] Joshua Schultz, Jun Ueda, A camera positioner driven by muscle-like actuation, in: 2012 4th IEEE RAS & EMBS International Conference on Biomedical Robotics and Biomechatronics (BioRob), IEEE, jun 2012, pp. 719–724.

[218] Joshua Schultz, Jun Ueda, Experimental verification of discrete switching vibration suppression, Mechatronics, IEEE/ASME Transactions on 17 (2) (2012) 298–308.

[219] Joshua Schultz, Jun Ueda, Nested piezoelectric cellular actuators for a biologically inspired camera positioning mechanism, IEEE Transactions on Robotics 29 (5) (2013) 1125–1138.

[220] Joshua Schultz, Jun Ueda, Two-port network models for compliant rhomboidal strain amplifiers, IEEE Transactions on Robotics 29 (1) (2013) 42–54.

[221] T.W. Secord, J. Ueda, H.H. Asada, Dynamic analysis of a high-bandwidth, large-strain, PZT cellular muscle actuator with layered strain amplification, in: Proceedings of 2008 IEEE International Conference on Robotics and Automation (ICRA 2008), 2008, pp. 761–766.

[222] Thomas W. Secord, H. Harry Asada, A variable stiffness actuator having tunable resonant frequencies, IEEE Transactions on Robotics 26 (6) (2010) 993–1005.

[223] K.A. Seffen, E. Toews, Hyperhelical actuators: coils and coiled-coils, in: 45th AIAA/ASME/ASCE/AHS/ASC Structures, Structural Dynamics and Materials Conference, apr 2004, pp. 19–22.

[224] Tomonori Seki, Minoru Sakata, Takuya Nakajima, Mikio Matsumoto, Thermal buckling actuator for micro relays, in: Solid State Sensors and Actuators, 1997. TRANSDUCERS'97 Chicago, 1997 International Conference on, vol. 2, IEEE, 1997, pp. 1153–1156.

[225] Brian Selden, Kyujin Cho, H. Harry Asada, Segmented shape memory alloy actuators using hysteresis loop control, Smart Materials and Structures 15 (2) (2006) 642–652.

[226] Mohsen Shahinpoor, Kwang J. Kim, H. Brett Schreyer, Artificial sarcomere and muscle made with conductive polyacrylonitrile (C-PAN) fiber bundles, in: Proceedings of SPIE, vol. 3987, 2000, p. 243.

[227] Jinjun Shan, Dong Sun, Dun Liu, Design for robust component synthesis vibration suppression of flexible structures with on–off actuators, IEEE Transactions on Robotics and Automation 20 (3) (jun 2004) 512–525.

[228] Qi Shan, Jiaya Jia, Aseem Agarwala, High-quality motion deblurring from a single image, ACM Transactions on Graphics (TOG) 27 (3) (2008) 73.

[229] Xiangrong Shen, Jianlong Zhang, Eric J. Barth, Michael Goldfarb, Nonlinear model-based control of pulse width modulated pneumatic servo systems, Journal of Dynamic Systems, Measurement, and Control 128 (3) (2006) 663.
[230] Min Shi, Jiang Yu Zheng, A slit scanning depth of route panorama from stationary blur, in: Computer Vision and Pattern Recognition, 2005. CVPR 2005. IEEE Computer Society Conference on, vol. 1, 2005, pp. 1047–1054.
[231] Pengfei Shi, John McPhee, Symbolic programming of a graph-theoretic approach to flexible multibody dynamics, Mechanics of Structures and Machines 30 (1) (2002) 123–154.
[232] W.Y. Shih, W.H. Shih, I.A. Aksay, Scaling analysis for the axial displacement and pressure of flextensional transducers, Journal of the American Ceramic Society 80 (5) (1997) 1073–1078.
[233] E.C.N. Silva, S. Nishiwaki, N. Kikuchi, Topology optimization design of flextensional actuators, Ultrasonics, Ferroelectrics and Frequency Control, IEEE Transactions on 47 (3) (2000) 657–671.
[234] G. Simmons, Y. Demiris, Optimal robot arm control using the minimum variance model, Journal of Robotic Systems 22 (11) (2005) 677–690.
[235] Neil C. Singer, Warren P. Seering, Preshaping command inputs to reduce system vibration, ASME Journal of Dynamic Systems, Measurement, and Control 112 (1990) 76–82.
[236] William Singhose, Erica Biediger, Hideto Okada, Saburo Matunaga, Experimental verification of real-time control for flexible systems with on–off actuators, ASME Journal of Dynamic Systems, Measurement, and Control 128 (2) (2006) 287–296.
[237] William Singhose, Bart W. Mills, Warren P. Seering, Closed form methods for generating on–off commands for undamped flexible spacecraft, Journal of Guidance, Control and Dynamics 22 (2) (1998) 378–382.
[238] William Singhose, Neil Singer, Warren Seering, Residual vibration reduction using vector diagrams to generate shaped inputs, ASME Journal of Mechanical Design 116 (2) (1994) 654–659.
[239] G. Song, V. Chaudhry, C. Batur, Precision tracking control of shape memory alloy actuators using neural networks and a sliding-mode based robust controller, Smart Materials and Structures 12 (2) (2003) 223.
[240] Gangbing Song, Nick V. Buck, Brij N. Agrawal, Spacecraft vibration reduction using pulse-width pulse-frequency modulated input shaper, in: AIAA Guidance, Navigation and Control Conference, AIAA, Baltimore, MD, 1998.
[241] A.V. Srinivasan, D.M. McFarland, H.A. Canistraro, E.K. Begg, Multiplexing embedded NiTiNOL actuators to obtain increased bandwidth in structural control, Journal of Intelligent Material Systems and Structures 8 (3) (1997) 202–214.
[242] A.V. Srinivasan, D. Michael McFarland, Smart Structures: Analysis and Design, Cambridge University Press, 2000.
[243] Stanley Wallace Jacob, Clarice Ashworth Francone, Walter J. Lossow, Structure and Function in Man, fifth edition, W.B. Saunders Co., 1982.
[244] Samuel C. Stanton, Clark C. McGehee, Brian P. Mann, Nonlinear dynamics for broadband energy harvesting: investigation of a bistable piezoelectric inertial generator, Physica D: Nonlinear Phenomena 239 (10) (2010) 640–653.
[245] R.B. Stein, E.R. Gossen, K.E. Jones, Neuronal variability: noise or part of the signal?, Nature Reviews Neuroscience 6 (5) (2005) 389–397.
[246] Erik Edward Steltz, Redesign of the micromechanical flying insect in a power density context, PhD thesis, University of California, Berkeley, may 2008.
[247] M.D. Stern, G. Pizarro, E. Rios, Local control model of excitation-contraction coupling in skeletal muscle, Journal of General Physiology 110 (4) (1997) 415.

[248] Gilbert Strang, Linear Algebra and Its Applications, third edition, Thomson Learning Inc., Toronto, 1988.

[249] Xiao Ping S. Su, Henry S. Yang, Design of compliant microleverage mechanisms, Sensors and Actuators, A: Physical 87 (3) (2001) 146–156.

[250] Engin Tanik, Volkan Parlaktas, A new type of compliant spatial four-bar (RSSR) mechanism, Mechanism and Machine Theory 46 (5) (2011) 593–606.

[251] Y. Tian, B. Shirinzadeh, D. Zhang, D. Zhong, Three flexure hinges for compliant mechanism designs based on dimensionless graph analysis, Precision Engineering 34 (2010) 92–100.

[252] Y. Tian, Bijan Shirinzadeh, D. Zhang, Y. Zhong, Closed-form compliance equations of filleted V-shaped flexure hinges for compliant mechanism design, Precision Engineering 34 (2010) 408–418.

[253] E. Todorov, Cosine tuning minimizes motor errors, Neural Computation 14 (6) (2002) 1233–1260.

[254] E. Todorov, Optimality principles in sensorimotor control, Nature Neuroscience 7 (9) (2004) 907–915.

[255] E. Todorov, Stochastic optimal control and estimation methods adapted to the noise characteristics of the sensorimotor system, Neural Computation 17 (5) (2005) 1084–1108.

[256] E. Todorov, W. Li, A generalized iterative LQG method for locally-optimal feedback control of constrained nonlinear stochastic systems, in: American Control Conference, 2005. Proceedings of the 2005, IEEE, 2005, pp. 300–306.

[257] Michael T. Tolley, Robert F. Shepherd, Bobak Mosadegh, Kevin C. Galloway, Michael Wehner, Michael Karpelson, Robert J. Wood, George M. Whitesides, A resilient, untethered soft robot, Soft Robotics 1 (3) (2014) 213–223.

[258] Juana Torres, H. Harry Asada, High-gain, high transmissibility PZT displacement amplification using a rolling-contact buckling mechanism and preload compensation springs, Robotics, IEEE Transactions on 30 (4) (2014) 781–791.

[259] Barry Trimmer, Bram Vanderborght, Yiğit Mengüç, Michael Tolley, Joshua Schultz, Soft robotics as an emerging academic field, Soft Robotics 2 (4) (2015) 131–134.

[260] Nikolaus F. Troje, Decomposing biological motion: a framework for analysis and synthesis of human gait patterns, Journal of Vision 2 (5) (jan 2002) 371–387.

[261] M.J.A. Tyreman, J.E. Molloy, Molecular motors: nature's nanomachines, IEE Proceedings – Nanobiotechnology 150 (3) (dec 2003) 95–102.

[262] K. Uchino, Piezoelectric Actuators and Ultrasonic Motors, Kluwer Academic Publishers, 1997.

[263] J. Ueda, L. Odhner, H.H. Asada, Broadcast feedback of stochastic cellular actuators inspired by biological muscle control, The International Journal of Robotics Research 26 (11–12) (nov 2007) 1251–1265.

[264] J. Ueda, T. Secord, H.H. Asada, Design of PZT cellular actuators with power-law strain amplification, in: Intelligent Robots and Systems, 2007. IROS 2007. IEEE/RSJ International Conference on, 2007, pp. 1160–1165.

[265] J. Ueda, T. Secord, H.H. Asada, Piezoelectric cellular actuators using nested rhombus multilayer mechanisms, in: 1st Annual Dynamic Systems and Control Conference (DSCC 2008), 2008.

[266] J. Ueda, T. Secord, H.H. Asada, Static lumped parameter model for nested PZT cellular actuators with exponential strain amplification mechanisms, in: IEEE International Conference on Robotics and Automation, 2008. ICRA 2008, 2008, pp. 3582–3587.

[267] J. Ueda, T.W. Secord, H.H. Asada, Large effective-strain piezoelectric actuators using nested cellular architecture with exponential strain amplification mechanisms, Mechatronics, IEEE/ASME Transactions on 15 (5) (2010) 770–782.
[268] Jun Ueda, Piezoelectrically actuated robotic end-effector with strain amplification mechanisms, in: Advanced Mechatronics and MEMS Devices, Springer, 2013, pp. 25–52.
[269] Jun Ueda, David MacNair, Ellenor Brown, Quantized control of compliant cellular actuator arrays for biological movement generation, in: ASME 2012 5th Annual Dynamic Systems and Control Conference joint with the JSME 2012 11th Motion and Vibration Conference, American Society of Mechanical Engineers, 2012, pp. 233–241.
[270] Jun Ueda, L. Odhnar, H. Asada, A broadcast-probability approach to the control of vast DOF cellular actuators, in: Proceedings of 2006 IEEE International Conference on Robotics and Automation (ICRA '06), may 2006, pp. 1456–1461.
[271] Jun Ueda, L. Odhner, Sang-Gook Kim, H.H. Asada, Distributed stochastic control of MEMS-PZT cellular actuators with broadcast feedback, in: The First IEEE/RAS-EMBS International Conference on Biomedical Robotics and Biomechatronics (BioRob 2006), feb 2006, pp. 272–277.
[272] Jun Ueda, Lael Odhner, H. Harry Asada, Broadcast feedback for stochastic cellular actuator systems consisting of nonuniform actuator units, in: Proceedings of 2007 IEEE International Conference on Robotics and Automation (ICRA '07), apr 2007, pp. 642–647.
[273] Jun Ueda, Lael Odhner, H. Harry Asada, Broadcast feedback of large-scale, distributed stochastic control systems inspired by biological muscle control, in: American Control Conference, 2007. ACC'07, IEEE, 2007, pp. 1317–1322.
[274] Jun Ueda, Tsuneo Yoshikawa, Mode-shape compensator for improving robustness of manipulator mounted on flexible base, Robotics and Automation, IEEE Transactions on 20 (2) (2004) 256–268.
[275] Jun Ueda, Tsuneo Yoshikawa, Robust arm configuration of manipulator mounted on flexible base, Robotics, IEEE Transactions on 20 (4) (2004) 781–789.
[276] Mitsunori Uemura, Katsuya Kanaoka, Sadao Kawamura, A new control method utilizing stiffness adjustment of mechanical elastic elements for serial link systems, in: Proceedings – IEEE International Conference on Robotics and Automation, 2007, pp. 1437–1442.
[277] Y. Uno, M. Kawato, R. Suzuki, Formation and control of optimal trajectory in human multijoint arm movement, Biological Cybernetics 61 (2) (1989) 89–101.
[278] Ronald Van Ham, Bram Vanderborght, Michael Van Damme, Bjorn Verrelst, Dirk Lefeber, MACCEPA, the mechanically adjustable compliance and controllable equilibrium position actuator: design and implementation in a biped robot, Robotics and Autonomous Systems 55 (10) (2007) 761–768.
[279] J. van Randeraat, R.E. Setterington (Eds.), Piezoelectric Ceramics, Mullard Limited, 1974.
[280] J.P. Van Zandwijk, M.F. Bobbert, J. Harlaar, A.L. Hof, From twitch to tetanus for human muscle: experimental data and model predictions for m. triceps surae, Biological Cybernetics 79 (2) (aug 1998) 121–130.
[281] Jan Peter Van Zandwijk, Maarten F. Bobbert, Guus C. Baan, Peter A. Huijing, From twitch to tetanus: performance of excitation dynamics optimized for a twitch in predicting tetanic muscle forces, Biological Cybernetics 75 (5) (nov 1996) 409–417.
[282] T. Villgrattner, H. Ulbrich, Design and control of a compact high-dynamic camera-orientation system, Mechatronics, IEEE/ASME Transactions on 16 (2) (2011) 221–231.
[283] Roman Vitushinsky, Sam Schmitz, Alfred Ludwig, Bistable thin-film shape memory actuators for applications in tactile displays, Microelectromechanical Systems, Journal of 18 (1) (2009) 186–194.

[284] Dana E. Vogtmann, Satyandra K. Gupta, Sarah Bergbreiter, Characterization and modeling of elastomeric joints in miniature compliant mechanisms, Journal of Mechanisms and Robotics 5 (nov 2013) 1–12.

[285] Gunter K. von Noorden, Emilio C. Campos, Binocular Vision and Ocular Motility: Theory and Management of Strabismus, 6th edition, Mosby, 2001.

[286] Oliver Whyte, Josef Sivic, Andrew Zisserman, Jean Ponce, Non-uniform deblurring for shaken images, International Journal of Computer Vision 98 (2) (2012) 168–186.

[287] B. Widrow, I. Kollár, Quantization Noise: Roundoff Error in Digital Computation, Signal Processing, Control, and Communications, Cambridge University Press, 2008.

[288] Levi Wood, Roger Kamm, Harry Asada, Stochastic modeling and identification of emergent behaviors of an endothelial cell population in angiogenic pattern formation, The International Journal of Robotics Research 30 (6) (2011) 659–677.

[289] Levi Wood, Jun Ueda, H. Harry Asada, Broadcast feedback with hysteresis loop control of stochastically behaving cellular units with application to cellular shape memory alloy actuators, in: ASME 2008 Dynamic Systems and Control Conference, American Society of Mechanical Engineers, 2008, pp. 425–432.

[290] Levi B. Wood, H. Harry Asada, Cellular stochastic control of the collective output of a class of distributed hysteretic systems, Journal of Dynamic Systems, Measurement, and Control 133 (6) (2011) 61011.

[291] Li Xu, Jiaya Jia, Two-phase kernel estimation for robust motion deblurring, in: Computer Vision–ECCV 2010, Springer, 2010, pp. 157–170.

[292] T.B. Xu, E.J. Siochi, L. Zuo, X. Jiang, J.H. Kang, Multistage force amplification of piezoelectric stacks. US Patent application. US9048759 B2, 2015.

[293] Pablo Valdivia y Alvarado, Kamal Youcef-Toumi, Design of machines with compliant bodies for biomimetic locomotion in liquid environments, Journal of Dynamic Systems, Measurement and Control, Transactions of the ASME 128 (1) (2006) 3–13.

[294] G.T. Yamaguchi, Dynamic Modeling of Musculoskeletal Motion, Kluwer Academic Publishers, 2001.

[295] J. Yan, R.J. Wood, S. Avadhanula, M. Sitti, R.S. Fearing, Towards flapping wing control for a micromechanical flying insect, in: Proceedings – IEEE International Conference on Robotics and Automation, vol. 4, 2001, pp. 3901–3908.

[296] Xuebo Yang, Xibin Cao, A new approach to autonomous rendezvous for spacecraft with limited impulsive thrust: based on switching control strategy, Aerospace Science and Technology 43 (2015) 454–462.

[297] Guo Ye, Wei Li, Yu-qiao Wang, Xue-feng Yang, Ling Yu, Kinematics analysis of bridge-type micro-displacement mechanism based on flexure hinge, in: Proceedings of the 2010 IEEE international conference on information and automation, jun 2010, pp. 66–70.

[298] Laurence R. Young, Lawrence Stark, Variable feedback experiments testing a sampled data model for eye tracking movements, IEEE Transactions on Human Factors in Electronics (1) (1963) 38–51.

[299] Elias Zakon, Mathematical Analysis I, The Trilla Group, 1975.

[300] Michael Zinn, Bernard Roth, Oussama Khatib, J.K. Salisbury, A new actuation approach for human friendly robot design, The International Journal of Robotics Research 23 (4) (apr 2004) 379–398.

Index

K

L

M

FSC
www.fsc.org
MIX
Papier | Fördert gute Waldnutzung
FSC® C083411

Kolibri 360 GmbH
Ferdinand-Jühlke-Straße 7
99095 Erfurt, Deutschland
produktsicherheit@kolibri360.de